XI International Symposium on Radiation from Relativistic Electrons in Periodic Structures (RREPS 2015)

AA006882

Journal of Physics: Conference Series Volume 732

Saint Petersburg, Russia
6 – 11 September 2015

Editors:

P. Karataev
A. Potylitsyn
M. Strikhanov
A. Tyukhtin

ISBN: 978-1-5108-2816-2
ISSN: 1742-6588

Printed from e-media with permission by:

Curran Associates, Inc.
57 Morehouse Lane
Red Hook, NY 12571

Some format issues inherent in the e-media version may also appear in this print version.

Copyright© (2015) by the Institute of Physics
All rights reserved. The material featured in this book is subject to
IOP copyright protection, unless otherwise indicated.

Printed by Curran Associates, Inc. (2016)

For permission requests, please contact the Institute of Physics
at the address below.

Institute of Physics
Dirac House, Temple Back
Bristol BS1 6BE UK

Phone: 44 1 17 929 7481
Fax: 44 1 17 920 0979

techtracking@iop.org

Additional copies of this publication are available from:

Curran Associates, Inc.
57 Morehouse Lane
Red Hook, NY 12571 USA
Phone: 845-758-0400
Fax: 845-758-2633
Email: curran@proceedings.com
Web: www.proceedings.com

Table of contents

Volume 732

XI International Symposium on Radiation from Relativistic Electrons in Periodic Structures (RREPS2015)
6–11 September 2015, Saint Petersburg, Russian Federation

Accepted papers received: 8 July 2016
Published online: 03 August 2016

Preface

Papers

General Aspects of Physical Phenomena and Processes Associated with Electromagnetic Radiation

Transition Radiation

Parametric X-Radiation

Diffraction Radiation, Smith-Purcell Effect

Coherent Bremsstrahlung and Channeling Radiation

THz sources and their applications

Journal of Physics: Conference Series **732** (2016) 011001 doi:10.1088/1742-6596/732/1/011001

Preface

These Proceedings are published as a recollection of contributions presented at the XI International Symposium on "Radiation from Relativistic Electrons in Periodic Structures" (RREPS-15), which was held in Saint Petersburg, September 6-11, 2015, Russian Federation.

RREPS-15 was co-organized by Saint-Petersburg State University, National Research Tomsk Polytechnic University, and National Research Nuclear University (MEPhI). The main goal of the symposium was to bring together the scientists from around the world who work on designs of new radiation sources and their applications.

There were 108 participants registered from 12 countries. The website of the symposium is available at http://rreps.tpu.ru/.

The scientific program of the symposium consisted of 6 sections and a satellite Workshop on Terahertz Radiation generation as a 7th section. All papers in these Proceedings refer to one from the following topics:

- Section 1: General Aspects of Physical Phenomena and Processes Associated with Electromagnetic Radiation
- Section 2: Transition Radiation
- Section 3: Parametric X- Radiation
- Section 4: Diffraction Radiation, Smith-Purcell Effect
- Section 5: Coherent Bremsstrahlung and Channeling Radiation
- Section 6: Monochromatic X-Ray and Gamma Beams Produced at Electron Accelerators
- Section 7: THz sources and their applications

The published papers cover nearly all "hot" topics of current interest on investigations of monochromatic and broadband radiation sources based on accelerators. The symposium discussed fundamental properties of all known mechanisms and their applications including Synchrotron Radiation, Transition Radiation, Diffraction Radiation, Cherenkov Radiation, Smith-Purcell Effect, Parametric X-ray and Channeling Radiation.

Content from this work may be used under the terms of the Creative Commons Attribution 3.0 licence. Any further distribution of this work must maintain attribution to the author(s) and the title of the work, journal citation and DOI.
Published under licence by IOP Publishing Ltd

We are extremely thankful to all authors for providing their valuable contributions for these Proceedings as well as the reviewers for their constructive recommendations and criticism aiding to improve the presented articles.

We are looking forward to welcoming all colleagues at the next Symposium of the biennial series RREPS-17, which will be hosted by DESY (Germany). We invite all researchers interested in the field including the authors of these Proceedings.

We thank Russian Foundation for Basic Research, National Research Nuclear University "MEPhI", Saint-Petersburg State University, Tomsk Polytechnic University and RASA centre in Tomsk for partial support.

Editors:
P. Karataev
John Adams Institute at Royal Holloway, University of London
A. Potylitsyn
National Research Tomsk Polytechnic University
M. Strikhanov
National Research Nuclear University MEPhI
A. Tyukhtin
Saint-Petersburg State University

Peer review statement

All papers published in this volume of *Journal of Physics: Conference Series* have been peer reviewed through processes administered by the proceedings Editors. Reviews were conducted by expert referees to the professional and scientific standards expected of a proceedings journal published by IOP Publishing.

Content from this work may be used under the terms of the Creative Commons Attribution 3.0 licence. Any further distribution of this work must maintain attribution to the author(s) and the title of the work, journal citation and DOI.
Published under licence by IOP Publishing Ltd

Transverse modulation of the positron beam density by using the laser standing wave

A A Babaev[1] **and S B Dabagov**[2,3,4]
[1]Tomsk Polytechnic University, Lenin Ave. 30, Tomsk 634050, Russia
[2]INFN Laboratori Nazionali di Frascati, via E. Fermi 40, Frascati 00044, Italy
[3]RAS P. N. Lebedev Physical Institute, Lenin Ave. 53, Moscow 119991, Russia
[4]NR Nuclear University MEPhI, Kashirskoe highway 31, Moscow 115409, Russia

E-mail: `krass58ad@mail.ru (A Babaev)`

Abstract. Recently it was shown that charged particles motion in the field of standing electromagnetic wave can undergo the features similar to the particles channeling in crystals. When a charged particle enters the channels formed by electromagnetic standing waves at a small angle to the node (anti-node) planes its motion represents namely the oscillations between two neighboring planes. The phenomenon is mostly known as channeling in a lattice of the standing waves. Obviously, this effect can be used to handle beams in accelerator physics, more general, for the beam shaping with the specific properties. The advantage of the plane wave channeling is the absence of inelastic scattering that takes place in a crystal. The possibility to re-distribute the current density of particles in the beam by means of the laser standing wave is demonstrated.

1. Introduction

The recent achievements in laser techniques open new areas for investigations in many fields of research. In particular, high intensity laser waves can find their applications in accelerator physics for the beam shaping: the beam deflection, focusing, micro-bunching etc (see, for example, in [1]). In [2] it was shown the standing laser wave could be also used for beam managing due to the channeling effect. The channeling in the laser standing wave has the similar nature as the famous channeling in crystals (see, for example, in [3,4] and references therein). As example of modern application of the channeling one can point out beam channeling in a bent crystal (see, for example, in [5,7] and references therein), when channeled projectiles follow the channel along bent crystallographic planes and, hence, they can be deflected outward the beam core. This scheme is suggested for beam collimation in modern hadron accelerators. In principle, the system of standing wave channels could be used instead of crystal channeling in all possible applications. The use of the laser standing wave has some evident advantages. Scattering of the beam by a crystal matter as well as the nuclear interactions between particles of the beam and crystal nuclei are also absent.

First, in the manuscript the short theoretical overview of the channeling in a standing wave is given. Further we investigate the evolution of the beam phase space to demonstrate the possibility of the beam focusing by the laser standing wave. After that the possible application of the focusing for the beam current re-distribution is analyzed. The simulations are carried out for 400 MeV positron beam (positron beam energy available at LNF INFN [7]). The laser has the wavelength $\lambda = 800$ nm and the intensity $I = 10^{17}$ W/cm^2. It will be shown such

Content from this work may be used under the terms of the Creative Commons Attribution 3.0 licence. Any further distribution of this work must maintain attribution to the author(s) and the title of the work, journal citation and DOI.
Published under licence by IOP Publishing Ltd

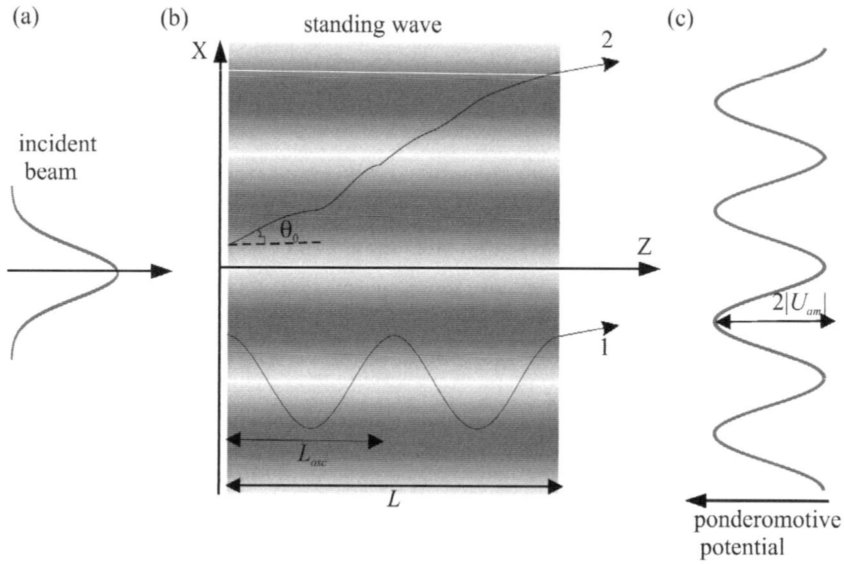

Figure 1. The geometry of computer experiment. The standing wave (b) is formed by two counterpropagating laser waves. The bunch of particles penetrates into the field along the Z-axis and particles enter the field at small glancing angle θ_0 to node planes (that are parallel to Z-axis and simultaneously are orthogonal to the figure plane). The ponderomotive potential forms cos-like structure in transverse X-direction with the wall $2|U_{\mathrm{am}}|$. Possible types of motion are: 1 — channeling, oscillations within single channel between node lines, 2 — quasichanneling, particle can cross channel borders.

intensity produces the potential barriers which are close to those in crystals at charged particles channeling. At the same time, the transverse space dimension of a laser channel, $a = \lambda/2$, is a few orders bigger than the width of a crystal channel.

2. Channeling in electromagnetic standing wave

Let consider the standing wave formed by two linear polarized counterpropagating (along the X-direction, see in figure 1) plane waves. The nodes (antinodes) of a standing wave form planes that are parallel to the YZ-plane. The bunch of relativistic charged particles moves by the way when the reference orbit coincides with the Z-direction. Particles penetrate into the wave field at small angles to the node (antinode) planes. In this case the particle motion could be described using averaged ponderomotive potential. Here we follow the work [2]. The ponderomotive forces of the standing laser wave acting on the projectile were considered also in [8,9]. The ponderomotive potential forms the channels along the Z-direction whereas in the XZ-cross-section the periodic cos-like potential takes place. The ponderomotive potential defines the smooth oscillating trajectory similar to the particle trajectory at crystal channeling. Namely, the channeled particle moves in the XZ-plane along the channel and oscillates between two neighboring node planes (or between antinode planes, this depends on the particle's velocity, see [2] for details). So, here we will not consider the motion in the Y-direction (orthogonal to the plane of figure 1). The small angle of particle incidence in the field with respect to the Z-axis is denoted by θ_0. The difference between real interactions and the interaction described by the averaged potential leads to small rapid oscillations along the smooth trajectory. These small oscillations are not considered in this work.

For the case of linear polarized counterpropagating waves the averaged potential has the form [2]

$$U(x) = -U_{\mathrm{am}} \cos(2kx), \tag{1}$$

where the potential amplitude is defined by the expression (here the projectile with the charge number ± 1 is assumed):

$$U_{\mathrm{am}} = \frac{4\pi\alpha\hbar I}{\omega^2 \gamma_z m}\left(2\beta_z^2 - 1\right). \tag{2}$$

In these expressions $v_z = \beta_z c$ is the longitudinal velocity of the particle (in the task considered here $v_z = \mathrm{const}$), c is the speed of light, $\gamma_z = (1-\beta_z^2)^{-1/2}$, m is the positron's mass, α is the fine structure constant, $\hbar$ is the Planck constant, $\omega = 2\pi c/\lambda$ is the angular frequency of the laser wave, $k = \omega/c$ is the wave number. The amplitude (2) defines the barrier $2|U_{\mathrm{am}}|$ of ponderomotive potential as shown in figure 1(c).

As in the case of channeling in a crystal, we can define the critical angle of projectile's penetration in the field

$$\theta_{\mathrm{lim}} = \sqrt{\frac{4|U_{\mathrm{am}}|}{\gamma\beta^2 mc^2}}$$

(we write the general expression which can be used at any energies). The channeling motion is impossible if $\theta_0 > \theta_{\mathrm{lim}}$. Nevertheless the particle's motion is still governed by averaged potential while θ_0 is of the order of a few θ_{lim}. This type of motion is known as the quasichanneling regime, when the particle can cross channel borders (see samples of trajectories in figure 1(b)). In general case the beam has the nonzero divergence, and both channeling and quasichanneling take place at the beam passage through the electromagnetic field. In this work we deal with non-divergent beam where all particles are captured in the channeling regime of motion.

The parameters of channeling for the case considered here are: $\gamma = 781$, $U_{\mathrm{am}} = 15.3$ eV, $\theta_{\mathrm{lim}} = 0.4$ mrad. Another important parameter is the oscillation length l_{osc} that is the length along the channel when a particle makes the complete transverse oscillation. The potential is not harmonic. Hence, this length depends on the initial particle transverse position in the channel (see in figure 2). For the aims of estimation it is convenient to evaluate $L_{\mathrm{osc}} = l_{\mathrm{osc}}(0)$ for a particle initially penetrating into the field along the Z-axis near the channel center. For 400 MeV positrons this estimation gives $L_{\mathrm{osc}} = 2.0$ mm. The figure 2 demonstrates the oscillation length l_{osc} increases rapidly when the initial transverse coordinate approaches $a/2$. Nevertheless, particles with initial transverse coordinates within the range $(-0.37a, 0.37a)$ (i.e. about of 70% of particles in the beam) have $l_{\mathrm{osc}} < 1.5L_{\mathrm{osc}}$. Therefore, the parameter L_{osc} is the characteristic parameter that defines the longitudinal dimension of the standing wave lattice for the observation of channeling-related effects. Namely, it defines the channel length when the effects of beam focusing could be observable. It should be noticed the dimensions of the lattice formed by the laser standing wave is a few order bigger than the crystal lattice. Hence, in principle, some thin features in the beam density re-distribution which are not observable by using the crystal channeling could be found with the standing wave lattice.

3. Evolution of the transverse phase space with the penetration depth

The idealized positron beam is considered in this section to demonstrate the main features of the standing wave channeling. The beam has no divergence, all particles penetrate into the field along the Z-axis. The transverse beam dimension coincides with the channel width a. figure 3 demonstrates the transverse phase space evolution in dependence on the penetration depth L. In figure 3 x is the transverse coordinate. The angle between velocity v and Z-axis is used instead of transverse velocity v_x as the parameter conjugate with the coordinate x. All particles are channeled.

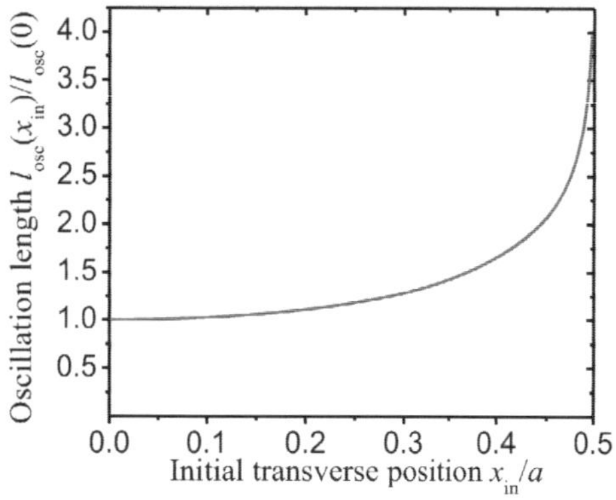

Figure 2. The dependence of oscillation length l_{osc} on initial transverse coordinate x_{in} for the case considered in the paper. The value $l_{osc}(0)$ is denoted L_{osc} through the text. The function l_{osc} tends to infinity at $x_{in} \to a/2$.

As one can see from figure 3 the beam shape could be controlled by the varying of the field length L. One of the interesting task of the accelerator physics is the possibility of beam focusing by a standing wave. If the potential will be harmonic, this beam is precisely focused at distances $L = 0.25L_{osc}$, $L = 0.75L_{osc}$, $L = 1.25L_{osc}$ etc. In the case of potential (1) the precise focusing is, of course, not possible. The best conditions for the beam focusing is at $L = 0.25L_{osc}$. After that, at $L = 0.75L_{osc}$, $L = 1.25L_{osc}$ and at successive "focused points" the defocusing takes place because of the potential inharmonicity.

In general, the evolution of the transverse phase space can be explained as follows. The initial beam is almost space-uniform and all particles have the same, zero, transverse velocity as shown in the panel (a). The phase points in the channel center and at the channel borders are stationary (see equation (1) and figure 1(c)). So, the line shown in the panel (a) is curving to keep its continuity and to keep the stationary points fixed while the beam propagates. During the beam motion the line is whirling around the channel center and it forms two-branch helix. For both the space and transverse velocity distributions this leads to the forming of peak in the center of distribution with the periodicity $L = 0.5L_{osc}$. The peak for the space distribution means the focusing penetration depth. The peak for the transverse velocity distribution means the minimal divergence at this moment. In general, when the beam is mostly focused, at this moment the beam demonstrates the maximal divergence (as in panel (b)), and vice versa, when the beam obtains the minimal divergence its transverse space tends to become uniform (see in panel (c)). The central peaks is smeared while the beam propagates. Finally, at $L \gg L_{osc}$ (the panel (d)) the beam shape does not depend on L. So, the best conditions for beam handling are provided at $L \leq L_{osc}$.

4. Transverse modulation of the beam current density

From figure 3(b) one can conclude there is a possibility to produce the kind of modulation in transverse beam cross-section if the beam is much wider than the channel width a. The channel width a defines the modulation dimension. Of course, this transverse modulation should take place also for crystal channeling, but the modulation dimension in that case is about of several Å, and it is too small to be observed. Moreover, scattering in the crystal target could prevent appearance of the modulation. On the contrary, for the case of laser standing wave the transverse modulation, perhaps, can be resolved by existing detectors; it depends on the laser wavelength λ used.

To simulate the transverse modulation the non-divergent Gaussian-shaped beam with the RMS width $\sigma_x = 5a$ was considered. The field length $L = 0.25L_{osc}$ was chosen to satisfy the best focusing condition for every channel. Results of simulations presented in figure 4 confirm the suggestion on the beam transverse modulation. Every channel gives the one peak of transverse

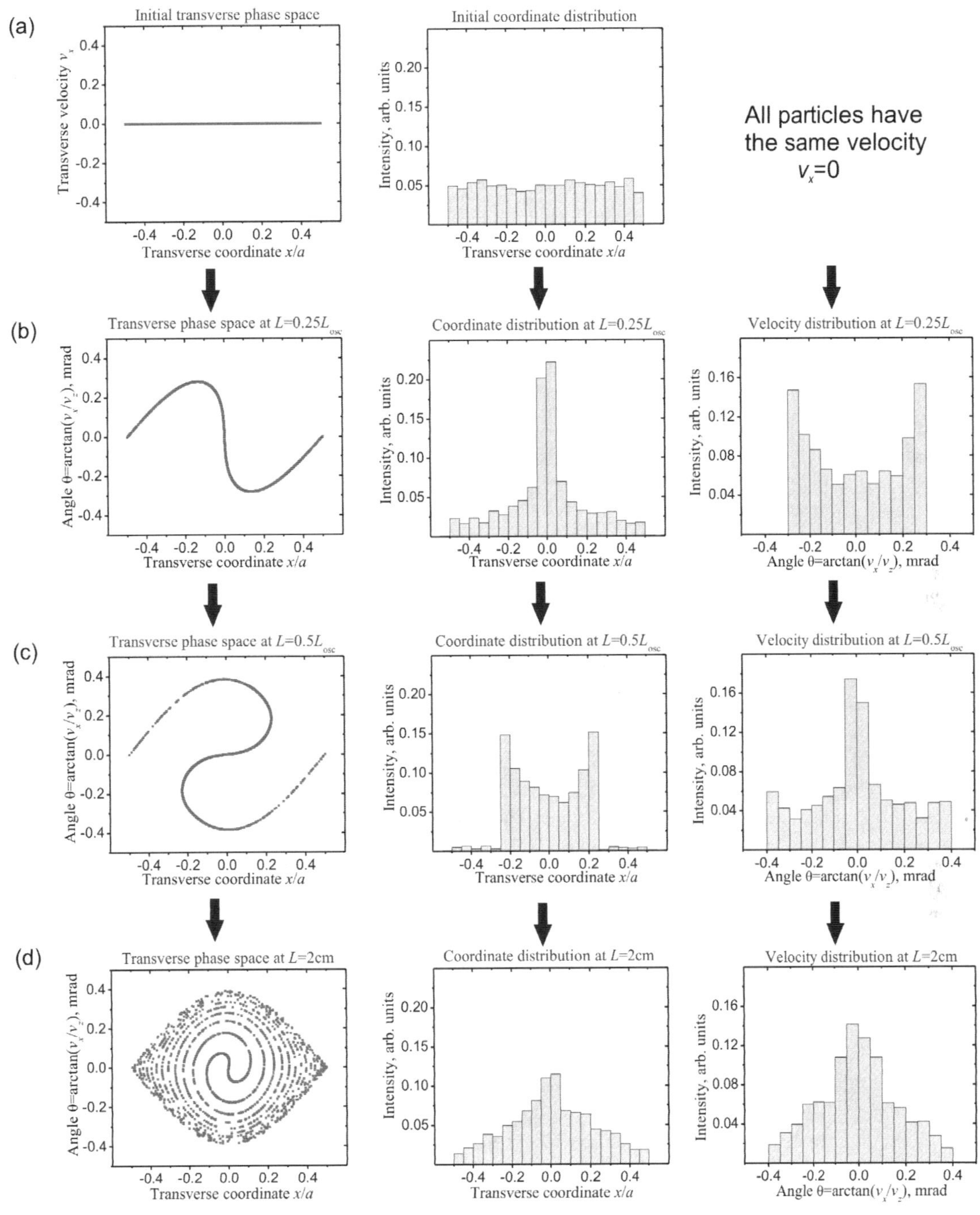

Figure 3. The transverse phase space evolution while the initially non-divergent beam propagates in the standing wave field. Panels demonstrate the phase space at different penetration depth L. The left pictures are the phase space, the central pictures are space distribution, the right pictures are the angular distributions. Following from the top picture to bottom one can trace the evolution of phase space, space distribution and angular distribution correspondingly. (a) Initial beam, (b) $L = 0.25L_{\rm osc}$, (c) $L = 0.5L_{\rm osc}$, (d) $L = 2$ cm.

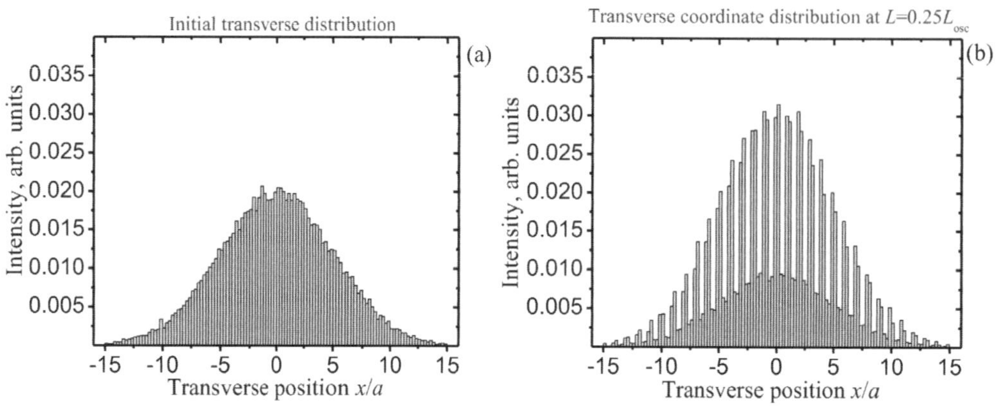

Figure 4. The appearance of transverse density modulation in the non-divergent beam that passages through the standing electromagnetic wave in the channeling regime: (a) the initial transverse beam profile, (b) the beam profile at $L = 0.25L_{\rm osc}$. The bin size corresponds to transverse dimension 100 nm.

beam density, so the beam splitting is evident. The full beam width remains the same as before the standing wave field. Hence, the peak beam intensity is higher than the maximal beam intensity before the field (the intensity increase is about of 50%). It should be mentioned the modulation might be quickly dissolved after the field because at $L = 0.25L_{\rm osc}$ the beam obtains a strong divergence (see in figure 3(b), this figure presents also the angular distribution of particles for the case considered in this section).

Now to obtain the transverse modulation of the beam the mask with the set of slits is used [10]. The mask leads to the significant particle loss (95% in [10], for example) whereas the suggested method simply re-distributes particles in the beam. The significant loss takes place because usually the distance between slits (1 mm in [10]) is several orders bigger than the slit width (50 um in [10]). The channel width in the scheme considered in the paper is defined by the laser wave length. It could be made significantly less than the slit width. The space between channels is absent. These features could make the plane wave channeling attractive for certain applications in comparison with masks. However, the plane wave channeling could increase the beam divergence, as shown in figure 3.

5. Conclusion

The main goal of this work was to demonstrate the possible applications of the beam channeling in the standing laser wave for the beam shaping tasks. It was pointed out that this effect can have some possible advantages in comparison with the crystal channeling. It was shown the potential wall of the standing wave lattice can get the same values as one of the crystal lattice. Hence, the laser standing wave channeling is effective such as the crystal channeling. Simultaneously, the channel width of the standing wave is few order bigger than the channel width of the crystal channel. Also, the scattering of particles by the solid target is absent in the case of the standing wave lattice. Therefore, the channeling in the standing wave can open new possibilities for the beam shaping that are inaccessible for the case of crystal channeling. Moreover, the use of the tunable laser sources enables the "dynamic" beam shaping during the bunch passage through the field.

This work deals with the light particles channeling as well as its application for the beam handling. The positron beam is considered here but the ponderomotive potential (1) does not depend on the charge sign. Hence, there is no difference in the motion of electron or positron beams at the same energies. The examples with simple model beams demonstrate the beam focusing as well as the transverse modulation of the beam current density. Nevertheless, the problem of particle's radiation influence [2] on the feasibility of light particles beam handling by the electromagnetic standing wave needs additional investigation.

Acknowledgments

The work is supported by the RF President Grant MK-5202.2015.2 and by the Ministry of Education and Science of RF in the frames of Competitiveness Growth Program of NRNU MEPhI, Agreement 02.A03.21.0005.

References

[1] Hemsing E, Stupakov G, Xiang D and Zholents A 2014 *Rev. Mod. Phys.* **86** 897
[2] Dabagov S B, Dik A V and Frolov E N 2015 *Phys. Rev. ST Accel. Beams* **18** 064002
[3] Ugerhoj U I 2005 *Rev. Mod. Phys.* **77** 1131
[4] Dabagov S B and Zhevago N K 2008 *Rivista Nuovo Cimento* **31** (9) 491
[5] Scandale W, Arduini G, Butcher M, Cerutti F, Garattini M, Gilardoni S, Lechner A, Losito R, Masi A, Mereghetti A *et al* 2015 *Phys. Lett.* B **748** 451
[6] Babaev A and Dabagov S B 2012 *Eur. J. Phys. Plus* **127** 62
[7] Bellucci S, Balasubramanian C, Grilli A, Micciulla F, Raco A, Popov A, Baranov V, Biryukov V, Chesnokov Yu and Maisheev V 2006 *Nucl. Instr. and Meth. in Phys. Res.* B **252** 3
[8] Pokrovsky A L and Kaplan A E 2005 *Phys. Rev.* A **72** 043401
[9] Smorenburg P W, Kanters J H M, Lassise A, Brussaard J H, Kamp L P J and Luiten O J 2011 *Phys. Rev.* A **83** 063810
[10] Sun Y-E, Piot P, Johnson A, Lumpkin A H, Maxwell T J, Ruan J and Thurman-Keup R 2010 *Phys. Rev. Lett.* **105** 234801

Cooling of ultrarelativistic β and μ particles by laser channels

A V Dik[1,3], **S B Dabagov**[1,3,4] **and E N Frolov**[1,2,3]

[1] LPI RAS, Moscow, Russia
[2] NR TPU, Tomsk, Russia
[3] NRNU MEPhI, Moscow, Russia
[4] LNF INFN, Frascati (Roma), Italy

E-mail: dik.alexey@gmail.com

Abstract. The descriprion of ultrarelativistic classical particles' movement in interference laser field formed by multichannel "sandwich" structures taking into account the radiative energy losses is present. The muon channeling case is described in detail. The critical angle for muon bound motion in the potential well of laser field is defined. The feasibility of beam cooling for charged particles due to radiation losses is shown.

1. Introduction

The investigation of charge particle dynamic in interference laser field presents great interest. The first article dedicated to this studies was published in the first half of 20th century [1]. Since that many papers describing charge particles interaction with interference electromagnetic field have been published. In particular, new applications of the effect of charge particle interaction with interference electromagnetic field were suggested in [2, 3, 4, 5, 6, 7, 8]. Theoretical investigation of interaction of charge particle with a standing electromagnetic wave were performed in [9, 10, 11]. A new point of view was suggested in the paper [12], in which for describing such interaction mathematical formalism of theory of charge particle channeling in crystals was used [13]. Using the mathematical apparatus of theory of charge particle channeling in crystals the effective potential of channeling was derived that allowed investigation of charge particle interaction in various geometries of interference laser field and some new features, such as the inversion of potential, to be studied in detail. Till now all investigations were made without taking into account the radiation losses of the particle, and only electrons were considered. The charge particle dynamics including the radiation losses, and the probability of muon channeling in laser field are covered in this paper.

2. Muon channeling in laser field

In [14] relativistic electron dynamics in the field of two interfering plane polarized laser waves was considered. Nonzero components of total field formed by the lasers can be written in the form

Content from this work may be used under the terms of the Creative Commons Attribution 3.0 licence. Any further distribution of this work must maintain attribution to the author(s) and the title of the work, journal citation and DOI.
Published under licence by IOP Publishing Ltd

$$\begin{aligned} E_x &= 2E_0 \sin\alpha \cos(k\cos\alpha x)\sin(\omega t - kz\sin\alpha), \\ E_z &= 2E_0 \cos\alpha \sin(k\cos\alpha x)\cos(\omega t - kz\sin\alpha), \\ H_y &= 2E_0 \cos(k\cos\alpha x)\sin(\omega t - kz\sin\alpha), \end{aligned} \tag{1}$$

where E_0 is the magnitude of electric field of a single laser wave, $\pi - 2\alpha$ is the angle between the wave vectors $\mid \mathbf{k} \mid = \omega_0/c$ of the lasers waves. The periodic channels can be formed for some specific conditions, and, in generally, they are capable to trap any charged particle. The effective potential describing interaction between the field and the charged particle of the mass m, the charge e, the longitudinal Lorentz-factor $\gamma_\parallel$ and the longitudinal velocity $\beta_\parallel$, in respect to the channel axis, which coincides with the Oz-axis, has the form

$$U_{eff}(x) = U_0(\beta_\parallel, \alpha)\cos(2kx\cos\alpha), \tag{2}$$

where

$$U_0(\beta_\parallel, \alpha) = \frac{e^2 E_0^2(-\cos(2\alpha) - 2\beta_\parallel \sin\alpha + \beta_\parallel^2(1+\cos^2\alpha))}{2\gamma_\parallel m \omega'^2}, \tag{3}$$

$$\omega' = \omega_0(1 - \beta_\parallel \sin\alpha) \tag{4}$$

As seen, the depth of a channel for muon less than for electron, with the same longitudinal Lorentz-factor, because of the muon mass is more than the electron one. Because of it, the only difference between muon and electron interactions with laser channels is quantitative and not qualitative. For example, let consider the critical angle of channeling defined by the following expression

$$\theta_{cr} = \frac{p_\perp^{max}}{p_\parallel} \approx \sqrt{\frac{2U_0(\beta_\parallel, \alpha)}{\gamma_\parallel mc^2}} \tag{5}$$

For muon in the field of a standing wave formed by the micrometer laser it can be written in a convenient form for the estimates

$$\theta_{cr}[rad] \approx 1,15 \cdot 10^{-10} \frac{\sqrt{I}}{\gamma_\parallel} \left[\frac{W}{cm^2}\right] \tag{6}$$

In the expression (6) the laser intensity is given in W/cm^2, while the critical angle measured in radians. In this way, for the muon with $\gamma_\parallel = 10^3$ and for the laser intensity $I = 10^{16} \div 10^{20}$ W/cm^2 the critical angle is $\theta_{cr} = 10$ μrad $\div$ 1 mrad. The ratio of muon and electron critical angles is the same as their masses $m^\mu/m^e \approx 200$. In [12] the estimation of the electrons critical concentration, which can be held by channel field, gives

$$n_0^e \leq \frac{2\gamma_\parallel I}{m\omega_0^2 c d_{ch}^2} \tag{7}$$

As one can conclude from expression (7) the difference in critical angles for muon and electron is defined by the mass, and the range of the laser intensity above the range of the critical concentration for muon is $10^{19} \leq n_0^\mu \leq 10^{23}$ cm^{-3}. It should be underlined, the expression (7) valid for non-disperse bunch. For a real bunch the critical concentration is typically less than (7).

3. Radiation losses

Let consider the motion of relativistic charge particle in external laser field. The equation of motion including energy losses is written in the following [15]:

$$mc\frac{du^i}{ds} = \frac{e}{c}F^{ik}u_k + g^i, \tag{8}$$

where F^{ik}, u_k, g^i, s are the electromagnetic field tensor, the 4-vector component of velocity, the 4-vector component of stopping force and the proper length of trajectory, respectively. Hereinafter, lower and upper indexes for 4-vectors correspond to covariance and contravariance components, meanwhile for the space vectors the upper index defines a power, unless it is otherwise specified. For the high energy particle the expression for stopping force can be rewritten in the next form (for convenience we use the 4-vector of momentum instead of the velocity $P^i = mcu^i$, and the 4-vector of momentum is upper-case)

$$g^i = \frac{2e^4}{3m^5c^8}P^i\left(F_{kl}P^l\right)\left(F^{km}P_m\right), \quad P^i = \left(\frac{E}{c}, \mathbf{p}\right), \tag{9}$$

where $E, \mathbf{p}$ are the energy and the space momentum of particle, respectively. Let pass from the proper length to the time, then (8) can be presented in the form

$$\frac{dP^i}{dt} = \frac{e}{c}F^{ik}\frac{P_k}{P^0} + \frac{mc^2}{P^0}g^i \tag{10}$$

Following the technique of our previous papers let present the particle motion as a sum of slow smooth trajectory $\bar{\mathbf{r}}$ and 4-vector of momentum $\bar{P}^i$, and fast oscillation trajectory $\mathbf{r}_\xi \ll \bar{\mathbf{r}}$ and momentum $P^i_\xi \ll \bar{P}^i$. Hereinafter the index ξ denominates the fast oscillation values. In this way for the fast oscillation motion in first approximation we can write

$$\frac{dP^i_\xi}{dt} = \frac{e}{c}F^{ik}\frac{\bar{P}_k}{\bar{P}^0} \tag{11}$$

and for the smooth motion

$$\frac{d\bar{P}^i}{dt} = \frac{e\bar{P}_k}{c\bar{P}^0}\overline{(\mathbf{r}_\xi, \nabla F^{ik})} + \frac{e}{c\bar{P}^0}\overline{F^{ik}\left(P^\xi_k - \frac{\bar{P}_k}{\bar{P}^0}P^0_\xi\right)} + \frac{mc^2}{\bar{P}^0}\bar{g}^i + \frac{mc^2}{\bar{P}^0}\overline{\delta g^i}, \tag{12}$$

where

$$g^i = \frac{2e^4}{3m^5c^8}\bar{P}^iF_{kl}\bar{P}^lF^{km}\bar{P}_m, \tag{13}$$

$$\delta g^i = \frac{2e^4}{3m^5c^8}\Big(P^i_\xi F_{kl}\bar{P}^lF^{km}\bar{P}_m + \bar{P}^i\left(\bar{P}^l\bar{P}_mF^{km}(\mathbf{r}_\xi,\nabla)F_{kl} + F_{kl}F^{km}\bar{P}_mP^l_\xi + \right.$$
$$\left. + F_{kl}\bar{P}^l\bar{P}_m(\mathbf{r}_\xi,\nabla)F^{km} + F_{kl}F^{km}\bar{P}^lP^\xi_m\right)\Big) + \frac{P^0_\xi}{\bar{P}^0}g^i \tag{14}$$

The g^i is nonzero in Eq. (12) because of momentum squared, therefore after averaging this doesn't equal zero, in generally. In the equation for fast oscillation we neglected influence of the energy losses as follows from the equations above derived. Till now we did not consider the period of averaging, which depends on longitudinal velocity in respect to direction of laser wave propagation [12, 14]. The period of fast oscillations is defined as $T = 2\pi/\omega'$, where ω' is the frequency of fast oscillations (4). Neglecting the losses the longitudinal velocity does not change. In reality that does not work. Nonetheless, we assume longitudinal velocity is constant during the time of one small oscillation. Let consider relativistic particle moving with high speed along the channel axis (Oz-axis), and low velocity along the transverse axis (Ox-axis), i.e. $P^0 \sim p_{\|} \gg p_x$. Then the component of 4-vector can be rewritten as follows

$$P^0 = \gamma mc = mc\sqrt{1 + \frac{\mathbf{p}^2}{m^2c^2}} \approx \bar{P}^0 + P^0_\xi, \tag{15}$$

$$P^1 = \bar{p}_x + p^\xi_x, \quad P^2 = 0, \quad P^3 = \bar{p}_{\|} + p^\xi_{\|}, \tag{16}$$

where

$$\bar{P}^0 = mc\sqrt{1 + \frac{\bar{p}^2_{\|}}{m^2c^2}} = \bar{\gamma}_{\|} mc,$$

$$P^0_\xi = \frac{\bar{p}_{\|} p^\xi_{\|}}{(\bar{P}^0)^2} = \frac{\bar{p}_{\|} p^\xi_{\|}}{mc\bar{\gamma}_{\|}}$$

Taking into account these approximations losses, we can find the following expressions for the components of space stopping force

$$f_x = -\frac{2e^2}{3m^4c^6}\left[\gamma^2_{\|} m^2 c p_x \omega'^2 \left(\frac{(eE_0)^2}{\gamma_{\|} m\omega_0^2} + 2U_{eff}\right) + \right.$$
$$\left. + \left(\frac{\partial U_{eff}}{\partial x} - 2\beta_{\|} p_x mc \frac{\omega_0}{\omega'} \frac{\sin\alpha - \beta_{\|}}{\cos(2\alpha)} \frac{\partial U_{eff,0}}{\partial x}\right) \frac{\partial U_{eff}}{\partial x}\right], \tag{17}$$

$$f_z = -\frac{2e^2}{3m^4c^6}\left[\gamma^2_{\|} m^2 c p_{\|} \omega'^2 \left(\frac{(eE_0)^2}{\gamma_{\|} m\omega_0^2} + 2U_{eff}\right) - \right.$$
$$-2\gamma_{\|} mc(1 + \beta^2_{\|}) \frac{\omega_0}{\omega'} \frac{\sin\alpha - \beta_{\|}}{\cos(2\alpha)} \frac{\partial U_{eff,0}}{\partial x} \frac{\partial U_{eff}}{\partial x} +$$
$$+\beta_{\|} \frac{(eE_0)^4}{2\omega'^2} \cos^2\alpha \left(1 + \frac{2m\omega_0^2 U_{eff,0}}{(eE_0)^2 \cos(2\alpha)}\right) \left(3\left(\frac{\omega'}{\omega_0}\right)^2 - \right.$$
$$\left.\left. + \left(\frac{\cos^2\alpha}{\gamma^2_{\|}} - 6\beta_{\|}\sin\alpha + 3(\sin^2\alpha + \beta^2_{\|})\right) \frac{2m\omega_0^2 U_{eff,0}}{(eE_0)^2\cos(2\alpha)}\right)\right], \tag{18}$$

where U_{eff} is the effective potential of relativistic particle (2), and $U_{eff,0}$ is the effective potential of nonrelativistic particle ($U_{eff,0} = U_{eff}(\beta_{\|} \to 0)$)

$$U_{eff,0} = -\frac{e^2E_0^2\cos(2\alpha)}{2m\omega_0^2}\cos(2kx\cos\alpha)$$

As seen from the expressions for the components of space stopping force (17) and (18), there is a significant contribution in the losses from both small oscillations under the influence of laser wave and channeling motion. The expression for stopping force includes both the effective potential and the derivative of effective potential. The losses take place when there are no channels. This happens due to the electron scattering in a laser field. The analysis of particle motion in the field of arbitrary geometry becomes difficult. Because of it we consider the standing wave geometry with $\alpha = 0$. In this case the expression for the stopping force is very simplified. Let consider the ultrarelativistic particle $\beta_{\|} \to 1$ in the field of laser channels and assume $p_{\|} \approx \gamma_{\|} mc$, then the equation of motion can be written in the following

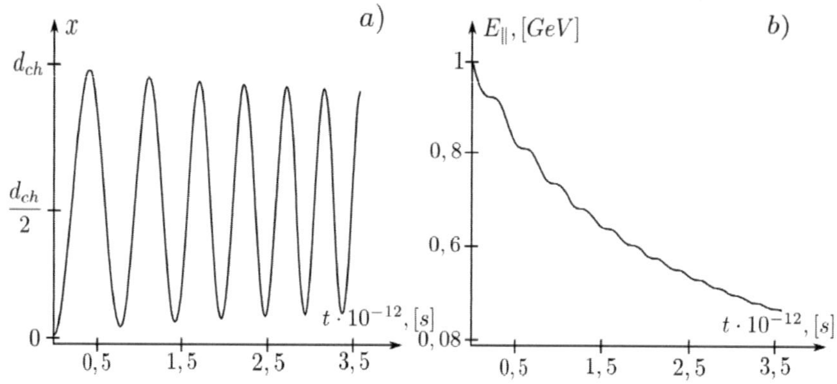

Figure 1. *a) The transverse coordinate x of electron versus time t; b) the longitudinal electron energy $E_{\parallel}$ versus time t in the field of micrometer wavelength laser with intensity of 10^{19} W/cm^2.*

$$\ddot{x} + \frac{e^6 E_0^4}{3\gamma_z m^5 c^7 \omega_0^2} \sin^2(2kx)\dot{x} + \frac{1}{\gamma_{\parallel} m}\frac{\partial U_{eff}}{\partial x} = 0, \tag{19}$$

$$\dot{\gamma}_{\parallel} = -\frac{2e^4 E_0^2 \gamma_{\parallel}^2}{3m^3 c^5}(1 + \cos(2kx)) \tag{20}$$

In the expression (20) the therms with ratio of effective potential and longitudinal energy are neglected at $U_{eff}/E_{\parallel} \to 0$. The solutions of the equations (19), (20) are shown in the figure 1. As seen, in such approximation there is no energy loss near the channel axis $x = \pi/(2k)$, while on the hill of potential the losses are maximal.

Hance, the maxima of the energy losses for electron (positron) and muon in the field of laser channels are given by

$$\left.\frac{dE_{\parallel}^{e}}{dz}\right|_{max} \approx 2,7 \cdot 10^{-16} I \gamma_{\parallel}^2, \quad \left[\frac{eV}{cm}\right] \tag{21}$$

$$\left.\frac{dE_{\parallel}^{\mu}}{dz}\right|_{max} \approx 7 \cdot 10^{-21} I \gamma_{\parallel}^2, \quad \left[\frac{eV}{cm}\right] \tag{22}$$

In the expressions above the laser intensity is expressed in W/cm^2. It worth mentioning that the (21) and (22) define energy losses per unit length at peaks of the effective potential, while a particle oscillating in the considered system moves from regions of high potential to regions of minimum potential. Hence, 1 GeV electron in the field of 10^{20} W/cm^2 intensity looses half of its energy at 0,3 mm length, while at the same length in the 10^{22} W/cm^2 intensity field 1 TeV muon looses only ~ 1 % of its energy.

Conclusions

The dynamics of classical charged particle in the field of laser channels including energy losses is considered. It is shown that the critical angle for muon is less than for electron as expected. The analytical expression for the stopping force influence on the charged particle in the channels of two plane polarized laser waves is obtained. A particular case of charged particle dynamics in a standing wave field is considered. The expression for the particle energy losses per length is derived. According to the equation (21) the electron energy losses per length unit are immense, while for muons (22) those are 5 orders less than for electron, but still essential. The use of the channels formed by lasers may become an effective mechanism for cooling and shaping ultrarelativistic charged particle beams.

References

[1] Kapitza P L, Dirac P A M 1933 *Math. Proc. Cambridge* **29** 297
[2] Fedorov M V, Oganesyan K B, Prokhorov A M 1988 *Appl. Phys. Lett.* **53** 353
[3] Balcou Ph 2010 *Eur. Phys. J.* **59** 525
[4] Andriyash I A, Balcou Ph, Tikhonchuk V T 2011 *Eur. Phys. J.* **65** 533
[5] Andriyash I A, d'Humi'eres E, Tikhonchuk V T, Balcou Ph 2012 *Phys. Rev. Lett.* B **109** 244802
[6] Andriyash I A, d'Humi'eres E, Tikhonchuk V T, Balcou Ph 2013 *J. Phys.: Conf. Ser.* B **414** 012008
[7] Andreev A V, Akhmanov S A 1991 *Zh. Eksp. Teor. Fiz., Pis'ma Red.* **53** 18
[8] Frolov E N and Dik A V, Dabagov S B 2015 *Nucl. Instr. Meth. Phys. Res.* B **355** 206
[9] Kaplan A E, Pokrovsky A L 2005 *Phys. Rev. Lett.* **95** 053601
[10] Pokrovsky A L, Kaplan A E 2005 *Phys. Rev.* A **72** 043401
[11] Smorenburg P W, Kanters J H M, Lassise A 2011 *Phys. Rev.* A **83** 063810
[12] Frolov E N and Dik A V, Dabagov S B, Artyomov K P 2014 *J. Phys.: Conf. Ser.* **552** 012035
[13] Lindhard J 1965 *K. Dan. Vidensk. Selsk. Mat.-Fys. Medd.* **34(14)** 1
[14] Dabagov S B, Dik A V, Frolov E N 2015 *Phys. Rev. ST Accel. Beams* **18** 064002
[15] Landau L D, Lifshitz E M 1988 *Field Theory* Moscow, Nauka 512

Influence of an Electric Field on the Propagation of a Photon in a Magnetic field

V M Katkov

Budker Institute of Nuclear Physics, Akademika Lavrentieva Ave. 11, Novosibirsk, 630090, Russian Federation

E-mail: katkov@inp.nsk.su

Abstract. In this work, a constant and uniform magnetic field is less than the Schwinger critical value. In turn, an additional constant and uniform electric field is taken much smaller than the magnetic field value. The propagation of a photon in this electromagnetic field is investigating. In particular, in the presence of a weak electric field, the root divergence is absent in the photon effective mass near the thresholds of pair creation. The effective mass of a real photon with a preset polarization is considered in the quantum energy region as well as in the quasiclassical one.

1. Introduction

The photon propagation in electromagnetic fields and the dispersive properties of the space region with magnetic fields is of very much interest. This propagation is accompanied by the photon conversion into a pair of charged particles when the transverse photon momentum is larger than the process threshold value $k_{\perp} > 2m$ (the system of units $\hbar = c = 1$ is used). In 1971 Adler [1] had calculated the photon polarization operator in a magnetic field using the proper-time technique developed by Schwinger [2] and Batalin and Shabad [3] had calculated this operator in an electromagnetic field using the Green function found by Schwinger [2]. In 1971 Batalin and Shabad [3] had calculated polarization operator in an electromagnetic field using the Green function found by Schwinger [2]. In 1975 the contribution of charged-particles loop in an electromagnetic field with n external photon lines had been calculated in [4]. For $n = 2$ the explicit expressions for the contribution of scalar and spinor particles to the polarization operator of photon were given in [4]. For the contribution of spinor particles obtained expressions coincide with the result of [3], but another form is used.

In this work, a constant and uniform magnetic field is less than the Schwinger critical value $H_0 = m^2/e = 4.41 \cdot 10^{13}$ G. In turn, an additional constant and uniform electric field is taken much smaller than the magnetic field value. In these fields, we consider the polarization operator on mass shell ($k^2 = 0$, the metric $ab = a^0b^0 - \mathbf{ab}$ is used) at arbitrary value of the photon energy ω.

2. General expressions

Our analysis is based on the general expression for the contribution of spinor particles to the polarization operator obtained in a diagonal form in [4] (see equations (3.19) and (3.33)). The

Content from this work may be used under the terms of the Creative Commons Attribution 3.0 licence. Any further distribution of this work must maintain attribution to the author(s) and the title of the work, journal citation and DOI.
Published under licence by IOP Publishing Ltd

eigenvalue κ_i of this operator on the mass shell ($k^2 = 0$) determines the effective mass of the real photon with the polarization e_i directed along the corresponding eigenvector:

$$\Pi^{\mu\nu} = -\sum_{i=2,3} \kappa_i \beta_i^\mu \beta_i^\nu, \qquad \beta_i\beta_j = -\delta_{ij}, \qquad \beta_i k = 0; \tag{1}$$

$$e_i^\mu = \frac{b_i^\mu}{\sqrt{-b_i^2}}, \qquad b_2^\mu = (Bk)^\mu + \frac{2\Omega_4}{\Omega}(Ck)^\mu, \tag{2}$$

$$b_3^\mu = (Ck)^\mu - \frac{2\Omega_4}{\Omega}(Bk)^\mu;$$

$$\kappa_2 = r\left(\Omega_2 - \frac{2\Omega_4^2}{\Omega}\right), \qquad \kappa_3 = r\left(\Omega_3 + \frac{2\Omega_4^2}{\Omega}\right), \tag{3}$$

$$\Omega = \Omega_3 - \Omega_2 + \sqrt{(\Omega_3 - \Omega_2)^2 + 4\Omega_4^2}, \qquad r = \frac{\omega^2 - k_3^2}{4m^2}.$$

The consideration realizes in the frame where electric $\mathbf{E}$ and magnetic $\mathbf{H}$ fields are parallel and directed along the axis 3. In this frame the tensor of electromagnetic field $F_{\mu\nu}$ and tensors $F^*_{\mu\nu}$, $B_{\mu\nu}$ and $C_{\mu\nu}$ have a form

$$\begin{aligned} &F_{\mu\nu} = C_{\mu\nu}E + B_{\mu\nu}H, \qquad F^*_{\mu\nu} = C_{\mu\nu}H - B_{\mu\nu}E, \qquad C_{\mu\nu} = g^0_\mu g^3_\nu - g^3_\mu g^0_\nu, \\ &B_{\mu\nu} = g^2_\mu g^1_\nu - g^1_\mu g^2_\nu; \qquad eE/m^2 = E/E_0 \equiv \nu, \qquad eH/m^2 = H/H_0 \equiv \mu; \end{aligned} \tag{4}$$

$$\Omega_i = \frac{\alpha m^2}{\pi}\mu \int_{-1}^{1} \mathrm{d}v \int_0^\infty f_i(v,x) \exp\left(\mathrm{i}\psi(v,x)\right) \mathrm{d}x. \tag{5}$$

Here

$$\begin{aligned} \frac{1}{\nu x} f_1 &= \frac{\cos(\mu x v)\cosh(\nu x v)}{\sin(\mu x)\sinh(\nu x)} - \frac{\cos(\mu x)\cosh(\nu x)\sin(\mu x v)\sinh(\nu x v)}{\sin^2(\mu x)\sinh^2(\nu x)}, \\ \frac{1}{\nu x} f_2 &= 2\frac{\cosh(\nu x)(\cos(\mu x) - \cos(\mu x v))}{\sinh(\nu x)\sin^3(\mu x)} + f_1, \\ \frac{1}{\nu x} f_3 &= 2\frac{\cos(\mu x)(\cosh(\nu x) - \cosh(\nu x v))}{\sin(\mu x)\sinh^3(\nu x)} - f_1, \\ \frac{1}{\nu x} f_4 &= \frac{\cos(\mu x)\cos(\mu x v) - 1}{\sin^2(\mu x)} \frac{\cosh(\nu x)\cosh(\nu x v) - 1}{\sinh^2(\nu x)} + \frac{\sin(\mu x v)\sinh(\nu x v)}{\sin(\mu x)\sinh(\nu x)}; \end{aligned} \tag{6}$$

$$\psi(v,x) = 2r\left(\frac{\cosh(\nu x) - \cosh(\nu x v)}{\nu\sinh(\nu x)} + \frac{\cos(\mu x) - \cos(\mu x v)}{\mu\sin(\mu x)}\right) - x. \tag{7}$$

Let us note that the integration contour in the equation (5) is passing slightly below the real axis.

After all calculations have been fulfilled we can return to a covariant form of the process description using the following expressions

$$\begin{aligned} &E^2, H^2 = \left(\mathcal{F}^2 + \mathcal{G}^2\right)^{1/2} \pm \mathcal{F}, \qquad \mathcal{F} = \left(\mathbf{E}^2 - \mathbf{H}^2\right)/2, \qquad \mathcal{G} = \mathbf{EH}, \\ &\left(C^2\right)_{\mu\nu} = \left(F^2_{\mu\nu} + H^2 g_{\mu\nu}\right) / \left(E^2 + H^2\right), \qquad \left(C^2\right)_{\mu\nu} - \left(B^2\right)_{\mu\nu} = g_{\mu\nu}. \end{aligned} \tag{8}$$

The real part of κ_i determines the refractive index n_i of photon with the polarization $e_i^\mu = \beta_i^\mu$, $i = 1, 2$:

$$\kappa_i = (m_i^{\mathrm{ef}})^2, \qquad n_i = 1 - \frac{\mathrm{Re}\,\kappa_i}{2\omega^2}. \tag{9}$$

At $r > 1$, the proper value of polarization operator κ_i includes the imaginary part, which determines the probability per unit length of pair production:

$$W_i = -\frac{1}{\omega}\mathrm{Im}\,\kappa_i. \tag{10}$$

Let us not that in considered case, we have $\mu \ll 1$, $\nu \ll \mu$. The high energy region, $r \gtrsim 1/\mu^2$, is contained in the region of the standard quasiclassical approximation (SQA) [5]. In SQA, the effective mass of a photon depends on the parameter κ only: $\kappa^2 = 4r(\mu^2+\nu^2) = -(Fk)^2/m^2H_0^2$. For $\kappa \gtrsim 1$ the influence of weak electric field on the polarization operator is small. Because of this we consider now the case of energies $r \ll 1/\mu^2$.

3. Region of intermediate photon energies

Choose a point x_0 in the following way: $\varphi'(x_0) = 0$, were

$$\varphi\left(x\right) = -\psi(0,x) = 2r\left(\frac{1}{\mu}\tan\frac{\mu x}{2} - \frac{1}{\nu}\tanh\frac{\nu x}{2}\right) + x. \tag{11}$$

Then, we have the following equation for x_0:

$$\tan^2\frac{\nu s}{2} + \tanh^2\frac{\mu s}{2} = \frac{1}{r}, \qquad x_0 = -\mathrm{i}s. \tag{12}$$

We represent the integral for Ω_i as

$$\Omega_i = \frac{\alpha m^2}{\pi}(a_i + b_i), \tag{13}$$

where

$$a_i = \mu\int\limits_{-1}^{1}\mathrm{d}v\int\limits_{0}^{x_0} f_i(v,x)\exp\left(\mathrm{i}\psi(v,x)\right)\mathrm{d}x, \tag{14}$$

$$b_i = \mu\int\limits_{-1}^{1}\mathrm{d}v\int\limits_{x_0}^{\infty} f_i(v,x)\exp\left(\mathrm{i}\psi(v,x)\right)\mathrm{d}x. \tag{15}$$

In the integral a_i in the equation (13), the small values $x \sim 1$ contribute. We calculate this integral expanding the entering functions over x, take into account that in the region under consideration, the condition $r\mu^2 \ll 1$ is fulfilled. Then in exponent, we keep the term $-x$ only and extend the integration over x to infinity. In the result of not complicated integration over v, we have:

$$a_2 = -\frac{16}{45}\left(\mu^2+\nu^2\right), \qquad a_3 = -\frac{28}{45}\left(\mu^2+\nu^2\right),$$
$$\kappa_2^a = -\frac{4\alpha m^2\kappa^2}{45\pi}, \qquad \kappa_3^a = -\frac{7\alpha m^2\kappa^2}{45\pi}, \qquad \kappa^2 = -\frac{(Fk)^2}{H_0^2m^2}. \tag{16}$$

These asymptotics are well known.

In the integral b_i equation (15), small values v contribute. Expanding entering functions over v and extending the integration over v to infinity, we get

$$b_i = \mu\int\limits_{-\infty}^{\infty}\mathrm{d}v\int\limits_{x_0}^{\infty} f_i(0,x)\exp\left[-\mathrm{i}\left(\varphi\left(x\right) + v^2\chi\left(x\right)\right)\right]\mathrm{d}x, \tag{17}$$

$$\chi\left(x\right) = rx^2\left(\frac{\nu}{\sinh(\nu x)} - \frac{\mu}{\sin(\mu x)}\right). \tag{18}$$

After the integration over v, one has

$$b_i = \mu\sqrt{\pi}\exp\left(-\mathrm{i}\frac{\pi}{4}\right)\int\limits_{x_0}^{\infty}\frac{f_i(0,x)}{\sqrt{\chi(x)}}\exp\left(-\mathrm{i}\varphi(x)\right)\mathrm{d}x. \tag{19}$$

The case $\mu^{-2} \gg r \gg \mu^{-2/5}$ was considered in [6] (see an equation (31)). The result has the following invariant form:

$$\kappa_2 = \frac{\alpha m^2\kappa}{8}\sqrt{\frac{3}{2}}\exp\left(-\frac{8}{3\kappa}+\frac{64\widetilde{\mathcal{F}}}{15\kappa^3}\right), \qquad W_3 = 2W_2, \tag{20}$$

$$\kappa = 4r\sqrt{\mu^2+\,nu^2}, \qquad \widetilde{\mathcal{F}} = \frac{\nu^2-\mu^2}{2}. \tag{21}$$

Leaving the main terms of the expansion in the parameter $\xi = \nu/\mu \ll 1$, we have

$$\kappa_2\left(\xi\right)/\kappa_2\left(0\right) = \exp\left[\frac{2\Delta}{3\sqrt{r}}\left(1+\frac{1}{r}\right)\right]. \tag{22}$$

4. The quantum energy region near the lower thresholds

We consider now the energy region where $r - 1 \ll 1$ when the moving of created particles is nonrelativistic. In this case, the equation (12) and its solutions are given by the following approximate equations

$$\frac{\xi^2 l^2}{16} \simeq \exp\left(-l\right)+\frac{1-r}{4}, \qquad l = \mu s, \qquad \xi = \frac{\nu}{\mu}; \tag{23}$$

$$l \simeq 2\ln\frac{4}{\xi\ln\left(4/\xi\right)}, \qquad |r-1| \ll \xi^2 l^2; \tag{24}$$

$$l \simeq \ln\frac{4}{r-1}, \qquad r-1 \gg \xi^2 l^2; \tag{25}$$

Leaving the main terms of the expansion, we obtain

$$\mathrm{i}\varphi(x) \simeq \beta(r) - \gamma\mathrm{e}^{-\mathrm{i}z} - \mathrm{i}zq + \mathrm{i}\Delta x^3/12, \tag{26}$$

$$x = \left(z-\mathrm{i}l\right)/\mu, \qquad q = \left(r-1\right)/\mu, \qquad \Delta = \xi^2/\mu, \tag{27}$$

$$\beta(r) = 2r/\mu - ql(r), \qquad \gamma = q + \Delta l^2/4, \tag{28}$$

$$\chi(x) \simeq x, \qquad f_{1,2,4}(0,x) \simeq 0, \qquad f_3(0,x) \simeq -\mathrm{i}. \tag{29}$$

In the region of lower thresholds, where the particles occupy not very high energy levels, we present the equation (19) for b_3 in the form

$$b_3 = -\mathrm{i}\sqrt{\pi\mu}\exp\left(-\mathrm{i}\frac{\pi}{4}\right)\exp\left(-\beta(r)\right)\int\limits_0^{\infty}\frac{\mathrm{d}z}{\sqrt{z-\mathrm{i}l}}\sum_{k=0}^{\infty}\frac{\gamma^k}{k!}\exp\left\{\mathrm{i}\left[\left(q-k\right)z-\frac{\Delta\left(z-\mathrm{i}l\right)^3}{12}\right]\right\}. \tag{30}$$

If $|\delta| \ll 1, \delta = q - n, \Delta \ll 1, \gamma \simeq n$, the large z contributes and we have after the change of variables

$$b_3 \simeq -\mathrm{i}\exp\left(-\mathrm{i}\frac{\pi}{4}\right)\sqrt{\mu\pi}\exp\left(-\beta(r)\right)\frac{n^n}{n!}\int\limits_0^{\infty}\frac{\mathrm{d}z}{\sqrt{z}}\exp\left[\mathrm{i}\left(\delta z-\Delta z^3/12\right)\right]. \tag{31}$$

After integration over z, we have the following approximate expressions

$$b_3 = 2\sqrt{\mu\pi}\frac{n^n}{n!}\exp\left(-\beta(r)\right) d\left(\delta, \Delta\right), \tag{32}$$

were

$$d\left(\delta, \Delta\right) = -\exp\left(\mathrm{i}\frac{\pi}{2}\vartheta(\delta)\right)\sqrt{\frac{\pi}{\delta}}, \qquad \Delta \ll |\delta|^3; \tag{33}$$

$$d\left(\delta, \Delta\right) = -\frac{1}{6}\exp\left(\mathrm{i}\frac{\pi}{6}\right)\Gamma\left(\frac{1}{6}\right)\left(\frac{12}{\Delta}\right)^{1/6}, \qquad |\delta| \ll \Delta^{1/3}, \tag{34}$$

where $\vartheta(z)$ is Heaviside function: $\vartheta(z) = 1$ for $z \geqslant 0$, $\vartheta(z) = 0$ for $z < 0$. The expression for κ_3 with the accepted accuracy can be rewritten in the following form:

$$\kappa_3^b \simeq -\alpha m^2\sqrt{\frac{\mu}{|\delta|}}\exp\left(\mathrm{i}\frac{\pi}{2}\vartheta(\delta)\right)\mathrm{e}^{-\zeta}\frac{(2\zeta)^n}{n!}, \qquad \Delta \ll |\delta|^3, \qquad \zeta = 2r/\mu. \tag{35}$$

$$\kappa_3^b \simeq -\frac{\sqrt{3}+\mathrm{i}}{12}\alpha m^2\sqrt{\frac{\mu}{\pi}}\Gamma\left(\frac{1}{6}\right)\left(\frac{12}{\Delta}\right)^{1/6}\mathrm{e}^{-\zeta}\frac{(2\zeta)^n}{n!}, \qquad |\delta| \ll \Delta^{1/3}, \tag{36}$$

In considered quantum case $n \sim 1$ the real part of κ_3^b is exponentially small compared to κ_3^a (see the equation (16)), but the imaginary part of the effective mass is given by the equation (35).

At $\Delta \gtrsim 1$, we have $\gamma \gg 1$, and the small z contributes to the integral in the equation (19). In this case, it is possible to use the method of stationary phase

$$\kappa_3^b \simeq -\mathrm{i}\alpha m^2\sqrt{\frac{\mu}{2\left(q+\Delta l^2/4\right)l}}\exp\left(-\beta(r)+\gamma+\Delta l^3/12\right). \tag{37}$$

5. Low energy region $\mathbf{r} - 1 < 0$

The case $1 - r \ll \nu^{2/3}$ was discussed above (see the equation (36)) where we have to put $n = 0$. For the case $1 \gg 1 - r \gg \nu^{2/3}$, the following expression

$$l \simeq \frac{2}{\xi}\sqrt{1-r}, \tag{38}$$

is valid and one can use the method of stationary phase. Then we have

$$\kappa_3^b \simeq -\mathrm{i}\frac{\alpha m^2\mu}{\sqrt{1-r}}\exp\left[-\frac{2r}{\mu}-\frac{4\left(1-r\right)^{3/2}}{3\nu}\right]. \tag{39}$$

This method is valid for lower photon energy when $\nu^{-2} \gg 1/r - 1 \gg \nu^{2/3}$. Then the approximate solution of the equation (12) has form

$$s \simeq \frac{2}{\nu}\arctan\sqrt{\frac{1}{r}-1}, \qquad \sqrt{-\frac{\chi\left(x_0\right)\varphi''\left(x_0\right)}{2}} \simeq \frac{\nu s}{2},$$
$$f_3\left(x_0\right) \simeq \mathrm{i}\frac{\nu s}{r\sqrt{r\left(1-r\right)}}, \qquad f_{1,2,4}(0,x) \simeq 0, \tag{40}$$

$$\kappa_3^b \simeq -\mathrm{i}\frac{\alpha m^2\mu}{\sqrt{r(1-r)}}\exp\left(-\frac{2r}{\mu}-\frac{2}{\nu}\arctan\sqrt{\frac{1}{r}-1}+\frac{2r}{\nu}\sqrt{\frac{1}{r}-1}\right). \tag{41}$$

For $1 - r \ll 1$, expanding the incoming in the equation (41) functions we have the equation (39). When the value r small enough $\nu^3 \ll r^{3/2} \ll \nu$, leaving the leading term of decomposition over r, we get

$$\kappa_3^b \simeq -\mathrm{i}\frac{\alpha m^2 \mu}{\sqrt{r}} \exp\left(-\frac{2r}{\mu} - \frac{\pi}{\nu} + \frac{4\sqrt{r}}{\nu}\right). \quad (42)$$

The imaginary part of this expression coincides with the equation (40) [6]. The ratio of this expression to the corresponding formula in a pure electric field has the form

$$\kappa_3^b/\kappa_3^b\,(\mu = 0) = \frac{\pi}{\xi} \exp\left[r\left(\frac{\pi}{\nu} - \frac{2}{\mu}\right)\right]. \quad (43)$$

The equations (42) and (43) are valid for the photon energy $r \gg \nu^2$. For the region of the photon energy $r \lesssim \nu^2$, we can use the results of [6].

6. Conclusion

We have considered the polarization operator of a photon in a constant magnetic fields in the presence of a weak electric field. The effective mass of a photon was calculated using three different overlapping approximation. In the region of SQA applicability, the created by a photon particles have ultrarelativistic energies. The role of fields in this case is to transfer the required transverse momentum and the electric field actions less than that of the magnetic field. At lower energies, the role of the electric field increases. It is necessary to note a special significance of a weak electric field $E = \xi H$ ($\xi \ll 1$) in the removal of the root divergence of the probability when the particles of pair are created on the Landau levels with the electron and positron momentum $p_3 = 0$ [5]. The frame is used where $k_3 = 0$.

Generally speaking, at $\xi \ll 1$ the formation time t_c of the process under consideration is $1/\mu$. Here we use units $\hbar = c = m = 1$. At this time the particles of creating pair gets the momentum $\delta p_3 \sim \xi$. If the value ξ^2 becomes larger than the distance apart Landau levels μ ($\Delta = \xi^2/\mu \gg 1$) all levels overlapped. Under this condition the divergence of the probability is vanished and the method of stationary phase is valid even in the energy region $r - 1 \lesssim \mu$, whereas that is inapplicable in the absence of electric field [5]. In the opposite case $\nu^2 \ll \mu^3$ for the small value of $p_3 \ll \sqrt{\mu}$, in the region where the influence of electric field is negligible, the formation time of the process t_f is $1/p_3^2$ and $\delta p_3 \sim \nu/p_3^2 \ll p_3$. It is follows from above that $\nu^{1/3} \ll p_3 \ll \sqrt{\mu}$. At this condition the value of discontinuity is $\sqrt{t_f/t_c} \sim \sqrt{\mu}/p_3$. For $\nu^{1/3} \gg p_3$ the time t_f is determined by the self-consistent equation $\delta\varepsilon^2 \sim 1/t_f \sim \nu^2 t_f^2$, $t_f \sim \nu^{-2/3}$ and the value of discontinuity becomes $\sqrt{\mu t_f} \sim (\mu^3/\nu^2)^{1/6}$ instead of $\sqrt{\mu}/p_3$.

In the region $\omega \lesssim 2m$ ($r \lesssim 1$) the energy transfer from electric field to the created particles becomes appreciable and for $\omega \ll m$ it determines the probability of the process mainly. At $\omega \ll eE/m$ ($\sqrt{r} \ll \nu$), the photon assistance in the pair creation comes to the end and the probability under consideration defines the probability of photon absorption by the particles created by electromagnetic fields. The influence of a magnetic field on the process is connected with the interaction of the magnetic moment of the created particles and magnetic field. This interaction, in particular, has appeared in the distinction of the pair creation probability by field for scalar and spinor particles [2].

Acknowledgments
The work was supported by the Ministry of Education and Science of the Russian Federation. The author is grateful to the Russian Foundation for Basic Research grant No. 15-02-02674 for partial support of the research.

References

[1] Adler S L 1971 *Ann. Phys.* **67** 599
[2] Schwinger J 1951 *Phys. Rev.* **82** 664
[3] Batalin I A and Shabad A E 1971 *Sov. Phys. - JETP* **33** 483
[4] Baier V N, Katkov V M and Strakhovenko V M 1975 *Sov. Phys. - JETP* **41** 198
[5] Baier V N and Katkov V M 2007 *Phys. Rev.* D **75** 073009
[6] Katkov V M 2012 *Sov. Phys. - JETP* **114** 226

Scattering of electromagnetic wave by dielectric cylinder in eikonal approximation

V V Syshchenko

Belgorod National Research University, Pobedy Street, 85, Belgorod 308015, Russian Federation

E-mail: syshch@bsu.edu.ru, syshch@yandex.ru

Abstract. The scattering of the plane electromagnetic wave on a spatially extended, fiber-lake target is considered. The formula for the scattering cross section is obtained using the approximation analogous to eikonal one in quantum mechanics.

1. Introduction

The problem of multiple scattering of the electromagnetic wave under oblique incidence on the system of parallel dielectric fibers was considered in [1]. The scattering on the single fiber had been described there in the limit of infinitely thin fiber. The wave scattering cross section in that case possesses the axial symmetry for small angle of incidence $\psi \ll 1$.

The scattering cross section by the cylindric fiber of finite radius is calculated in the present article using eikonal approximation. The possibility of substantial axial anisotropy of the scattered radiation is demonstrated.

2. APPROXIMATIONS IN RADIATION SCATTERING THEORY

The equations for the electric field of a monochromatic wave $\mathbf{E}(\mathbf{r})e^{-i\omega t}$ in non-uniform medium

$$\left(\Delta + \frac{\omega^2}{c^2}\,\varepsilon\right)\mathbf{E} = \mathrm{grad}\,\mathrm{div}\,\mathbf{E}, \tag{1}$$

$$\mathrm{div}\,(\varepsilon\,\mathbf{E}) = 0 \tag{2}$$

(where Δ is the Laplasian operator and $\varepsilon(\mathbf{r})$ is the dielectric permittivity of the medium for the given frequency ω) could be easily derived from Maxwell equations (see, e.g. [2, §68]). The solution of equations (1), (2) for the case of a spatially localized target in vacuum is convenient to find as the superposition

$$\mathbf{E}(\mathbf{r}) = \mathbf{E}^{(0)}(\mathbf{r}) + \mathbf{E}^{(1)}(\mathbf{r}), \tag{3}$$

where $\mathbf{E}^{(0)}$ is the electric field of the incident wave that satisfies the equation

$$\left(\Delta + \frac{\omega^2}{c^2}\right)\mathbf{E}^{(0)} = 0. \tag{4}$$

Hence the field $\mathbf{E}^{(1)}$ in (3) has to be treated as the field of the scattered radiation.

Content from this work may be used under the terms of the Creative Commons Attribution 3.0 licence. Any further distribution of this work must maintain attribution to the author(s) and the title of the work, journal citation and DOI.
Published under licence by IOP Publishing Ltd

Equation (1) can be written using (2), (3) and (4) in the form

$$\left(\Delta+\frac{\omega^2}{c^2}\right)\mathbf{E}^{(1)}=\frac{\omega^2}{c^2}(1-\varepsilon)\mathbf{E}+\text{grad div}\,((1-\varepsilon)\mathbf{E}). \tag{5}$$

The last equation could be presented in the integral form,

$$\mathbf{E}^{(1)}(\mathbf{r})=\int G(\mathbf{r}-\mathbf{r}')\left\{\frac{\omega^2}{c^2}(1-\varepsilon(\mathbf{r}'))\mathbf{E}(\mathbf{r}')+\text{grad div}\,((1-\varepsilon(\mathbf{r}'))\mathbf{E}(\mathbf{r}'))\right\}d^3r', \tag{6}$$

where $G(\mathbf{r}-\mathbf{r}')$ is the Green function of equation (5),

$$G(\mathbf{r}-\mathbf{r}')=\int\frac{e^{i\boldsymbol{\kappa}(\mathbf{r}-\mathbf{r}')}}{(\omega/c)^2-\kappa^2+i0}\frac{d^3\kappa}{(2\pi)^3}. \tag{7}$$

The asymptotic of that function on large distances r from the domain where $\varepsilon(\mathbf{r})\neq 1$,

$$G(\mathbf{r}-\mathbf{r}')\big|_{r\to\infty}\to-\frac{1}{4\pi}\frac{e^{ik_f r}}{r}e^{-i\mathbf{k}_f\mathbf{r}'}, \tag{8}$$

where $\mathbf{k}_f=(\omega/c)\mathbf{r}/r$ is the wave vector of the scattered wave, is needed to find the field of the scattered radiation. Substituting the last formula into (6) and integrating the second term by parts, we obtain

$$\mathbf{E}^{(scattered)}(\mathbf{r})=\mathbf{E}^{(1)}(\mathbf{r})\Big|_{r\to\infty}=-\frac{1}{4\pi}\frac{e^{ik_f r}}{r}\left(\frac{\omega^2}{c^2}\mathbf{I}-\mathbf{k}_f(\mathbf{k}_f\cdot\mathbf{I})\right), \tag{9}$$

where

$$\mathbf{I}=\int(1-\varepsilon(\mathbf{r}))\,\mathbf{E}(\mathbf{r})\,e^{-i\mathbf{k}_f\mathbf{r}}d^3r. \tag{10}$$

We see that the integrand in (10) would be nonzero only in the domain where the dielectric permittivity is not equal to unit. The origin of the scattered radiation from the motion of the electrons in the medium excited by the incident wave becomes especially clear.

The integral form (6) of the field equation (5) that leads to (9), (10) is useful for construction of various approximate solutions. The simplest case is

$$|1-\varepsilon|\ll 1, \tag{11}$$

when the difference $(1-\varepsilon)$ in (10) could be treated as a small perturbation, hence the whole electric field in the target $\mathbf{E}(\mathbf{r})$ in (10) could be replaced by the field

$$\mathbf{E}^{(0)}=\mathbf{e}_i e^{i\mathbf{k}_i\mathbf{r}} \tag{12}$$

of the incident wave (where $\mathbf{k}_i$ is the incident wave vector, $|\mathbf{k}_i|=|\mathbf{k}_f|$, $\mathbf{e}_i$ is its polarization vector, and the wave amplitude has no matter for computation of the scattering cross section). This approximation is equivalent to Born approximation in the quantum theory of scattering (see, e.g., [3]). According to that analogy, the approximation is valid under

$$|1-\varepsilon|\left(\frac{a\omega}{c}\right)^2\ll 1 \tag{13}$$

(where a is the characteristic size of the spatial domain with $\varepsilon\neq 1$) for $a\omega/c\ll 1$ and under

$$|1-\varepsilon|\frac{a\omega}{c}\ll 1 \tag{14}$$

for $a\omega/c \gg 1$.

The situation when (11) is valid, but (14) is not, needs another approach. The analogy to the quantum theory of scattering suggests the eikonal approximation in this case. The eikonal approximation means that the solution of equations (1), (2) has to be found in the form (see also [4])

$$\mathbf{E}(\mathbf{r}) = e^{i\mathbf{k}_i\mathbf{r}}\mathbf{\Phi}(\mathbf{r}) = e^{i(\omega/c)z}\mathbf{\Phi}(\mathbf{r}), \tag{15}$$

where the direction of the wave incidence on the target $\mathbf{k}_i$ is chosen as the z axis direction and the function $\mathbf{\Phi}(\mathbf{r})$ is supposed to change itself in space slow enough to neglect its second derivations in (1). Hence the solution of equation (1) that coincides with the non-disturbed field of the incident wave (12) under $z \to -\infty$ has the form

$$\mathbf{E}(\mathbf{r}) = \mathbf{e}_i \exp\left\{ i\frac{\omega}{c}\left[z - \frac{1}{2}\int_{-\infty}^{z} (1-\varepsilon(\mathbf{r}))\, dz' \right] \right\}. \tag{16}$$

The value $\mathbf{I}$ (10) with the field (16) is equal to

$$\mathbf{I} = \mathbf{e}_i \int e^{-i\mathbf{k}_{f\perp}\boldsymbol{\rho}}\, d^2\rho \int_{-\infty}^{\infty} (1-\varepsilon(\boldsymbol{\rho},z)) \exp\left\{ i\left[\left(\frac{\omega}{c} - k_{f_z}\right) z - \frac{\omega}{2c}\int_{-\infty}^{z} (1-\varepsilon(\boldsymbol{\rho},z'))\, dz' \right] \right\} dz. \tag{17}$$

Calculating the average energy flux of the scattered wave (7) to the solid angle $d\Omega$ and dividing it by the average intensity of the incident wave we obtain the scattering cross section

$$\frac{d\sigma}{d\Omega} = \frac{\omega^2}{(4\pi)^2c^2}\left|\mathbf{k}_f \times \mathbf{I}\right|^2, \tag{18}$$

3. THE SCATTERING ON THE DIELECTRIC CYLINDER

The formulae (17) and (18) describe the radiation scattering by the target of arbitrary structure. Consider now the simplest case of uniform cylinder of the radius a and the length $L \to \infty$ as the target (figure 1), when the cylinder's axis makes the angle ψ with the direction of the wave incidence $\mathbf{k}_i$. The integrals in (17) could be calculated after rotation of the coordinate axes to coincidence of the new axis z' with the cylinder's axis; the result could be written in the form

$$\mathbf{I} = -i\mathbf{e}_i(1-\varepsilon)\frac{2\pi \sin\psi\ \delta(k_{i_\parallel} - k_{f_\parallel})}{\frac{\omega}{c} - k_{f_z} - \frac{\omega}{2c}(1-\varepsilon)} \times \tag{19}$$

$$\times \left\{ \int_{-a}^{a} e^{i\left[-k_{f_y}y + \left(\frac{\omega}{c}\varepsilon - k_{f_z}\right)\frac{\sqrt{a^2-y^2}}{\sin\psi}\right]} dy - \int_{-a}^{a} e^{i\left[-k_{f_y}y - \left(\frac{\omega}{c} - k_{f_z}\right)\frac{\sqrt{a^2-y^2}}{\sin\psi}\right]} dy \right\}.$$

The presence of the δ-function expresses the equality of the components of $\mathbf{k}_i$ and $\mathbf{k}_f$ that parallel to the cylinder's axis, that means the azimuthal character of the scattering: since the absolute values of the wave vectors $\mathbf{k}_i$ and $\mathbf{k}_f$ are also equal, the scattered radiation will be directed along the surface of the cone with the axis along the cylinder and the half-opening angle equal to the incidence angle ψ.

The azimuthal character of the scattering permits clear interpretation as a manifestation of Cherenkov mechanism. Indeed, the incident wave produces a perturbation in the medium that moves along the fiber with the phase velocity $v = \omega/(k_i)_\parallel = c/\cos\psi > c$. This superluminal motion generates the radiation analogous to Cherenkov one. The half-opening angle of the Cherenkov cone for this case $\theta = \arccos(c/v)$ is just equal to ψ (figure 2).

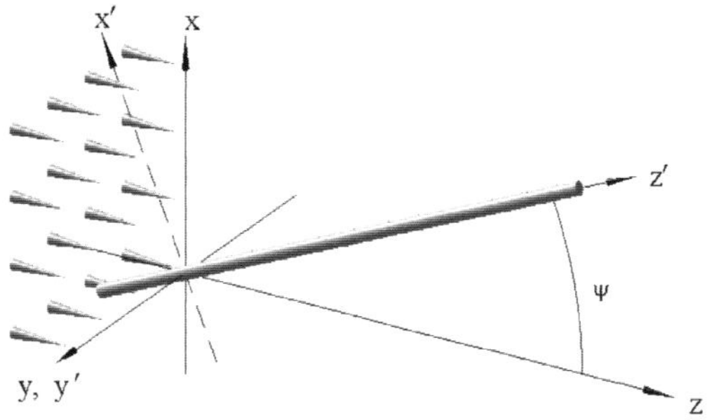

Figure 1. The radiation incidence along the z direction on the dielectric cylinder tilted by the angle ψ.

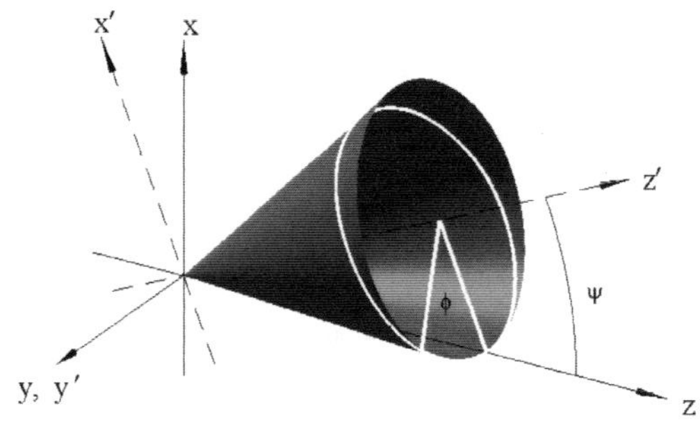

Figure 2. Direction diagram for the radiation scattered by the cylinder on figure 1. White curve represents the relative intensity of the scattered radiation in the limit of infinitely thin dielectric fiber (precise result [1]), the cone represents the azimuthally symmetric scattering. The angle of incidence is $\psi = 0.4$ radian.

Substitution of (19) into (18) gives (using the rule of δ-function squaring, $[\delta(k_{i_\parallel} - k_{f_\parallel})]^2 = \delta(k_{i_\parallel} - k_{f_\parallel}) \cdot L/2\pi$) the cross section (averaged over polarizations of the incident wave):

$$\frac{d\sigma}{d\Omega} = L\delta(k_{i_\parallel} - k_{f_\parallel})F(\phi), \tag{20}$$

where the function $F(\phi)$ describes the angular dependence of the cross section; this dependence is determined by the single parameter ϕ (the azimuthal angle in the plane perpendicular to the cylinder's axis) due to the azimuthal character of the scattering given by the delta-function in (20):

$$F(\phi) = \frac{1}{4\pi}\left(\frac{a\omega}{c}\right)^2 (1-\varepsilon)^2(1+\cos\theta)^2 \left(\frac{\sin\psi}{\frac{1+\varepsilon}{2} - \cos\theta}\right)^2 \times \tag{21}$$

$$\times \left\{ \left[\int_0^1 \cos\left(\frac{a\omega}{c} n_y \xi\right) \left[\cos\left(\frac{a\omega}{c}\frac{\varepsilon - \cos\theta}{\sin\psi}\sqrt{1-\xi^2}\right) - \cos\left(\frac{a\omega}{c}\frac{1-\cos\theta}{\sin\psi}\sqrt{1-\xi^2}\right)\right] dy \right]^2 + \right.$$

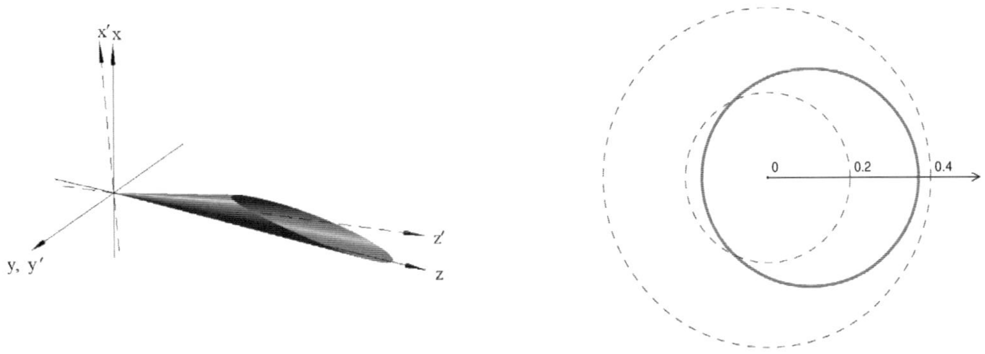

Figure 3. Plots of the function $F(\phi)$ (21) for $\varepsilon = 1.1$, $\psi = 0.1$, $a\omega/c = 50\sin\psi$ as the 3-dimensional direction diagram (left panel) and the polar plot (right panel; the polar axis corresponds to $\phi = 0$ that is the direction of the incident radiation).

$$+\frac{\pi^2}{4}\left[\frac{\varepsilon-\cos\theta}{\sqrt{n_y^2\sin^2\psi+(\varepsilon-\cos\theta)^2}}J_1\left(\frac{a\omega}{c\sin\psi}\sqrt{n_y^2\sin^2\psi+(\varepsilon-\cos\theta)^2}\right)-\right.$$

$$\left.-\frac{1-\cos\theta}{\sqrt{n_y^2\sin^2\psi+(1-\cos\theta)^2}}J_1\left(\frac{a\omega}{c\sin\psi}\sqrt{n_y^2\sin^2\psi+(1-\cos\theta)^2}\right)\right]^2\Bigg\}.$$

Indeed, both values in (21) that depend on the scattering direction, the angle θ between $\mathbf{k}_f$ and the z axis, and the projection n_y of the unit vector $(c/\omega)\mathbf{k}_f$ on the y axis, could be expressed via the azimuthal angle ϕ:

$$\cos\theta = (c/\omega)(k_f)_z = \cos\left[2\arcsin\left(\sin\psi\sin\frac{|\phi|}{2}\right)\right], \tag{22}$$

$$n_y = (c/\omega)(k_f)_y = \sin\psi\sin\phi, \tag{23}$$

where the angle ϕ is measured from the direction $(\mathbf{k}_i)_\perp$ (see figure 2).

The polar plots of the function $F(\phi)$ for some specific values of the parameters a, ω, ε and ψ are presented in figures 3 and 4. We see the substantial azimuthal asymmetry of the scattering cross section for large $a\omega/c$ (that is for large ratio of the cylinder's radius to the radiation wavelength). Similar behavior had been demonstrated earlier for the transition radiation of fast particles on the targets with cylindrical symmetry in Born [5] and eikonal [4] approximations.

4. Conclusion

The scattering of the electromagnetic wave under its oblique incidence on the linear extended target is considered. The formula for the scattering cross section is obtained using eikonal approximation. The azimuthal character of the scattering is demonstrated as well as the axial asymmetry around the target axis for the finite angle of incidence and the finite target thickness. The results could be used for improving the kinetic theory of propagation of the radiation through the system of parallel fibers [1].

Acknowledgments

This work was supported by the Russian Science Foundation Grant (project N 15-12-10019).

Figure 4. Plots of the function $F(\phi)$ (21) for $\varepsilon = 1.2$, $\psi = 0.1$, $a\omega/c = 500\sin\psi$ as the 3-dimensional direction diagram (left panel) and the polar plot (right panel; the polar axis corresponds to $\phi = 0$ that is the direction of the incident radiation).

References

[1] Shul'ga N F and Syshchenko V V 2014 *Problems of Atomic Science and Technjlogy* **93** (5) 115
[2] Landau L D and Lifshitz E M 1960 *Electrodynamics of Continuous Media* (Oxford: Pergamon Press)
[3] Landau L D and Lifshitz E M 1977 *Quantum Mechanics: Non-Relativistic Theory* (Oxford: Pergamon Press)
[4] Shul'ga N F and Syshchenko V V 2006 *Advanced Radiation Sources and Applications* ed H Wiedemann (Springer Netherlands) pp 129–148
[5] Shul'ga N F and Syshchenko V V 2003 *Phys. Lett.* A **313** 307

Coherent Light induced in Optical Fiber by a Charged Particle

Xavier Artru[1,2] **and Cédric Ray**[1,3]

[1] Université de Lyon, Université Lyon 1, F-69622 Villeurbanne, France
[2] Institut de Physique Nucléaire de Lyon, CNRS-IN2P3.
[3] Institut Lumière Matière, Université Lyon 1.

E-mail: `x.artru@ipnl.in2p3.fr`

Abstract. Coherent light production in an optical fiber by a charged particle (named PIGL, for *particle-induced guided light*) is reviewed. From the microscopic point of view, light is emitted by transient electric dipoles induced in the fiber medium by the Coulomb field of the particle. The phenomenon can also considered as the capture of virtual photons of the particle field by the fiber. Two types of captures are distinguished. Type-I takes place in a uniform part of the fiber; then the photon keeps its longitudinal momentum p_z. Type-II takes place near an end or in a non-uniform part of the fiber; then p_z is not conserved. Type-I PIGL is not affected by background lights external to the fiber. At grazing incidence it becomes nearly monochromatic. Its circular polarization depends on the angular momentum of the particle about the fiber and on the relative velocity between the particle and the guided wave. A general formula for the yield of Type-II radiation, based on the reciprocity theorem, is proposed. This radiation can be assisted by metallic objects stuck to the fiber, via plasmon excitation. A periodic structure leads to a *guided Smith-Purcell* radiation. Applications of PIGL in beam diagnostics are considered.

In memory of Karo Asatour Ispirian

1. Introduction

The field of a charged particle passing *through* or *near* an optical fiber induces a transient electric dipole moments of the atoms (see figure 1). Then these dipoles emit radiation coherently. Part of this radiation is channeled by the fiber [1, 2, 3, 4] and we call it *particle-induced guided light* (PIGL). It must not be confused with the incoherent radiation produced in scintillating fibers used as particle detectors. When the fiber radius is much larger than the wavelength, a description in terms of transition- or Cherenkov radiation is appropriate. This is the case for DIRC particle detectors [5, 6, 7]. Here we consider a bare fiber of radius a comparable to the wavelengths, therefore carrying one or a few number of modes. The number of produced photons per particle is too small for individual particle detection, but may be large enough for beam diagnostics. The fiber should have little effect on the beam emittance.

PIGL can also be considered as made of virtual photons from the particle field captured by the fiber. Two types of capture are distinguished.

- Type I : The particle passes *near* or *through* a uniform part of the fiber, far from an end. Then the capture conserves the longitudinal momentum of the photon, p_z.
- Type II : The particle passes near or through an end of the fiber or an added structure (e.g., an indentation or a metallic object touching or close to the fiber). Then p_z is not conserved.

Content from this work may be used under the terms of the Creative Commons Attribution 3.0 licence. Any further distribution of this work must maintain attribution to the author(s) and the title of the work, journal citation and DOI.
Published under licence by IOP Publishing Ltd

A first investigation of PIGL has been undertaken at Tomsk Polytechnic University [8]. The "fiber" was a dielectric rod of radius $a = 5$ mm and the wavelength domain was 3 - 30 mm. These lengths are about 10^3 times larger than in the optical domain, but the number of PIGL photons is invariant under the rescaling $a \to c\,a$, $\lambda \to c\,\lambda$. In the mm range a very large number of photons is needed to build a detectable signal. This was achieved using the coherent effect of all the electrons in mm size bunches. The set-up configuration shown in [8] is for type-II PIGL.

2. Spectrum of type-I PIGL

We consider a fiber of radius a and uniform index $n = \sqrt{\varepsilon}$ and denote by $\{\mathcal{M}, \omega\}$ the propagation modes. ω is the photon energy[1] and $\mathcal{M} = \{M, n_r, \tau, \sigma\}$ gathers the discrete variables: $M = L_z + S_z$ is the total angular momentum of the photon about the fiber axis $\hat{\mathbf{z}}$; n_r is the radial quantum number; $\sigma = \mathrm{sign}(p)$ indicates the direction of propagation, p being the longitudinal momentum of the photon. τ indicates the mode type: τ=EH for "transverse electric" $[E_z(a) < H_z(a)]$, τ=HE for "transverse magnetic" $[E_z(a) > H_z(a)]$. p and ω are linked by a dispersion relation:

$$\omega = \omega_{\mathcal{M}}(p) \;\leftrightarrow\; p = p_{\mathcal{M}}(\omega)\,. \tag{1}$$

The real electric field of the mode is[2] $\mathbf{E}_{\rm real} = 2\,\mathrm{Re}\{\vec{\mathcal{E}}_{\{\mathcal{M},\omega\}}(\mathbf{X},t)\,e^{i\eta}\}$, where

$$\vec{\mathcal{E}}_{\{\mathcal{M},\omega\}}(\mathbf{X},t) = \mathbf{E}_{\{\mathcal{M},\omega\}}(\mathbf{r})\,\exp(ipz - i\omega t) \tag{2}$$

(more precisely $\vec{\mathcal{A}} = -i\vec{\mathcal{E}}/\omega$) is the "photon wave function" and η is a *classical* phase determining the nodes of $\mathbf{E}_{\rm real}$. The transverse coordinate $\mathbf{r} = (x, y)$ is measured from the fiber axis. The dispersion relation and the analytic expressions of the proper fields in a uniform fiber can be found in textbooks [9, 10] or in [1, 3]. Figure 1 shows the phase velocity $v_{\rm ph}(\omega) = \omega/p$ and the group velocity $v_{\rm g}(\omega) = d\omega/dp$ of the lowest mode HE_{11} ($|M|$=1, n_r=1) for n=1.41 (fused silica). The photon spectrum produced in mode $\mathcal{M}$ by an electron of trajectory $\mathbf{X}(t)$ is given by

$$\frac{\omega\, d\mathcal{N}_{\mathcal{M}}}{d\omega} = \frac{1}{2\pi|P_{\mathcal{M}}(\omega)|}\left| e\int_{\rm traj.} d\mathbf{X}(t)\cdot\vec{\mathcal{E}}^{*}_{\{\mathcal{M},\omega\}}(\mathbf{X},t)\right|^2 , \tag{3}$$

where $P_{\mathcal{M}}(\omega)$ is the energy flow of the real field:

$$P_{\mathcal{M}}(\omega) = 2\int d^2\mathbf{r}\;\mathrm{Re}\left\{\mathbf{E}^{*}_{\{\mathcal{M},\omega\}}(\mathbf{r}) \times \mathbf{B}_{\{\mathcal{M},\omega\}}(\mathbf{r})\right\}_z . \tag{4}$$

[1] We use relativistic quantum units units: $\hbar = c = \varepsilon_0 = \mu_0 = 1$. $\lambda\!\!\!^{-} \equiv 1/\omega$. The Gauss law is written $\nabla\cdot\mathbf{E} = \rho$, not $4\pi\rho$. $e^2/(4\pi) = \alpha \simeq 1/137$.

[2] the factor 2 is a question of convention

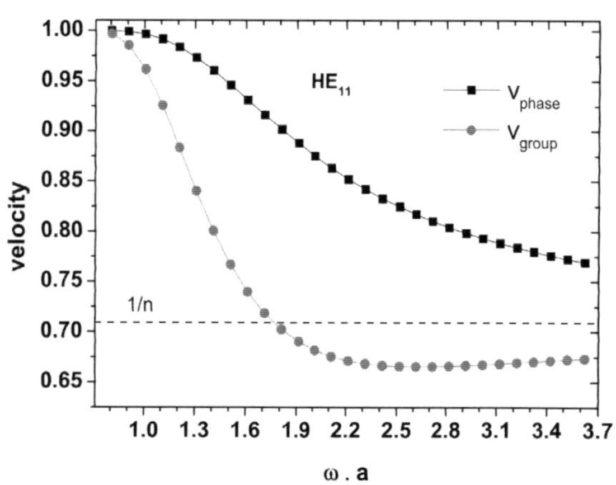

Figure 1. Phase velocity (squares) and group velocity (balls) of the mode HE_{11} as a function of ωa.

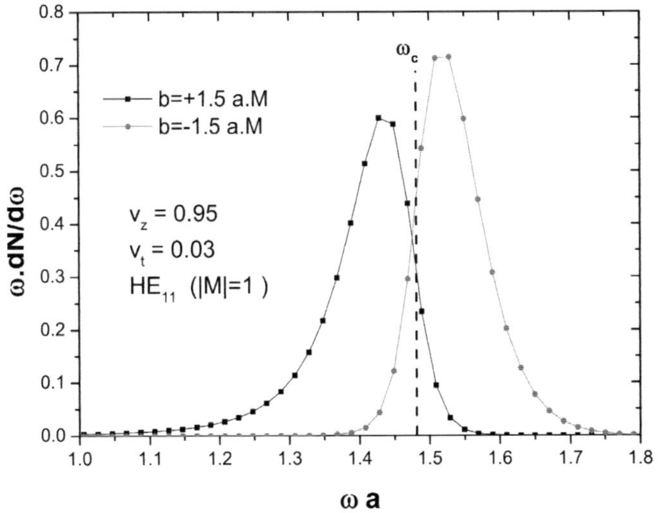

Figure 2. Photon spectra in the HE_{11}, $M=\pm 1$ modes, for a particle not touching the fiber: $|b| = 1.5\,a$. Squares: bM positive; balls: bM negative. The vertical line is at ω_C defined by Eq.(6). The fiber index is n=1.41.

Some spectra for the HE_{11} mod are presented in Refs.[1, 3] (with an error by a factor 2 in [1]). Figure 2 shows a HE_{11} spectrum for the "passing near" case ($b > a$) and the two helicities. The geometrical parameters are defined in figure 3.

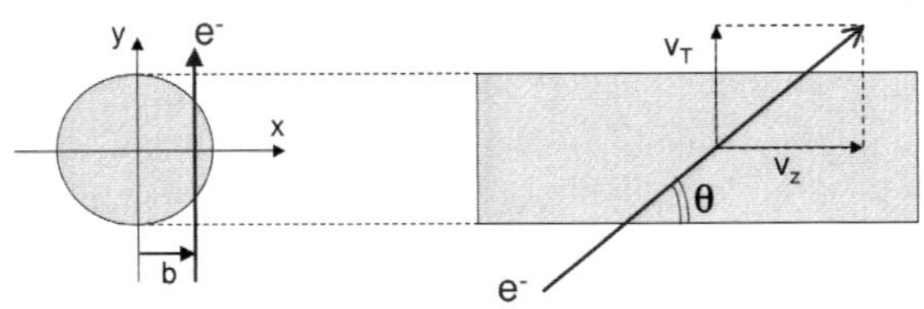

Figure 3. Geometrical parameters of the trajectory. Left side: front view; right side: side view.

Angular drag. In figure 2 one can observe a correlation between the angular momentum of the photon, M, and that of the electron, $J_e = \gamma m_e\,[\mathbf{b} \times \mathbf{v}] \cdot \hat{\mathbf{z}}$:

$$\langle M \rangle J_e \begin{cases} > 0 \text{ if } v_{ph} > v_z \quad (\omega < \omega_C), \\ < 0 \text{ if } v_{ph} < v_z \quad (\omega > \omega_C), \end{cases} \tag{5}$$

where ω_C is the frequency such that the electron "surfes" on the wave :

$$v_{ph}(\mathcal{M}, \omega_C) = v_z(particle)\,. \tag{6}$$

Strictly speeking, inequality (5) is obtained for $\mathbf{v}_T \ll v_z$ [4]. This "angular drag" effect can tell on which side of the fiber a microbeam is passing. An analogous - but opposite - correlation exists in helical undulator radiation, where $\langle M \rangle J_e$ is positive if v_z(photon) $> v_z$(electron), i.e., $\theta < 1/\gamma$, negative in the opposite case.

The small angle limit. If the crossing angle θ is small, the radiation in each mode becomes nearly monochromatic about a "Cherenkov peak" $\omega_C(v_z)$ given by (6). The number of photons under the peak is

$$\mathcal{N}_{\mathcal{M}} \simeq \frac{4\pi\alpha}{\omega_C\, P_{\mathcal{M}}(\omega_C)}\, |1/v_g(\omega_C) - 1/v_z|^{-1} \int_{\text{traj.}} dz(t)\, \left|\hat{\mathbf{z}} \cdot \mathbf{E}_{\{\mathcal{M},\omega_C\}}(\mathbf{r}(t))\right|^2 . \tag{7}$$

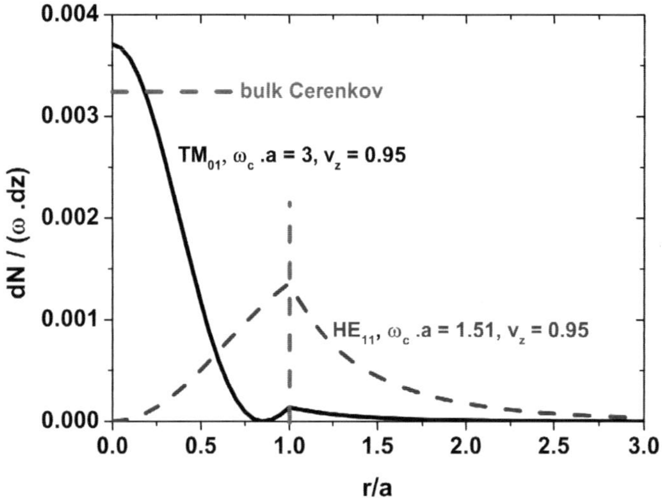

Figure 4. Rate of emitted photon along an electron trajectory at small angle with the fiber, as a function of r. The parameters are v_z=0.95 and $n_{\rm fiber}$=1.41. Peak frequencies: $\omega_{\rm C}$=1.51 a^{-1} for the $\rm HE_{11}$ mode, $\omega_{\rm C}$=3 a^{-1} for the $\rm TM_{01}$ ($\equiv$ $\rm EH_{01}$) mode. The line "bulk Cherenkov" is the spectrum of Cherenkov photon $d\mathcal{N}/d\omega/dz = \alpha\left[1-(nv)^{-2}\right]$ in infinite medium.

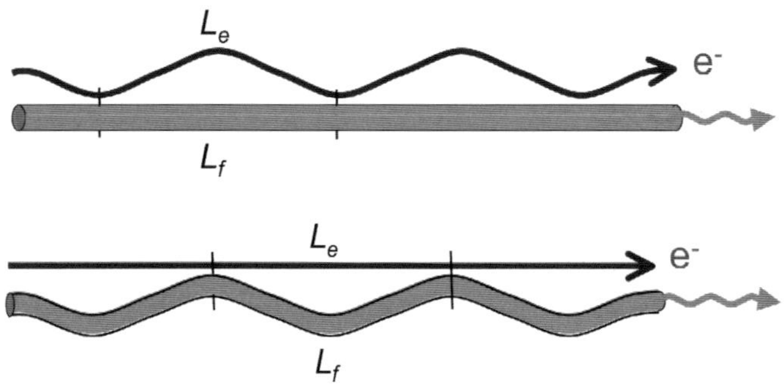

Figure 5. Two types of "fiber undulator".

This formula also applies when the fiber and/or the electron trajectory are weakly curved and at small angle, (x, y, z) being curvilinear coordinates following the fiber axis. Eq.(7) tells that the photon number increases linearly with the time spent by the particle in the region where the longitudinal field of the mode is important. Figure 4 plots $d\mathcal{N}_{\mathcal{M}}/(\omega_{\rm C}\, dz)$ from Eq.(7) as a function of r for the two lowest modes $\rm HE_{11}$ and $\rm TM_{01} \equiv EH_{01}$, at $v_{\rm ph} = v_z$=0.95. This quantity is the number of emitted photons per length $\lambda\!\!\!^{-} \equiv \omega_{\rm C}^{-1}$ of trajectory. $\mathcal{N}_{\mathcal{M}}$ behaves like $1/\theta$ and may exceed unity in spite of the $\alpha = 1/137$ factor. A rough estimation of the peak width is

$$\Delta\omega \sim \left|v_z/v_{\rm g}(\omega_{\rm C}) - 1\right|^{-1} v_{\rm T}/\max(a, |b|). \tag{8}$$

The exact peak position depends on the sign of $MJ_{\rm e}$, as shown in figure 2. The shift relative to $\omega_{\rm C}$ is comparable to $\Delta\omega$. Strict monochromaticity is obtained when the trajectory is parallel to the fiber. This kind of Cherenkov effect was studied in Refs. [11, 12].

Fiber undulator. If the fiber and/or the particle trajectory are undulated as in figure 5, the peak at $\omega \simeq \omega_{\rm C}$ is split in subpeaks such that

$$\Phi = \omega\ (L_{\rm f}/v_{\rm ph} - L_{\rm e}/v) = 2\pi \times \text{integer}\,, \tag{9}$$

$L_{\rm f}$ and $L_{\rm e}$ being the fiber length and particle path in one period. For a helical undulation a *Berry phase* [13] should be added to the third member of Eq.(9), but it is small for small bending angle.

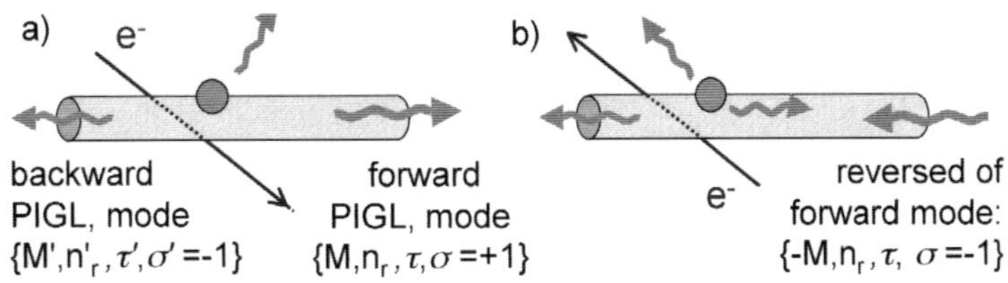

Figure 6. a) waves produced in a type-II PIGL experiment (here using a metallic ball). b) waves in the "inverse PIGL" process for the reciprocity theorem leading to Eq. (10).

3. Type-II PIGL

Figures 6a and 7 present different schemes of type-II photon capture. The PIGL yield may be obtained with a generalization of Eq.(3),

$$\frac{\omega\, d\mathcal{N}_{\mathcal{M}}}{d\omega} = \frac{1}{2\pi |P_{\mathcal{M}}(\omega)|} \left| e \int_{\mathrm{traj.}} d\mathbf{X}(-t) \cdot \vec{\mathcal{E}}_{\{\,\mathrm{rev}\mathcal{M},\omega\},\,\mathrm{in}}(\mathbf{X}(-t), t) \right|^2 , \tag{10}$$

suggested by the reciprocity theorem (or time reversal symmetry), like Eq.(2.33) of [14]. $\mathbf{X}(-t)$ is the time-reversed electron trajectory, $\mathrm{rev}\mathcal{M} = \{-M, n_r, \tau, -\sigma\}$ is the time-reversed mode and "in" refers to the *scattering* wave, solution of the macroscopic Maxwell equations *without the electron*, coming via the fiber *from* the detector direction. This wave contains a reflected part, a transmitted part and an outgoing unguided part indicated by the arrows in figure 6b.

For the schemes of Figs. 6a) and 7a,b,c) the Equivalent Photon method [15, 16] offers an alternative, more direct but less rigourous approach [3, 4]. The photon capture in scheme 7a) is more "adiabatic" than in scheme 7b), hence more efficient. Besides, its angular acceptance is narrower, therefore the background from ambient light is reduced. In schemes 6a), and 7c,d) the particle field excites first a plasmon resonance [17, 18] in the ball(s); then the plasmon

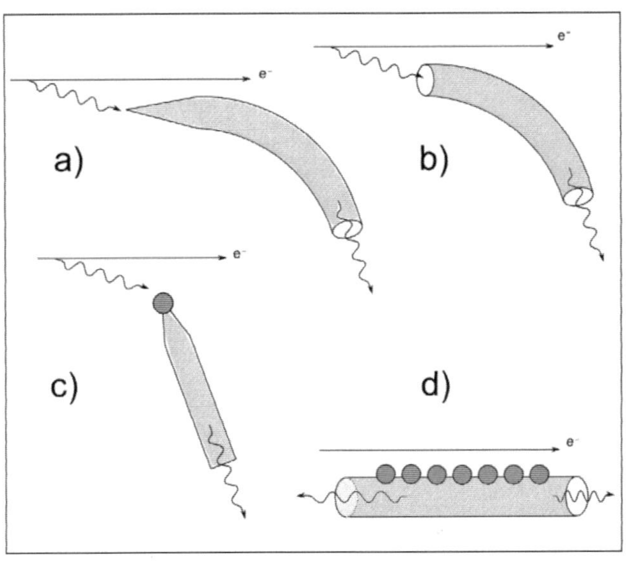

Figure 7. Different schemes of type-II PIGL: a) conical end; b) sharp-cut end; c) metallic ball at one end; d) regularly spaced metallic balls (figure taken from Ref.[3]).

is evacuated with a probability $\mathcal{P}_f$ as a photon in the fiber. It yields a quasi-monochromatic spectrum peaked at $\omega_{\rm plasmon} \simeq (4\pi\alpha\, n_c/3m_e)^{1/2}$ which is in the ultraviolet domain.

Guided Smith-Purcell light. In the scheme 7d) the interferences between balls produce peaks at ω such that

$$\Phi \equiv (1/v - 1/v_{\rm ph})\,\omega l = 2\pi \times \text{integer}\,, \tag{11}$$

l being the ball spacing and $\text{sign}(v_{\rm ph}) = +1$ and -1 for forward and backward PIGL respectively. The forward one has the highest frequency. To optimize the photon yield one must choose l such that an interference peak coincides with $\omega_{\rm plasmon}$. In the "kinematical" approach,

$$(d\mathcal{N}_{\mathcal{M}}/d\omega)_{N\,\rm balls} = (d\mathcal{N}_{\mathcal{M}}/d\omega)_{1\,\rm ball} \times \left[\sin^2(N\Phi/2)/\sin^2(\Phi/2)\right]. \tag{12}$$

Re-scattering and *shadowing* [19] by the balls should be taken into account using a "dynamical" theory like in [20]. This type of PIGL, which we call *Guided Smith-Purcell* radiation (GSP), is in competition with the external Smith-Purcell radiation (SP) from the balls, where $1/v_{\rm ph}$ in (11) is replaced by $\cos\theta_{\rm photon}$. If $(1/v + 1/|v_{\rm ph}|)\,\omega l = 2\pi/l$, only backward GSP and no ordinary SP is be emitted, since $|v_{\rm ph}| < 1$. This "exclusivity" reinforces the intensity of backward GSP.

4. Application to beam diagnostics

- Type-I PIGL at large crossing angle can be used for a transverse tomography of the beam with a resolution $\sim 2a$. No background is caused by real photons coming from upstream sources, e.g., synchrotron radiation from upstream bending magnets or transition radiation from an accelerator exit window. Due to p_z conservation, real photons are not captured but only *scattered* by the fiber. This is an advantage over Optical Transition Radiation and Optical Diffraction Radiation. For the same reason the resolution power of type-I PIGL is not affected by the large transverse tail ($\sim \gamma\lambda$) of the virtual photon cloud. Indeed, the virtual photons of this tail are "almost real".
- In microbeam diagnostics the "angular drag" of the circular polarization could tell on which side of the fiber the microbeam is passing.
- In a "fiber undulator" or in Guided Smith-Purcell effect, the width of the interference peaks and the peak/background ratio are sensitive respectively to the velocity and angular spreads of the beam.
- Type-II PIGL at a fiber end can provide directly a 2-dimensional transverse scan of the beam density. It can be assisted by a metallic ball of the type presented in [21, 22] or a SNOM probe. However, like OTR and ODR, it receives the background from upstream radiation sources and is sensitive to the transverse tail of the virtual photon cloud.
- PIGL of type I or II may be used to measure the time profile of the electron bunch. Note that in figure 1 the *group velocity* is almost constant for $\omega \geq 2/a$. Therefore in this frequency domain the signal shape is approximately conserved during the propagation to the detector.

5. Conclusion

We have reviewed the basic mechanism and some important aspects of light generation in monomode or few-mode optical fibers by charged particles (PIGL) : Cherenkov peak for grazing incidence, angular drag, absence of background in type-I, interference peaks for periodic structures or deformations, possibility of using a plasmon resonance. For the photon yield we have given a generalization of the type-I PIGL formula to the type-II case.

Whether or not PIGL can be successfully applied to beam diagnostics is too early to predict. Theoretical and experimental works must be pursued, in particular on the possibility to get a large enough signal, on the heat deposit in the fiber and on the radiation hardness. At least we can say that the narrowness of monomode fibers without cladding allows a minimal perturbation of the beam and the flexibility of fibers is certainly a great advantage over traditional optical systems.

If the fiber is bent to a radius below $R_{\rm cr} \sim \lambda\ (1-v_{\rm ph}^2)^{-3/2}$ light escapes from it by a tunneling process. The bending loss was investigated in [8]. In Ref. [23] a parallel is made between this effect and synchrotron radiation.

References

[1] Artru X and Ray C 2007, Photon production by charged particles in narrow optical fibers, *Proc. SPIE* **6634** ed Dabagov S B

[2] X. Artru and C. Ray 2008, Interference and shadow effects in the production of light by charged particles in optical fibers, *Nucl. Instr. Methods in Phys. Research* **266** 3725

[3] Artru X and Ray C 2012, Radiation Induced by Charged Particles in Optical Fibers, in *Selected Topics on Optical Fiber Technology* (InTech Open Access Company) ed Yasin M et al

[4] Artru X and Ray C 2013, Light induced by charged particles in optical fibers, *Nucl. Instr. Methods in Phys. Research* B **309** 162

[5] Coyle P et al 1994, *Nucl. Instr. Methods in Phys. Research* A **343** 292

[6] Contin A, De Salvo R, Gorodetzky P, Helleboid J M, Johnson K F, Juillot P, Lazic D and Lundin M 1994, *Nucl. Instr. Methods in Phys. Research* A **367** 271

[7] Janata E 2002, *Nucl. Instr. Methods in Phys. Research* A **493** 1

[8] Naumenko G, Potylitsyn A, Bleko V and Soboleva V 2015, Coherent radiation of relativistic electrons in dielectric fibers, *Nucl. Instr. Methods in Phys. Research* B **335** 125-128

[9] Okoshi T 1982, *Optical Fibers*, Academic Press

[10] Pocholle J P 1985, *L'Optique Guidée Monomode et ses Applications*, ed MASSON

[11] Bogdankevich L S and Bolotovskii B M 1957, *J. Exp. Theoret. Phys.* **32**, 1421 [*Sov. Phys. JETP* **5**, 1157]

[12] Zhevago N K and Glebov V I 1993, *Nucl. Instr. Methods* A **331** 592; 1997 *Zh. Exp. Teor. Fiz.* **111** 466

[13] Tomita A and Chiao R 1986, *Phys. Rev. Lett.* **57** 937

[14] Rullhusen P, Artru X and Dhez P 1998, *Novel Radiation Sources Using Relativistic Electrons*, World Scientific

[15] Jackson JD 1962, *Classical Electrodynamics*, John Wiley & Sons, Inc.

[16] Panofsky W K H and Phillips M 1962, *Classical Electricity and Magnetism*, Addison-Wesley, Inc.

[17] Zhevago N K 1991, *Europhys. Lett.* **15** 277

[18] Yamamoto N, Araya K and García de Abajo F J 2001, *Phys. Rev.* B **64** 205419

[19] Naumenko G, Artru X, Potylitsyn A, Popov Y, Sukhikh L and Shevelev M 2010, *J. Phys.: Conf. Series* **236** 012004

[20] García de Abajo F J 1999, *Phys. Rev. Lett.* **82** 2776; García de Abajo F J 2000, *Phys. Rev.* E **61** 5743

[21] Kalkbrenner T, Ramstein M, Mlynek J and Sandoghdar V 2001, *J. Microsc.* **202** 72

[22] Anger P, Bharadwaj P and Novotny L 2006, *Phys. Rev. Lett.* **96** 113002

[23] Artru X and Ray C 2015, Common characteristics of synchrotron radiation and light leaking from a bent optical fiber (these proceedings)

Coherent Cherenkov radiation as an intense THz source

V Bleko[1], P Karataev[2], A Konkov[3], K Kruchinin[2], G Naumenko[3], A Potylitsyn[3] and T Vaughan[2]

[1] RASA Center in Tomsk, Tomsk Polytechnic University, Lenin Ave. 30, Tomsk, 634050, Russian Federation
[2] John Adams Institute at Royal Holloway, University of London, Egham, TW20 0EX, UK
[3] Tomsk Polytechnic University, Lenin Ave. 30, Tomsk, 634050, Russian Federation

E-mail: bleko_vitold@mail.ru

Abstract. Diffraction and Cherenkov radiation of relativistic electrons from a dielectric target has been proposed as mechanism for production of intense terahertz (THz) radiation. The use of an extremely short high-energy electron beam of a 4th generation light source (X-ray free electron laser) appears to be very promising. A moderate power from the electron beam can be extracted and converted into THz radiation with nearly zero absorption losses. The initial experiment on THz observation will be performed at CLARA/VELA FEL test facility in the UK to demonstrate the principle to a wider community and to develop the radiator prototype. In this paper, we present our theoretical predictions (based on the approach of polarization currents), which provides the basis for interpreting the future experimental measurements. We will also present our hardware design and discuss a plan of the future experiment.

1. Introduction

The THz frequency of electromagnetic radiation always ranges from 10 – 0.1 THz and corresponds to the wavelength range from 30 μm to 3 mm [1]. Due to its ability to excite interatomic oscillations in many substances, THz radiation has numerous practical applications from spectroscopy of chemical and biological objects to fulfilling military or civilian security purposes [2–4].

However, there are difficulties in developing efficient methods for generating coherent THz radiation. Modern methods used in optical and infrared regions are not applicable in this case [5]. To generate the radiation in the frequency range form 0.1 – 1.6 THz, Gunn diodes, GaAs based Schottky Barrier Diodes and Backward Wave Oscillators (BWO) are widely used. They provide a CW radiation with continuous frequency adjustment [6] and power of order of several microwatts for diodes to a few watts for BWOs. All above-mentioned devices are low power and narrowband, which are the main disadvantages of them. The high frequency range is covered by new quantum cascade laser technology and gas lasers emitting radiation from 1.6 to 2.5 THz. Typical optical to THz conversion is very low (about 10% for a single stage [7]). The output power falls from 1 Watt at 1 THz to below 0.1 Watt at 3 THz with even newer distributed or travelling photomixer designs [8]. One should pay particular attention to Free Electron Lasers [9, 10], which, despite some of their obvious advantages, have a limited range of tasks because of their huge sizes.

Content from this work may be used under the terms of the Creative Commons Attribution 3.0 licence. Any further distribution of this work must maintain attribution to the author(s) and the title of the work, journal citation and DOI.
Published under licence by IOP Publishing Ltd

As is generally known, when a relativistic charged particle passes by any non-uniform medium with a distance comparable to or shorter than the parameter $\gamma\lambda/2\pi$ (here γ is the charged particle Lorentz-factor and λ is the radiation wavelength), its field polarizes the medium atoms [11]. The excited atoms start oscillating (representing currents changing in time) and become sources of EM radiation. Depending on the geometry of the medium, it can be treated as diffraction radiation (DR) from a metal surface [12] or Cherenkov radiation (ChR) from a transparent medium. If the radiation wavelength is comparable to or longer than the bunch length, all electrons emit radiation more or less in time stimulating each other's emission. As a result the photon yield is proportional to the square of the number of particles in the bunch; in the other words, the radiation is emitted coherently [13].

DR is emitted whenever a uniformly and rectilinearly moving charge passes in the vicinity of an optical inhomogeneity (target) [14] with impact parameter, h (being the shortest distance between the target and the electron beam), satisfying the following condition:

$$h \leq \frac{\gamma\lambda}{2\pi}. \tag{1}$$

Assuming 14 GeV beam energy of the LCLS accelerator [9] and 1 THz (0.3 mm) frequency any inhomogeneity (different vacuum pipe diameter, flanges, viewports, etc.) paced within 1.3 m radius away from the beam will radiate at THz frequencies. An optimal radiator might extract a moderate power from the beam and convert it into a THz radiation beam for utilization by THz user community.

ChR is generated when a charged particle speed is higher than the speed of light inside a medium [15]. Its direction is determined by the following relation:

$$\cos\theta_{\mathrm{Ch}} = \frac{v_{\mathrm{light}}}{v_{\mathrm{particle}}} = \frac{c/n}{c\beta} = \frac{1}{n\beta}. \tag{2}$$

However, the particle itself does not have to go through the substance. The energy loss due to ChR and DR is very small in comparison to the total power of the beam. Such generation geometry is very convenient. It excludes the ionization losses due to direct interaction with the target; and, therefore, it enables a parallel installation of several generating station for the same beam.

When generating ChR the particle also generates pure DR from all surfaces of the target. In [16–20] the authors have tried to unify the theory and merge different radiation types together. The method for the unified theory was called as polarization currents approach (PCA) [20]. It should be mentioned that the authors [21–24] have partially verified the model experimentally.

In this paper, we propose a scheme to generated intense THz radiation beams using the ChR mechanism at CLARA (Compact Linear Accelerator for Research and Application) facility in Daresbury Laboratory (UK) with intermediate energy of 50 MeV. The paper will review the theoretical background, overview the hardware status and experimental plans.

2. Theoretical background

In this paper we used polarization current method enabling us to calculate polarization radiation characteristics for conditions close to the experimental ones. The method is valid for description of polarization radiation generated by short charged particle bunches passing by a target with finite outer dimensions and known dielectric permittivity (see figure 1).

One should note that in such geometry ChR and DR will be generated simultaneously [21]. The DR intensity depends on the prism sides AC and CB (see figure 1), ChR is generated from the side AC. Changing the length of the side one may change the radiation intensity.

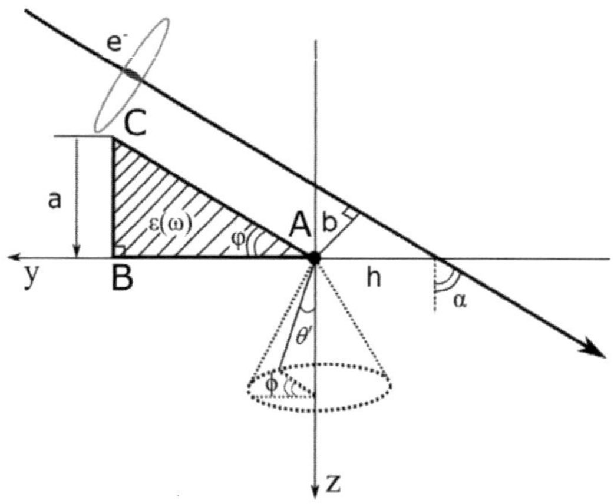

Figure 1. Scheme of polarization radiation generated by a charged particle moving rectilinearly in the vicinity of a dielectric prism.

To calculate the spectral-angular distribution of the ChR in vacuum we use the model derived in [19] (see the equation (16)):

$$\frac{d^2W}{d\omega d\Omega} = \frac{e^2}{4\pi^2 c}\frac{\beta^2\cos^2(\theta-\alpha)}{K^2|P|^2}\left|\frac{\varepsilon(\omega)-1}{\varepsilon(\omega)}\right|^2 \times$$
$$\left|1 - \frac{P\exp\left(i\frac{\omega}{\beta c}a\coth\varphi\right) + \Sigma\coth\varphi\exp\left(-ia\frac{\omega}{\beta c}P\right)}{P+\Sigma\coth\varphi}\right|^2 \times$$
$$\frac{\exp\left[-2\frac{\omega}{\gamma\beta c}(h+a\coth\varphi)K\cos\alpha\right]}{1-\beta^2\cos^2(\theta-\alpha)+\beta^2\sin^2\alpha\left(1-\sin^2(\theta-\alpha)\sin^2\phi\right)+2\beta\sin\alpha\sin(\theta-\alpha)\cos\phi}\times$$
$$\left[\left|\frac{\varepsilon(\omega)}{\varepsilon(\omega)\cos(\theta-\alpha)+U}\right|^2\left|\cos\alpha\left(\gamma^{-1}\sin(\theta-\alpha)-iKU\cos\phi\right)+\right. \right. \tag{3}$$
$$\left. + \sin\alpha\left(iK\sin(\theta-\alpha)+\gamma^{-1}U\cos\phi\right)-\gamma\beta U\sin(\theta-\alpha)\sin^2\phi\right|^2 +$$
$$\gamma^2\sin^2\phi\left|\frac{\sqrt{\varepsilon(\omega)}}{\cos(\theta-\alpha)+U}\right|^2\left(\sin^2(\theta-\alpha)+|U|^2\right)\times$$
$$\left.\left[1-\beta^2\cos^2(\theta-\alpha)+2\beta\gamma^{-2}\sin\alpha\sin(\theta-\alpha)\cos\phi-\gamma^{-2}\sin^2\alpha\left(K^2-\gamma^{-2}\right)\right]\right].$$

where we used the following notations:

$$U = \sqrt{\varepsilon(\omega)-\sin^2(\theta-\alpha)},$$
$$P = \cos\alpha - \beta U + i\gamma^{-1}K\sin\alpha,$$
$$\Sigma = \sin\alpha + \beta\sin(\theta-\alpha)\cos\phi - i\gamma^{-1}K\cos\alpha,$$
$$K = \sqrt{1+(\gamma\beta\sin(\theta-\alpha)\sin\phi)^2}.$$

In this equation ω is the frequency of the emitted radiation; α is the charged particle incident angle; θ and ϕ are the polar and azimuthal observation angles respectively; a is the side BC

(see figure 1); $h = b/\cos\alpha$ is the impact parameter. The radiation angle θ is measured from the trajectory of a charged particle and is connected with the angle θ' that defined relative to the z-axis of the Cartesian coordinates system (see figure 1) by the simple relation: $\theta' = \theta - \alpha$.

The equation (3) takes into account both DR and ChR types. The following pole in the denominator is responsible for the ChR:

$$\left|\cos\alpha - \beta\sqrt{\varepsilon(\omega) - \sin^2(\theta - \alpha)} + i\gamma^{-1}K\sin\alpha\right| \to 0. \quad (4)$$

The equation (4) is somewhat different from classical relation (2), defining the direction of the ChR propagation. It should be noted that, when $\alpha \to 0$, the equation (4) coincides with classical Cherenkov radiation condition represented in vacuum.

The fact that the equation (3) contains poles responsible for ChR. It enables us to control the direction of the Cherenkov radiation photon yield by changing the prism angle φ in the range $0 < \varphi < \pi/2$. This aspect might significantly simplify an experimental setup eliminating the need to use additional optical elements for radiation focusing or extraction.

The spectral-angular distributions of polarization radiation appearing when a fast charged particle passes by parallel to the dielectric prism ($\alpha = \pi/4$) are illustrated in the figure 2. As a target a PTFE (Teflon) triangular prism with the opening angle of $\pi/2$ is used. The calculation parameters were chosen for the linear electron accelerator CLARA [25] and listed up in the table 1.

Table 1. CLARA beam parameters.

Beam Energy	55 MeV
Beam Charge	100 pC
r.m.s. Beam Size (x/y)	50/200 μm
Bunch Length (r.m.s.)	0.15 ps

ChR maximum intensity corresponds to the angle of $\theta_{\rm Ch} = 43°$ with respect to the particle trajectory determined by the equation (4).

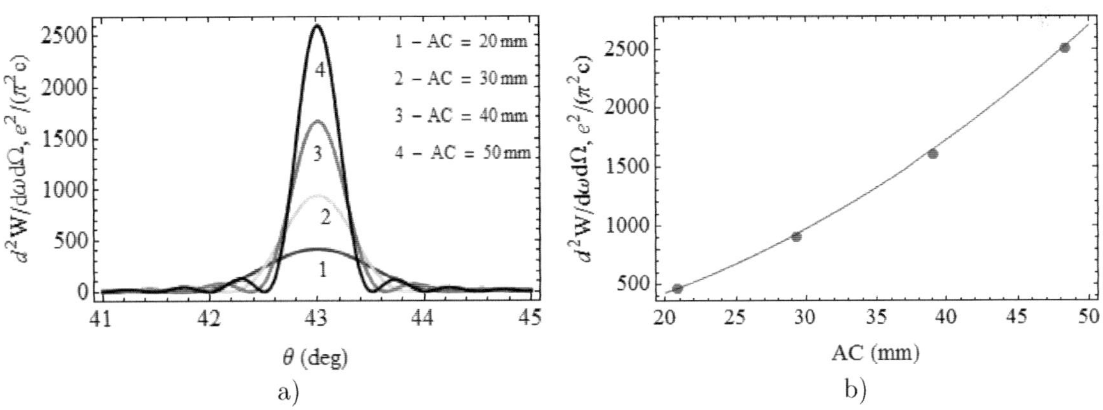

Figure 2. Angular distribution of the polarization radiation intensity for different AC side length (a-left). Dependence of the angular distribution maximum intensity versus AC side length (b-right). The calculation parameters are: $\varepsilon(\omega) = 1.9$ (PTFE); $\gamma = 97$; $\lambda = 0.3$ mm; $b = 5$ mm; $\alpha = 45°$; $\varphi = 45°$; $\phi = 0°$.

The radiation intensity in the maximum of Cherenkov radiation distribution grows quadratically with increase of interaction length of the particle with the target as expected, because the interaction surface increases. Increasing the AC side length from 20 mm to 50 mm we can significantly increase the photon yield.

The next step is to consider the intensity in maximum as a function of the impact parameter b. Figure 3 illustrates the dependences calculated for the same parameters as figure 2 and has the same color identities.

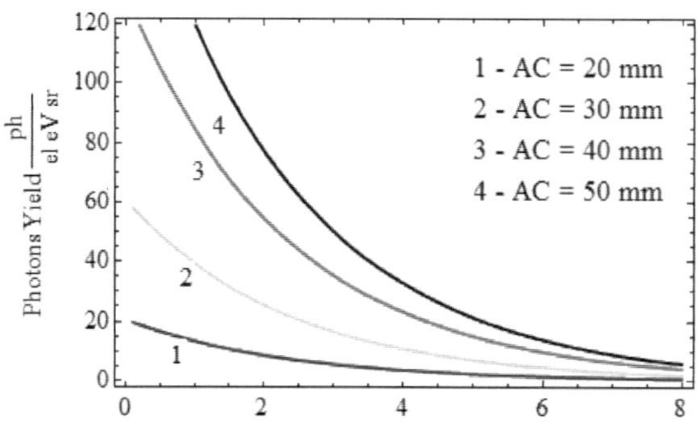

Figure 3. Dependences of ChR intensity in maximum as a function of the impact parameter calculated for different AC side lengths.

It is clear from the figure 3 that reducing the impact parameter the polarization radiation intensity increases exponentially for all AC side lengths. The analysis of the dependences enables us to choose an optimal impact parameter, will minimize the influence of the target on the beam and provide essential intensity.

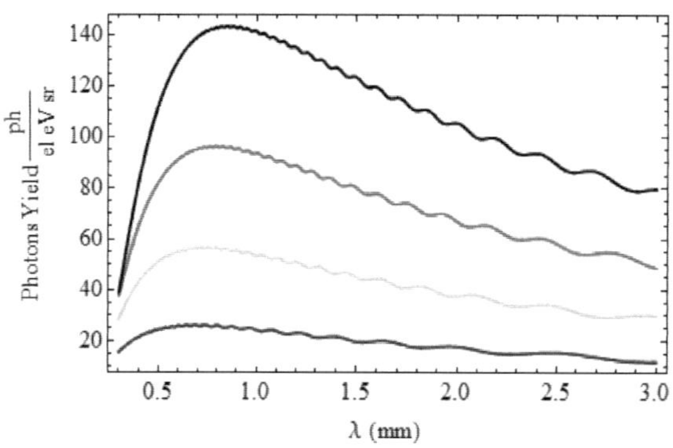

Figure 4. ChR spectra in angular distribution maximum calculated for different AC side lengths.

To evaluate the average photon yield per electron, we need to calculate the ChR photon spectrum using the equation (3). Figure 4 illustrates the calculation results in maximum of the angular distribution ($\theta_{\mathrm{Ch}} = 43°$) for the same parameters as previous figures. One may see that the maximum of the ChR photon yield is in the THz spectral range ($\lambda = 0.5 - 1$ mm).

Finnaly a number of CChR photons can be defined as

$$N = N_e^2 \int_{\Delta\omega} \frac{\mathrm{d}\omega}{\hbar\omega} \exp\left[-2\left(\frac{\omega}{\beta c}\sigma_z\right)^2\right] \int_{\Delta\Omega} \frac{\mathrm{d}^2 W}{\mathrm{d}\omega \mathrm{d}\Omega} \mathrm{d}\Omega, \tag{5}$$

where N_e is the number of electrons per bunch, $\hbar = h/2\pi$ is the reduced Planck constant. The exponential factor with the bunch length σ_z in the equation (5) is describe the longitudinal bunch form factor for the ChR [26].

Assuming 10^9 bunch population, 10^{-3} sr solid angle ($\Delta\theta = 0.5°$ and $\Delta\phi = 10°$ around the ChR maximum), 1 meV photon energy interval around maximum, and the AC side length of 50 mm, we have about 1.4×10^{14} photons per electron, corresponding to approximately 150 kW of peak power. Such a number of photons and the peak power are high enough to perform spectrometry experiments for chemical and biological samples. We intend to study the phenomenon spectral-angular characteristics and absolute power in order to propose the new radiator for the state of the art X-ray FELs as an auxiliary advantageous equipment.

3. Future plan and experimental setup

The experimental setup has been developed such that the direct interaction of the electron beam and the target can be controlled, and, if necessary, excluded. Non-invasive nature of the ChR gives an additional advantage as the beam parameters are kept almost the same as the initial ones. Therefore the beam can later be used by other experimental groups during the entire beam time. To realise the plan, we created an experimental setup schematically illustrated in the figure 5. Together with ChR the electron beam will generate forward DR, which will propagate along the electron beam trajectory but not in the direction of ChR.

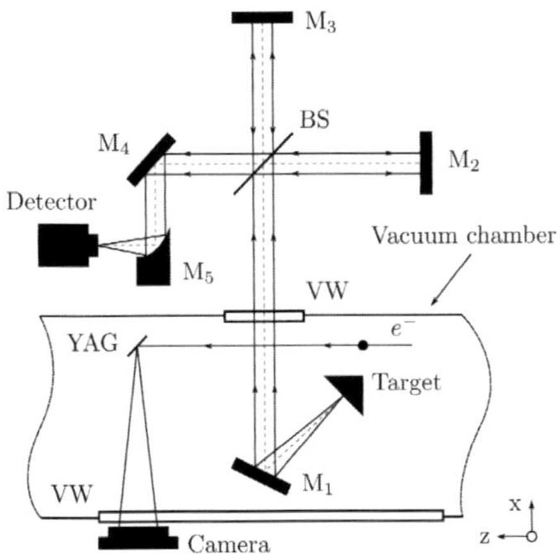

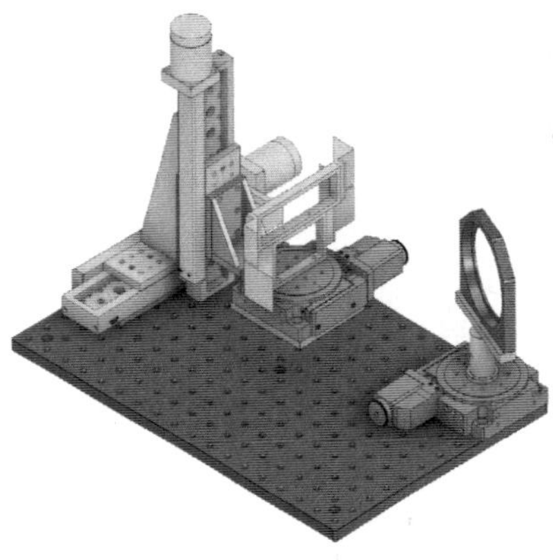

Figure 5. a) Principle scheme of the experiment; b) CAD drawing of the in-vacuum target and mirror M_1 manipulation system.

In the experiment we plan to use four different target with different AC side lengths to be able to compare the quadratic dependence of the ChR photon yield as a function of the interaction area with the theoretical predictions shown in the figure 2b. Geometrically all targets are the same, just the AC base length is 20, 30, 40, and 50 mm. The targets are mounted on a rotation/translation stage (see figure 5b) enabling to change the targets with respect to the beam. Horizontal translation also enables optimal impact parameter adjustment. A YAG screen placed downstream the target on a linear translation stage enables us to observe and optimise the electron beam shape and position. The image of the beam is observed in real time by the CCD camera installed outside the vacuum chamber. It shows the target position with respect to the beam.

The ChR induced by the electron beam Coulomb field propagates towards the rotating mirror M_1 and, then, which reflects the radiation out of the vacuum chamber through the fused silica vacuum window with the diameter of 150 mm. The mirror M_1 is mounted on the rotation stage allowing the angular distribution of the radiation to be scanned.

In order to measure the spectrum of the generated radiation the Michelson interferometer will be used. The interferometer consists of a stationary mirror M_2, movable mirror M_3 and a beam splitter BS. After the interferometer the radiation is directed by the mirror M_4 to the off-axis parabolic mirror M_5 which focuses the beam of radiation into the detector. A pyroelectric and Golay Cell detectors are used to detect the power of the radiation.

4. Conclusion

In the paper we have proposed an experiment on generation of intense THz radiation using non-invasive method based on ChR. We have presented the theoretical analysis of the ChR spectral-angular characteristics for the parameters of the CLARA facility being constructed at Daresbury Laboratory (UK). We have optimised the experimental conditions for the experiment, developed experimental hardware and prepared computer codes for future comparison with the experimental results.

Non-invasive nature of the ChR enables to put several radiator on the same beam to generate intense THz for several experiments in parallel. The spectral characteristics of the radiation will be applied for longitudinal bunch profile studies with femtosecond resolution when CLARA facility is commissioned.

Acknowledgments

The work was partly supported by the program "Nauka" of the Russian Ministry of Education and Science within the grant no. 3.709.2014/K, ADG-TX Leverhulme Trust Foundation International Network grant (no. IN-2015-012) and the Competitiveness Enhancement Program of Tomsk Polytechnic University. We would also like to acknowledge the Russian Foundation for Basic Research (grant no. 14-02-31642-mol_a).

References

[1] Siegel P H 2002 *IEEE Trans. on Microwave Theory and Techniques* **50** 910
[2] Smye S W, Chamberlain J M, Fitzgerald A J and Berry E 2001 *Physics in Medicine and Biology* **46** R101
[3] Kampfrath T, Tanaka K and Nelson K A 2013 *Nature Photonics* **7** 680
[4] Dudovich N, Oron D and Silberberg Y 2002 *Nature* **418** 512
[5] Van der Weide D 2003 *Opt. Photonics News* **14** 48
[6] Gorshunov B, Volkov A, Spektor I, Prokhorov A *et al* 2005 *Int. J. of Infrared and Millimeter Waves* **26** 1217
[7] Verghese S, McIntosh K A, and Brown E R 1997 *IEEE Trans. Microwave Theory Tech* **45** 1301
[8] Matsuura S, Blake G A, Wyss R A, Pearson J C *et al* 1999 *Appl. Phys. Lett.* **74** 2872
[9] LCLS Technical Design Report 2002 (*SLAC*) R-593 http://www-ssrl.slac.stanford.edu/lcls/cdr/
[10] XFEL Technical Design Report 2006 (*DESY*) 097 http://xfel.desy.de
[11] Ryazanov M I and Tilinin I S 1976 *Sov. Phys. - JETP* **44** 1092
[12] Karataev P, Araki S, Hamatsu R, Hayano H *et al* 2004 *Phys. Rev. Lett.* **93** 244802
[13] McMillan E M 1945 *Phys. Rev.* **68** 144
[14] Bolotovskii B M and Galstyan E A 2000 *Phys.-Usp.* **43** 755
[15] Jelley J V 1958 *Čerenkov radiation and its applications* (London: Pergamon Press) p 331
[16] Karlovets D V and Potylitsyn A P 2009 *JETP Lett.* **90** 368
[17] Karlovets D V 2011 *J. Exp. Theor. Phys.* **113** 27
[18] Kruchinin K O and Karlovets D V 2012 *Russ. Phys. J.* **55** 9
[19] Shevelev M V and Konkov A S 2014 *J. Exp. Theor. Phys.* **118** 501
[20] Shevelev M V, Konkov A S and Aryshev A S 2015 *Phys. Rev.* A **92** 053851
[21] Naumenko G A, Potylitsyn A P, Shevelev M V and Popov Y A 2011 *JETP Lett.* **94** 258
[22] Shevelev M, Naumenko G, Potylitsyn A and Popov Yu 2012 *J. Phys: Conf. Ser.* **357** 012020
[23] Naumenko G, Potylitsyn A, Shevelev M, Bleko V *et al* 2014 *J. Phys: Conf. Ser.* **517** 012004
[24] Bleko V V, Konkov A S and Soboleva V V 2015 *Nucl. Instrum. Methods* B **355** 129
[25] CLARA Conceptual Design Report 2013 (*STFC*) http://www.stfc.ac.uk/files/science-publications/clara-conceptual-design-report-2013/
[26] Naumenko G A 2015 *Adv. Mat. Res.* **1084** 138

Coherent radiation of relativistic electrons in wire metamaterial

V Soboleva, G Naumenko, V Bleko

Tomsk Polytechnic University, Lenina str. 2, Tomsk, 634050, Russia

E-mail: sobolevaveronica@mail.ru

Abstract. We present in this work the experimental investigation of the interaction of relativistic electron field with wire metamaterial. The measurements of the spectral-angular characteristics of coherent radiation were done in millimeter wavelength region in far-field zone at the relativistic electron beam with energy of 6 MeV. Used target represent the right triangular prism that consist of periodic placed copper wires. We showed that bunched electron beam passing near wire metamaterial prism generates coherent Cherenkov radiation. Spectral-angular characteristics of radiation from the wire target were compared with the characteristics of Cherenkov radiation generated in similar experimental conditions in a dielectric target (Teflon prism) that has the same form and sizes.

1. Introduction

The different types of metamaterials have been intensively studied in recent years [1–4]. Metamaterials represent composite periodic structures that consist of scattering elements whose dimensions are much smaller than wavelength of the electromagnetic field. The examples of such media can be left-handed materials (LHM), wire metamaterials and other various artificial structures whose electromagnetic properties are not observed in natural materials.

The processes of radiation generation in metamaterials are of large interest to researchers because the radiation generated in such media has very unusual properties. For example, Cherenkov radiation that is generated in the result of interaction of a relativistic charged particle with LHM is propagated in backward semi-sphere. It is so-called backward Cherenkov radiation [5–8]. In the case when a charged particle interacts with a wire medium, radiation is also generated. The properties of radiation generated in wire metamaterials were theoretical studied in paper [9–11]. One of the motivations for this study is that in work [11] have been shown the possibility to boost the Cherenkov emission by several orders of magnitude in such composite structures in comparison to natural media.

In this paper, we present the results of experimental investigation of relativistic electron field with wire metamaterial and dielectric target in similar experimental conditions.

2. Experimental setup and technique

The experiments were carried out at the extracted electron beam of microtron of Tomsk Polytechnic University, with the following beam parameters: electron energy was 6 MeV ($\gamma = 12$), bunch period – 380 ps, macropulse duration varied from 3 to 5 μs, bunch population $N_e = 6 \cdot 10^8$, bunch length $\sigma_b \approx 1.9 \div 2.4$ mm, dimensions of the extracted electron beam 4×2 mm^2, angular divergence of the

Content from this work may be used under the terms of the Creative Commons Attribution 3.0 licence. Any further distribution of this work must maintain attribution to the author(s) and the title of the work, journal citation and DOI.
Published under licence by IOP Publishing Ltd

extracted beam $\sigma = 0.08$ rad. For such bunch length and bunch population the radiation with wavelength $\lambda > 8$ mm is coherent and radiation intensity increases by 8 orders. This fact allows using a room-temperature detector. Used for the measurements detector was produced at Tomsk Scientific Research Institute of Semiconductor Devices on the basis of broadband antenna with high frequency diode built into the center of antenna. This detector makes it possible to register electromagnetic radiation in the wavelength range from 3 to 30 mm.

The design of the studied targets is presented in figure 1. Both of them represent the right triangular prisms that have the same form. Wire target consist of 4 layers placed at the distance 10 mm. The diameter of copper wire is 0.5 mm, the wire period is 1 mm. The length of the both cathetuses is 170 mm. Material of the second target is a Teflon with refraction index $n = 1.41$. The height of Teflon prism is 74 mm. The length of cathetuses is 175 mm.

a) 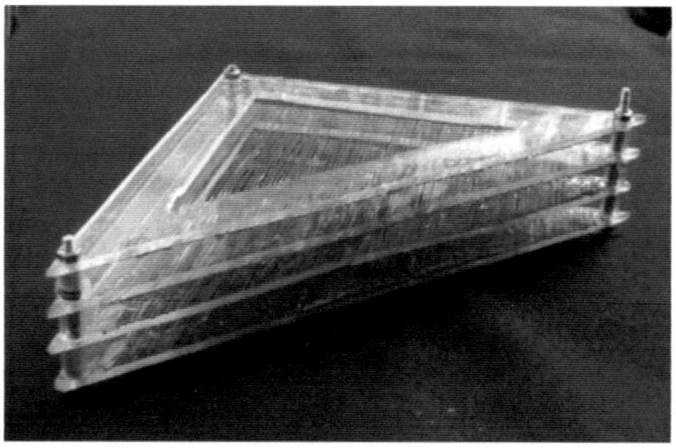 b) 

Figure 1. Design of the targets: a) Wire target; b) Teflon target.

The scheme of the measurements is shown in figure 2. The studied targets were placed close to the electron beam trajectory at the distance 10 mm (impact parameter h). The distance from the target to the extraction window was 300 mm. To exclude the effect of pre-wave zone [12] we used the detector located in focus of the parabolic mirror. Both of them are placed on the rod that rotates around the axis, which passes through the center of the emitting surface. In this case, the measured angular characteristics of radiation coincide with the characteristics measured in far-field zone [13].

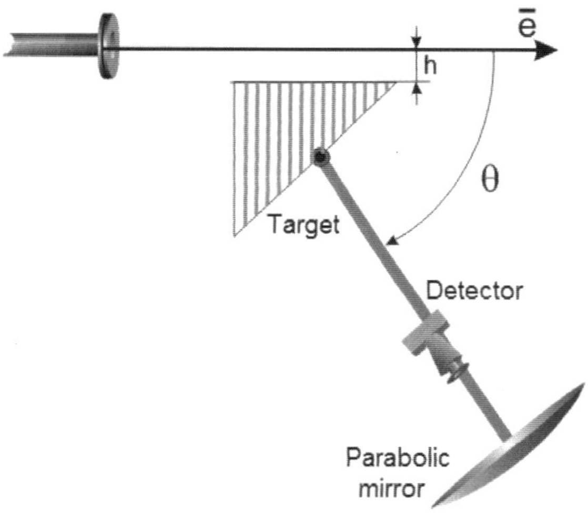

Figure 2. Scheme for the measuring of angular characteristics of radiation.

To show that the observed effect is caused by availability of periodic wire structure (but not target frame) was made a target that represents just a frame of wire target (see figure 3). The measurement of angular distribution of radiation from this target was carried out in the same geometry of experiment (see figure 2).

Figure 3. Design of the wire target frame.

3. Measurements

In experiment was measured the horizontal polarization of radiation. Measured angular distributions of radiation from studied targets are given in figure 4.

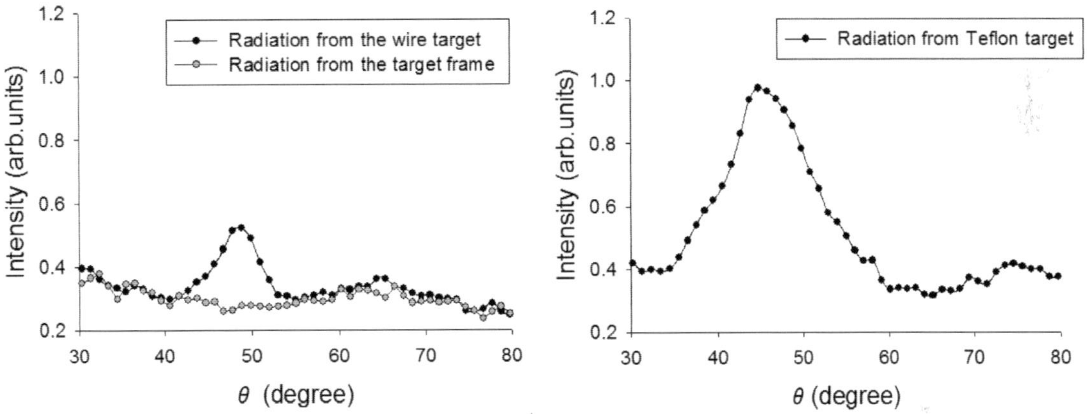

a) b)

Figure 4. Angular distribution of coherent radiation from: a) Wire target (black points) and target frame (open circles); b) Teflon prism.

As can be seen from obtained dependences (see figure 4(a)) radiation from the wire target is observed at the angle $\theta = 48^\circ$. This radiation is the result of interaction of relativistic electron field with wire structure of the target. The frame of wire target does not contribute to observed effect in the considered angle range. Radiation from Teflon prism is observed at the angle $\theta = 45^\circ$ (see figure 4(b)). Observation angle of radiation corresponds to the condition for Cherenkov radiation: $\cos\theta = \frac{1}{n \cdot \beta}$ (where β is a velocity of charge in terms of the light velocity, n is a refraction index).
As one can see the intensity of Cherenkov radiation from the Teflon prism is more by three times than radiation intensity from wire target. It is caused by that the studied wire target consists of only four layers. However, the increasing of the layers number by an order or more can provide the radiation intensity much more than intensity of radiation from Teflon target of the same dimensions.

In figure 5 are shown the spectra of radiation measured using interferometer with wave front division on the basis of two mirrors, one of which is movable [14].

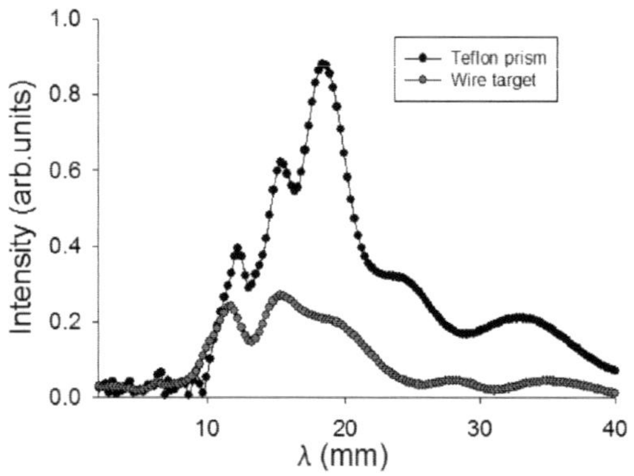

Figure 5. The spectra of radiation from the wire target (gray curve) and Teflon prism (dark curve).

As can be seen from the obtained dependences, the spectra of radiation from wire and Teflon targets have the similar spectral distributions. The radiation intensity from the wire target is smaller than intensity from the Teflon target because we used only four layers of wires.

4. Conclusion

We have shown that bunched electron beam passing near a wire metamaterial prism generates the coherent radiation. Spectral-angular characteristics of radiation from the wire prism were compared with the characteristics of Cherenkov radiation generated in the same experimental conditions in Teflon target that has the same form and similar dimensions. For the future we plan the experimental investigation of the spectral-angular characteristics of radiation in wire metamaterial that has the same period of the wire in the horizontal and vertical planes.

Acknowledgment

This work was supported by grant of Ministry of Education and Science of Russian Federation № 3761.

References

[1] Grbic A, Eleftheriades G V 2004 *Phys. Rev. Lett.* **92** 117403
[2] Lee H, Xiong Y, Fang N 2005 *New Journal of Physics* **7** 255
[3] Pendry J B, Schurig D, Smith D R 2006 *Science* **312** 1780
[4] Kildishev A V, Shalaev V M 2011 *Phys. Usp.* **54** 53
[5] Galyamin S, Tyukhtin A 2010 *Phys. Rev. B* **81** 235134
[6] Tyukhtin A V, Galyamin S N, Belonogaya E S 2010 *Proceedings of IPAC10* **3** 1068
[7] Sheng Xi, Chen H, Tao Jiang 2009 *Phys. Rev. Lett.* **103** 190801
[8] Duan Z Y, Wu B I, Xi S *et al* 2009 *Prog. Electromagn. Res.* **90** 75
[9] Tyukhtin A V, Vorobev V V 2014 *Phys. Rev. E* **89** (1) 013202
[10] Vorobev V V, Tyukhtin A V 2012 *Phys. Rev. Lett.* **108** 184801
[11] Fernandes D E, Maslovski S I, Silverina M G 2012 *Phys. Rev. B* **85** 155107
[12] Verzilov V A 2000 *Phys. Lett. A* **273** 135
[13] Kalinin B N, Naumenko G A, Potylitsyn A P *et al* 2006 *JETP Letters* **84** (3) 110
[14] Naumenko G A, Potylitsyn A P, Shevelev M V *et al* 2009 *Russian Physics Journal* **11** (2) 254

Common characteristics of synchrotron radiation and light leaking from a bent optical fiber

Artru X[1,2] and Ray C[1,3]

[1] Université de Lyon, Université Lyon 1, F-69622 Villeurbanne, France
[2] Institut de Physique Nucléaire de Lyon, CNRS-IN2P3.
[3] Institut Lumière Matière, Université Lyon 1.

E-mail: x.artru@ipnl.in2p3.fr

Abstract. Light leaking from a bent optical fiber shares many properties with synchrotron radiation : in ray optics, both lights are emitted tangentially to a *light cylinder* ; in wave optics, the emission mechanism involves a *tunnel effect.* The angular distributions of these two radiations are studied in parallel and found to be similar. The same is done for the impact parameter distributions. The latter show interference fringes of the Airy function type. The far field escaped from the fiber is calculated with the *Volume Current Method.* An optical system observing the impact parameter profile is proposed.

1. Introduction

Synchrotron radiation (SR) is the best known type of radiation emitted by a relativistic electron. Its spectrum and angular distribution is well calculated in classical electrodynamics for most practical purpose. However, its distribution in *impact parameter* space [1, 2, 3, 4], which is relevant in beam profile measurement, is not so well known. Furthermore, the most consistent classical equation of motion including radiation reaction, the so-called Abraham-Lorentz-Dirac (ALD) equation is plagued with unphysical *run-away* solutions. Notwithstanding this difficulty, a link has been established between the impact parameter of an emitted photon, a "side-slip" of the electron and the Schott term of the ALD equation [5, 6].

An heuristic explanation of SR is the following: the whole Coulomb field of the electron cannot follow the curved motion of the latter. Its part outside the "light cylinder" (a concept used in pulsar physics) of radius[1] R/v would go faster than light. It detaches and becomes free radiation. Light escaping a bent optical fiber (LEBF) can be explained is the same way (see figure 1), replacing the Coulomb field by the evanescent wave surrounding the fiber and v by the phase velocity $v_{\rm ph}$ of the internal wave.

In this paper we pursue this analogy quantitatively. We assume that light circulates in the fiber in a definite mode $\mathcal{M}$ (see Ref. [7]) and has a definite frequency ω_0. In Section 2 we compare the far-field angular distributions of both radiations. In Section 3 we analyze the fields in the impact parameter b parallel to the curvature plane and propose an optical system for observing the b-distribution.

[1] R is the orbit radius of the electron or the bending radius of the fiber. In our units the speed of light is 1.

Content from this work may be used under the terms of the Creative Commons Attribution 3.0 licence. Any further distribution of this work must maintain attribution to the author(s) and the title of the work, journal citation and DOI.
Published under licence by IOP Publishing Ltd

Angular momentum considerations. In SR, the photon angular momentum J_Y about the orbit axis (see figure 2) can be expressed in two ways [4]

$$J_Y = \hbar\omega\, R/v_e\,, \quad (1)$$
$$J_Y = R_{\rm phot}\,\hbar\omega\,\cos\psi\,, \quad (2)$$

$R_{\rm phot}$ is the distance between the light ray (in ray optics) and the orbit axis and ψ is the angle between the ray and the orbit plane. For a photon escaping a fiber bent about the Y-axis we have the same formulae, but replacing v_e by the phase velocity $v_{\rm ph}$. From (1-2) we obtain

$$R_{\rm phot} = R/(v\cos\psi) \quad (3)$$

with $v = v_e$ or $v_{\rm ph}$. We see that the light ray does not come from the electron trajectory (SR case) or from the fiber (LEBF case), but is tangent to a cylinder of radius $R_{\rm phot} > R$. The bundle of rays form a *caustic*. The *classical impact parameter* of the photon is

$$b_{\rm cl} \equiv R_{\rm phot} - R \simeq (\gamma^{-2} + \psi^2)\, R/2 \quad (4)$$

where ψ is the angle between the ray and the curvature plane and $\gamma = (1-v)^{-1/2}$ with $v = v_e$ or $v_{\rm ph}$. We have assumed $\psi \ll 1$ and $\gamma \gg 1$, so that $b_{\rm cl} \ll R$. The finite impact parameter explains the frequency cutoff of SR or the critical bending radius of LEBF. $b_{\rm cl}$ must be less than the transverse size of the electron Coulomb field (for SR) or of the evanescent wave (for LEBF). These are both $\sim \gamma/\omega$. We obtain therefore formally the same condition for a significant SR or LEBF yield :

$$\omega\, R \lesssim \gamma^3\,. \quad (5)$$

This condition can also be obtained by considering the emission of the photon as a *tunneling* through the centrifugal barrier [8].

Electron side-slipping [5, 6]. In the SR case, the electron "side-slips" in the direction of the force exerted by the field. The displacement

$$b_e = b_{\rm cl}\,\hbar\omega/(\gamma m_e) \quad (6)$$

is such that, just after the emission, the centre-of-inertia of the *photon + electron* system prolongates the initial electron trajectory. $b_{\rm cl}$ may be measurable with a thin enough electron beam, but $b_e \sim \lambda\!\!\!^{-}_c \simeq 400$ fm is practically not visible. Nevertheless, the global side-slipping due to many photon emissions in the $\hbar \to 0$ limit is responsible for the *Schott term* in $d^3X/d\tau^3$ of the ALD equation.

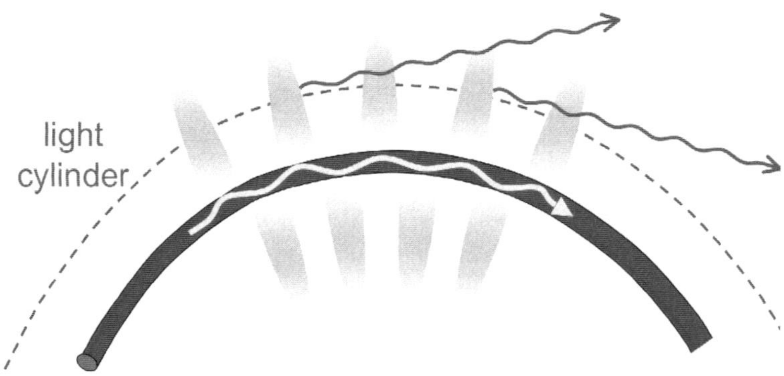

Figure 1. Escape of light from a bent optical fiber. The shaded areas represent the evanescent wave. A similar figure can be drawn for synchrotron radiation.

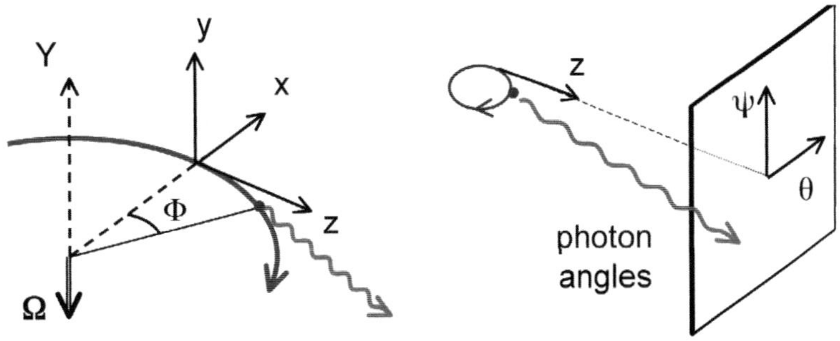

Figure 2. Coordinate system for synchrotron radiation or LEBF. The Y-axis is the bending axis.

2. SR and LEBF amplitudes

Let us treat in parallel SR and LEBF. The electron trajectory or the fiber is bent in the "horizontal" (x, z) plane along the circle of radius R centered at $(-R, 0, 0)$ as shown in figure 2. The angular velocity Ω of the electron or the wave about the Y axis is negative. The right of figure 2 specifies the spherical angular coordinate θ and ψ of the emitted photon momentum, $\mathbf{K} = \omega(\cos\psi\sin\theta, \sin\psi, \cos\psi\cos\theta)$. We take as linear polarization vectors $\hat{\mathbf{e}}_\theta(\mathbf{K}) = (\cos\theta, 0, -\sin\theta)$ and $\hat{\mathbf{e}}_\psi(\mathbf{K}) = (-\sin\psi\sin\theta, \cos\psi, -\sin\psi\cos\theta)$.

The current sources. SR is emitted by the current density

$$\mathbf{j}(t, \mathbf{X}) = -e\,\mathbf{v}_\mathrm{e}(t)\,\delta^3[\mathbf{X} - \mathbf{X}_\mathrm{e}(t)]\,, \tag{7}$$

where $\mathbf{X}_\mathrm{e}(t) = (R\cos\phi - R, 0, R\sin\phi)$, $\phi = vt/R$ and $\mathbf{v}_\mathrm{e} = (-\sin\phi, 0, \cos\phi)$.

LEBF can be considered as emitted by the *polarization current* $\mathbf{j}_\mathrm{pol}$ induced by the channeled wave in the fiber medium. For the mode $\{\mathcal{M}, \omega\}$ defined in [7] the current density is

$$\mathbf{j}_\mathrm{pol}(\mathbf{X}, t) = 2\,\mathrm{Re}\{\mathbf{j}_{\{\mathcal{M},\omega\}}(\mathbf{r})\,\exp(ipz - i\omega t + i\eta)\}\,, \tag{8}$$

η being a "classical" phase. *Macroscopic* Maxwell equations in medium give $\mathbf{j}_\mathrm{pol} = (\varepsilon - 1)\,\partial_t\mathbf{E}_\mathrm{real}$, whence

$$\mathbf{j}_{\{\mathcal{M},\omega\}}(\mathbf{r}) = -i\omega\,[\varepsilon(\mathbf{r}) - 1]\,\mathbf{E}_{\{\mathcal{M},\omega\}}(\mathbf{r})\,. \tag{9}$$

$\mathbf{j}_{\{\mathcal{M},\omega\}}$ is complex, like $\mathbf{E}_{\{\mathcal{M},\omega\}}$ and $\vec{\mathcal{E}}_{\{\mathcal{M},\omega\}}$ of Ref. [7]. In a straight fiber the current density $\mathbf{j}_\mathrm{pol}$ is in uniform translation motion at speed v_ph, therefore does not radiate. In a bent fiber it emits synchrotron-like radiation, the retarded field obeying the *microscopic* Maxwell equations

$$\begin{array}{ll} \nabla\times\mathbf{B} - \partial_t\mathbf{E} = \mathbf{j}_\mathrm{pol}\,, & \nabla\cdot\mathbf{E} = \rho_\mathrm{pol}\,, \\ \nabla\times\mathbf{E} + \partial_t\mathbf{B} = 0\,, & \nabla\cdot\mathbf{B} = 0\,. \end{array} \tag{10}$$

The *Volume Current Method* [9, 10] consists in taking the $\mathbf{j}_\mathrm{pol}$ of the *straight tangent fiber* in (10), replacing the variable z in (8) by the curvilinear abscissa along the fiber.

2.1. Angular distributions

The radiation field emitted by the current density $\mathbf{j}$ is expanded in free plane waves:

$$\mathbf{E}_\mathrm{rad}(t, \mathbf{r}) = \int \frac{d^3\mathbf{K}}{(2\pi)^3}\,f(\omega)\;\mathrm{Re}\left\{\tilde{\mathbf{E}}(\mathbf{K})\;e^{i\mathbf{K}\cdot\mathbf{X} - i\omega t}\right\}\,, \tag{11}$$

$$f(\omega)\,\tilde{\mathbf{E}}(\mathbf{K}) = -\int dt\, d^3\mathbf{X}\ \exp(i\omega t - i\mathbf{K}\cdot\mathbf{X})\,\mathbf{j}_\perp(t,\mathbf{X})\,, \tag{12}$$

where $\omega = |\mathbf{K}|$, $f(\omega) = 1$ for a polychromatic source (SR case), $f(\omega) = 2\pi\delta(\omega-\omega_0)$ for a monochromatic source (LEBF case); $\mathbf{j}_\perp$ is the component of $\mathbf{j}$ orthogonal to $\mathbf{K}$. In SR one measures the **spectral**-angular distribution,

$$\frac{dW}{d\omega\, d\Omega} = \frac{\omega^2}{16\pi^3}\sum\nolimits_\mu |\hat{\mathbf{e}}^*_\mu(\mathbf{K})\cdot\tilde{\mathbf{E}}(\mathbf{K})|^2\,, \tag{13}$$

and in LEBF, the angular **power** distribution,

$$\frac{dW}{dt\, d\Omega} = \frac{\omega_0^2}{8\pi^2}\sum\nolimits_\mu |\hat{\mathbf{e}}^*_\mu(\mathbf{K})\cdot\tilde{\mathbf{E}}(\mathbf{K})|^2\,; \tag{14}$$

$\mu = \theta$ or ψ denotes the linear polarization axis.

Synchrotron case. We consider the radiation from one single pass of orbital motion. From (7) and (12),

$$\tilde{\mathbf{E}}(\mathbf{K}) = eR\int_{-\pi}^{\pi} d\phi\ \exp\{iKR(\phi/v_{\rm e} - \cos\psi\ \sin\phi)\}\,[\sin\phi\,\hat{\mathbf{e}}_\theta + \sin\psi\ \cos\phi\,\hat{\mathbf{e}}_\psi]\,. \tag{15}$$

Using the ultra-relativistic approximations $\gamma \gg 1,\ \ \psi \ll 1$ we obtain

$$\tilde{\mathbf{E}}(\mathbf{K}) = e\,\theta_0 R\left[-i\theta_0\,\mathcal{A}'(\xi)\,\hat{\mathbf{e}}_\theta + \psi\,\mathcal{A}(\xi)\,\hat{\mathbf{e}}_\psi\right]\,e^{i\zeta}\,, \tag{16}$$

where

$$\theta_0 = (\omega R)^{-1/3}\,,\quad \xi = (\gamma^{-2}+\psi^2)\,\theta_0^{-2}/2\,, \tag{17}$$

$$\zeta = \omega R\,(\cos\psi\sin\theta - \theta/v) \tag{18}$$

and $\mathcal{A}(\xi)$ is a re-scaled Airy function

$$\mathcal{A}(\xi) = 2^{4/3}\pi\ \mathrm{Ai}\left(2^{1/3}\xi\right) = \int_{-\infty}^{+\infty} d\tau\ \exp(i\xi\tau + i\tau^3/6)\,. \tag{19}$$

Optical fiber case. We will assume that the fiber is sufficiently thin so that $v \equiv v_{\rm ph}$ is close to 1. Then we can use the same "ultrarelativistic" approximations as in SR and obtain

$$\begin{aligned}\tilde{\mathbf{E}}(\mathbf{K}) \;&=\; -\theta_0 R\Big\{\left[\tilde{\mathbf{j}}_x\,\mathcal{A}(\xi) + i\,\tilde{\mathbf{j}}_z\,\theta_0\,\mathcal{A}'(\xi)\right]\,\hat{\mathbf{e}}_\theta \\ &\quad+\;\left[\tilde{\mathbf{j}}_y - \psi\,\tilde{\mathbf{j}}_z\right]\mathcal{A}(\xi)\,\hat{\mathbf{e}}_\psi\Big\}\;e^{i\zeta}\end{aligned} \tag{20}$$

where

$$\tilde{\mathbf{j}} = \int dx\,dy\ \exp(-iK_y\,y)\,\mathbf{j}_{\{\mathcal{M},\omega\}}(x,y) \tag{21}$$

is calculated for a straight fiber and we use again the definitions (17-18).

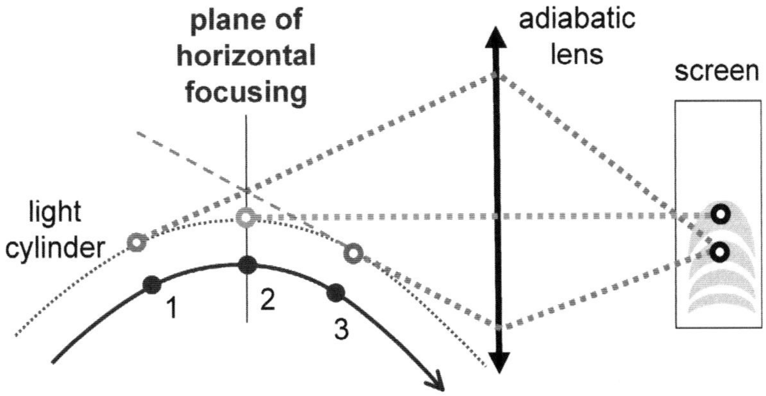

Figure 3. Optical device reproducing the hybrid impact parameter distribution of SR or LEBF on a screen. For the horizontal (resp. vertical) coordinate, the adiabatic lens is focused on the $z = 0$ (resp. $-\infty$) plane (the "vertical" axis $\hat{\mathbf{y}}$ is out of the figure plane). The hollow dots on the light cylinder are the classical emission points when the electron is in positions 1, 2 and 3. Rays emitted in 1 and 3 reach the same point on the screen, therefore interfere.

2.2. Impact parameter representation

We collect the light with an optical device, shown in figure 3, centered on the z axis and of aperture O such that $\theta_0 \ll O \ll 1$. Thus $\hat{\mathbf{e}}_\theta(\mathbf{K}) \simeq \hat{\mathbf{x}}$ and $\hat{\mathbf{e}}_\psi(\mathbf{K}) \simeq \hat{\mathbf{y}}$. We are interested only in the component b of the impact parameter parallel to the curvature plane, so we choose an astigmatic device horizontally focused at $z = 0$ and vertically focused at $z = -\infty$. We thus observe the partial Fourier transform

$$\check{\mathbf{E}}(\omega, x, K_y) = \int \frac{dK_x}{2\pi} \exp(iK_x x)\, \tilde{\mathbf{E}}(\mathbf{K}) \tag{22}$$

$$\simeq [\omega\theta_0/(2\pi)]\, \tilde{\mathbf{E}}(0, K_y, \omega)\, \mathcal{A}(\chi)\,, \tag{23}$$

with $\chi = [b_{\rm cl} - x]/b_0$ and $b_0 = R\,\theta_0^2$. The factor $\mathcal{A}(\chi)$ comes from the Fourier transform of $e^{i\zeta}$. According to the Parseval-Plancherel formula one passes from the K_x-distribution to the x-distribution by the substitution $|\tilde{\mathbf{E}}|^2 \to |\check{\mathbf{E}}|^2$ and $dK_x/(2\pi) \to dx$. We obtain thus the hybrid (x, ψ) distribution

$$\frac{dW}{dx\, d\psi\, (d\omega \text{ or } dt)} = \frac{\omega}{2\pi}\, \theta_0^2\, \mathcal{A}^2(\chi)\, \frac{dW}{d\theta\, d\psi\, (d\omega \text{ or } dt)}\,, \tag{24}$$

which applies both to SR (with $d\omega$) and LEBF (with dt). The simplicity of this result comes from the uniform rotation of the source, which gathers the θ dependence in the phase factor $e^{i\zeta}$.

The Airy fringes. The oscillating Airy function in (24) gives the crescent-shaped fringes on the screen. These can be explained by the interference between waves emitted from points of the light cylinder which are symmetric about the (x, y) plane, like the hollow points above 1 and 3 of figure 3. The same oscillations are present in the full (x, y) impact parameter distribution (see figure A5 of [2]). Note that the fringes are characteristic of the impact parameter representation of a *caustic* wave [11]. It is therefore possible to gather the fringes in *one single peak* in x, with an optical system which transforms the caustic into a true focus, for instance a curved mirror with a cubic deformation. This single peak would be more convenient and give a better resolution for beam profile measurements.

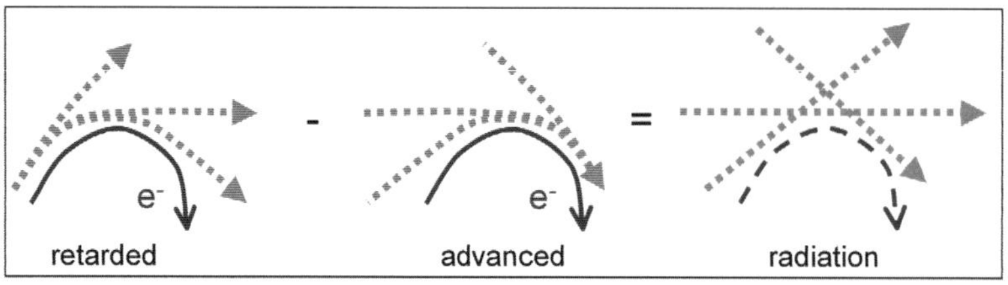

Figure 4. Relation between the fields $\mathbf{E}_{\rm rad}$, $\mathbf{E}_{\rm ret}$ and $\mathbf{E}_{\rm adv}$.

"Far-field" conditions. The x- or (x, y)-distribution of $\mathbf{E}_{\rm rad}$ cannot be measured with a probe put in the $z = 0$ plane, close to the electron beam or to the fiber : one would record the intensity of the *retarded* field $\mathbf{E}_{\rm ret}$ (of the electron for SR, of the polarization current for LEBF). Due to the identity $\mathbf{E}_{\rm rad} = \mathbf{E}_{\rm ret} - \mathbf{E}_{\rm adv}$ pictured in figure 4 the error would be equal to the *advanced* field $\mathbf{E}_{\rm adv}$. If, for the purpose of deflecting the light toward a focusing optics, one puts a mirror close to the electron beam, one adds a parasitic *diffraction radiation* by the mirror edge. We must therefore put any collecting system far enough from the source. $\mathbf{E}_{\rm adv}$ is still there, but propagating at large angle to the optical axis, as can be guessed from figure 4; therefore it cannot reach the image screen.

3. Conclusion

The strong similarity between synchrotron radiation (SR) and light escaping a bent fiber (LEBF) allows a deeper understanding of both processes. For example the cutoff frequency of SR and the critical curvature of a fiber are both derived from angular momentum considerations and reveal a tunneling mechanism. The Airy-type fringes of the impact parameter distribution, yet observed neither in SR nor in LEBF, could provide a tool for beam position monitoring. The principle of the optical system could be cheaply tested with LEBF.

References

[1] Hofmann A and Méot F 1982, *Nucl. Inst. Meth.* A **203** 483-493
[2] Andersson Å1997, Electron beam profile measurements and emitance manipulation at the MAX-laboratory, *doctoral dissertation*, Lund University
[3] Chubar O, Elleaume P, Snirigev A, 1999, *Nucl. Inst. Meth. in Phys. Research* A **435** 495
[4] Artru X 2006, Impact parameter profile of synchrotron radiation, *Advanced Radiation Sources and Applications*, ed Wiedemann H, NATO Science Series II **199** 387 ; arXiv:hep-ph/0503182
[5] Artru X and Bignon G 2002, A semiclassical approach to the radiation damping force, in *Electron-Photon Interaction in Dense Media*, ed Wiedemann H, *NATO Science series* II **49** pp. 85-90
[6] Artru X, Bignon G and Qasmi T 2001, Side-slipping of a radiating particle, *Problems of Atomic Science and Technology* **6** 98 ; arXiv:physics/0208005
[7] Artru X and Ray C 2015, Coherent light induced in optical fiber by a charged particle, this proceedings.
[8] Artru X 2015, Classical and quantum phenomenology in radiation by relativistic electrons in matter or in external fields, *Nucl. Inst. Meth. in Phys. Research* B **355** 11
[9] White I A 1979, Radiation from bends in optical waveguides: the volume-current method, *IEE Journal on Microwaves, Optics and Acoustics* 186.
[10] Kuznetsov M and Haus H 1983, Radiation loss in dielectric waveguide structures by the volume current method, *Quantum Electronics, IEEE* **10** 1505
[11] Landau L and Lifshitz E 1951, *The Classical Theory of Fields*, Addisson-Wesley

The study of undulator radiation of transversal oscillator

A S Kotanjyan and A A Saharian

Yerevan State University, 1 Alex Manoogian Street, 0025 Yerevan, Armenia
Institute of Applied Problems in Physics, 25 Nersessian Street, 0014 Yerevan, Armenia

E-mail: anna.kotanjyan@ysu.am, saharian@ysu.am

Abstract. We study the radiation intensity for a charged transversal oscillator moving around a dielectric cylinder. Similar to the case of coaxial circular motion, under certain conditions for the parameters of the trajectory and dielectric cylinder, strong narrow peaks appear in the angular distribution of the radiation intensity in the exterior medium for a given harmonic. Comparing to the case of a uniform coaxial circular motion, for a transversal oscillator new peaks may appear. The oscillations serve as a mechanism for the control of the spectral-angular characteristics of the radiation intensity.

1. Introduction

The radiation processes in medium from various types of sources have been widely discussed in the literature (see, for example, [1]-[4]). The presence of medium can essentially change the radiation features and gives rise to new types of phenomena. Well-known examples are Cherenkov, transition and diffraction radiations. Moreover, the operation of a number of devices assigned to production of electromagnetic radiation is based on the interaction of high-energy particles with materials.

By taking into account extensive applications of synchrotron radiation in a wide variety of experiments and in many disciplines, it is of interest to investigate the influence of a medium on the spectral and angular distributions of the synchrotron emission. This study is also important with respect to some astrophysical problems. The synchrotron radiation from a charged particle circulating in a homogeneous medium was considered in [5]-[13], where it has been shown that the interference between the synchrotron and Cherenkov radiations leads to remarkable effects. New interesting features arise in inhomogeneous media. In particular, the investigation of the radiation from a charged particle circulating around or inside a dielectric cylinder immersed in a homogeneous medium has shown [14]-[18] that, under the Cherenkov condition for the material of the cylinder and for the velocity of the particle image on the cylinder surface, there are narrow peaks in the angular distribution of the number of quanta emitted into the exterior space. For some values of the parameters the density of the number of quanta in these peaks exceeds the corresponding quantity for the radiation in vacuum by several orders. The electromagnetic fields and the radiation intensity for a charged particle moving along an arbitrary closed orbit around a dielectric cylinder immersed into a homogeneous medium have been investigated in [19].

In the present paper, we aim to study the sensitivity of the peaks with respect to distortions of the trajectory from circular one and possible new features in the radiation process. As an

Content from this work may be used under the terms of the Creative Commons Attribution 3.0 licence. Any further distribution of this work must maintain attribution to the author(s) and the title of the work, journal citation and DOI.
Published under licence by IOP Publishing Ltd

example of a charge motion, we consider transversal oscillations with respect to mean circular motion. The paper is organized as follows. In the next section we describe the charge motion under consideration and present the expression for the spectral-angular density of the radiation intensity. The features of the radiation and some numerical results are discussed in section 3. The section 4 concludes the main results.

2. Radiation intensity from a transversal oscillator

Consider a point charge q moving around a dielectric cylinder with radius ρ_c and with the dielectric permittivity ε_0. We assume that the cylinder is immersed in a homogeneous medium with the dielectric permittivity ε_1 (magnetic permeabilities for both the cylinder and surrounding medium will be taken to be unit). In cylindrical coordinates (ρ, ϕ, z), with the z-axis along the cylinder axis, the trajectory of the charge is given by the functions $(\rho_e(t), \phi_e(t), z_e(t))$, $\rho_e(t) > \rho_c$. We will assume that along the radial direction the charge makes an oscillatory motion around a circular trajectory with the radius $\rho_0 > \rho_c$ described by

$$\rho_e(t) = \rho_0 \left[1 + a \cos(n_0 \omega_0 t)\right], \tag{1}$$

where n_0 is a positive integer and a is the relative amplitude of the oscillations. Here $\omega_0 = 2\pi/T$ with T being the period of the transversal motion. Hence, n_0 gives the number of transversal oscillations of the charge during one period of the rotation. From the condition $\rho_e(t) > \rho_c$ it follows that $a < 1 - \rho_c/\rho_0$. For the motion along the cylinder axis z we will take the uniform one with the constant velocity $v_\|$: $z_e(t) = v_\| t$.

In order to fix the function $\phi_e(t)$ for the azimuthal motion, we consider a special case with the constant energy of the charge: $\mathcal{E} = \text{const}$. This is the case when the charge moves under a static magnetic field only. For the components of the charge velocity along the radial and azimuthal directions one has $v_\rho = d\rho_e(t)/dt$, $v_\phi = \rho_e(t) d\phi_e(t)/dt$. The function $\phi_e(t)$ is found from the condition $v^2 = v_\perp^2 + v_\|^2 = \text{const}$, where $v_\perp = \sqrt{v_\rho^2 + v_\phi^2}$ is the velocity of the transversal (with respect to the cylinder axis) motion. After the integration, this gives

$$\phi_e(t) = \frac{1}{\rho_0} \int_0^t dt \frac{\sqrt{v_\perp^2 - a^2 n_0^2 v_{\perp 0}^2 \sin^2(n_0 \omega_0 t)}}{1 + a \cos(n_0 \omega_0 t)}, \tag{2}$$

where $v_{\perp 0} = \omega_0 \rho_0$. Note that $v_{\perp 0}$ is the velocity of the motion along the circle with radius ρ_0 having the same period T. In particular, we can take $v_\perp = v_{\perp 0}$. In this case, we should additionally require that $a n_0 < 1$ and, in the limit $a \to 0$, the motion under consideration reduces to a uniform circular motion with the period T. For the components of the velocity one finds

$$v_\rho = -n_0 a v_{\perp 0} \sin(n_0 \omega_0 t), \; v_\phi = \sqrt{v_\perp^2 - v_\rho^2}. \tag{3}$$

The transversal oscillations can be generated by using magnetic structures similar to those employed in planar undulators (see, for example, [20]).

In what follows, we will denote by θ the angle between the wave vector of the radiated photon and the cylinder axis. The frequency of the radiation, $|\omega_n|$, on a given harmonic $n = 1, 2, \ldots$, is given by the expression

$$\omega_n = \frac{n \omega_0}{1 - \beta_{1\|} \cos\theta} \tag{4}$$

with the notation $\beta_{1\|} = v_\| \sqrt{\varepsilon_1}/c$. For the general case of the charge trajectory $(\rho_e(t), \phi_e(t), v_\| t)$, at large distances from the cylinder, the angular density of the radiation intensity on the harmonic n, averaged over the period T, is presented as [19]

$$\frac{dI_n}{d\Omega} = \frac{q^2}{4\pi^3 c} \frac{n^2\omega_0^2\sqrt{\varepsilon_1}}{\left|1-\beta_{1\|}\cos\theta\right|^3} \sum_{m=-\infty}^{+\infty} \left[\left|\sum_{p=\pm1} pD_{m,n}^{(p)}\right|^2 + \left|\sum_{p=\pm1} D_{m,n}^{(p)}\right|^2 \cos^2\theta \right], \tag{5}$$

where $d\Omega = \sin\theta d\theta d\varphi$ is the solid angle element,

$$\begin{aligned} D_{m,n}^{(p)} &= \frac{\pi}{2ic}\left[J_{m+p,m}(n,\lambda_1) - H_{m+p,m}(n,\lambda_1)\frac{W(J_{m+p},J_{m+p})}{W(J_{m+p},H_{m+p})}\right] \\ &+\frac{i\pi\lambda_1}{2ck_z}\left[J_{m,m,z}(n,\lambda_1) - H_{m,m,z}(n,\lambda_1)\frac{W(J_m,J_m)}{W(J_m,H_m)}\right] \\ &+\frac{pJ_m(\lambda_0\rho_c)J_{m+p}(\lambda_0\rho_c)}{c\rho_c\alpha_m W(J_{m+p},H_{m+p})}\left[k_z\frac{H_{m,m,z}(n,\lambda_1)}{W(J_m,H_m)} + \frac{\lambda_0}{2}\sum_{l=\pm1}\frac{H_{m+l,m}(n,\lambda_1)}{W(J_{m+l},H_{m+l})}\right], \end{aligned} \tag{6}$$

and

$$\lambda_0 = \frac{\omega_n}{c}\sqrt{\varepsilon_0 - \varepsilon_1\cos^2\theta},\ \lambda_1 = \frac{\omega_n}{c}\sqrt{\varepsilon_1}\sin\theta,\ k_z = \frac{\omega_n}{c}\sqrt{\varepsilon_1}\cos\theta. \tag{7}$$

Here and below, $J_m(x)$ and $H_m(x)$ are the Bessel and Hankel (of the first kind) functions, and

$$W(J_{m+p},F_{m+p}) = J_m(\lambda_0\rho_1)\frac{\partial F_m(\lambda_1\rho_1)}{\partial\rho_1} - F_m(\lambda_1\rho_1)\frac{\partial J_m(\lambda_0\rho_1)}{\partial\rho_1}, \tag{8}$$

where $F = J$ and $F = H$. Other notations are defined by the relations

$$\begin{aligned} F_{m',m,l}(n,\lambda_1) &= \frac{1}{T}\int_0^T dt\, v_l(t)F_{m'}(\lambda_1\rho_e(t))e^{-im\phi_e(t)+in\omega_0 t}, \\ F_{m+p,m}(n,\lambda_1) &= F_{m+p,m,\phi}(n,\lambda_1) - ipF_{m+p,m,\rho}(n,\lambda_1), \end{aligned} \tag{9}$$

where $l = \rho, \phi, z$ and

$$\alpha_m = \frac{\varepsilon_0}{\varepsilon_1 - \varepsilon_0} - \frac{\lambda_0}{2}J_m(\lambda_0\rho_1)\sum_{l=\pm1}\frac{lH_{m+l}(\lambda_1\rho_1)}{W(J_{m+l},H_{m+l})}. \tag{10}$$

The equation $\alpha_m = 0$ determines the eigenmodes of the dielectric cylinder for the electromagnetic field. The information on the charge motion is encoded in the functions $F_{m',m,l}(n,\lambda_1)$ with $m' = m, m\pm1$.

3. Features of the radiation

First let us consider the special case of the charge motion along a circular helix. In this case one has $v_l(t) = \text{const}$, $\rho_e(t) = \rho_0$, $\phi_e(t) = \omega_0 t$ and, hence, $F_{m',m,l}(n,\lambda_1) = v_l F_{m'}(\lambda_1\rho_0)\delta_{mn}$. The only nonzero contribution to the series over m in (5) comes from the term with $m = n$ and we obtain the result of [17] (for the radiation emitted by a charged particle moving along a circular helical orbit inside a dielectric cylinder see [16]). By using Debye's asymptotic expansions for the Bessel and Neumann functions with large values of the order, in [17] it has been shown that, under the condition $|\lambda_1|\rho_c < n$, at points where the real part of the function α_n, given by formula (10), is equal to zero, the contribution of the imaginary part of this function into the coefficients $D_{n,n}^{(p)}$ in the expression of the radiation intensity can be exponentially large for large values n. This gives rise to strong narrow peaks in the angular distribution for the radiation intensity at a given harmonic n. The condition for the real part of the function α_n to be zero is obtained from the equation determining the eigenmodes for the dielectric cylinder by the replacement $H_n \to Y_n$, where $Y_n(x)$ is the Neumann function. This equation has no solutions for $\lambda_0^2 < 0$, which is possible only for $\varepsilon_0 < \varepsilon_1$. Hence, the peaks do not appear for the case $\lambda_0^2 < 0$ which corresponds to the angular region $\cos^2\theta > \varepsilon_0/\varepsilon_1$. As necessary conditions for the presence of the strong narrow peaks in the angular distribution for the radiation intensity one has $\varepsilon_0 > \varepsilon_1$, $\widetilde{v}\sqrt{\varepsilon_0}/c > 1$, were $\widetilde{v} = \sqrt{v_\parallel^2 + \omega_0^2\rho_c^2}$ is the velocity of the charge image on the cylinder surface. The second condition is the Cherenkov condition for the velocity of the charge image on the cylinder surface and dielectric permittivity of the cylinder.

In figure 1, by the full curves we have plotted the angular density of the number of the radiated quanta per period T,

$$\frac{dN_n}{d\Omega} = \frac{T}{\hbar\,|\omega_n|}\frac{dI_n}{d\Omega}, \tag{11}$$

as a function of the angle θ, in the case of a circular motion along the orbit with $\rho_c/\rho_0 = 0.7$ around a dielectric cylinder with permittivity $\varepsilon_0/\varepsilon_1 = 3.75$. The numbers near the curves correspond to the values of the harmonic n ($n = 6, 8$ for figure 1(a) and $n = 10, 12$ for figure 1(b)). The dashed curves correspond to the radiation in a homogeneous medium with permittivity ε_1. For the other parameters we have taken $\beta_{1\parallel} = 0.4$, $\beta_{1\perp} = v_\perp\sqrt{\varepsilon_1}/c = 0.9$, $v_{\perp 0} = v_\perp$. For the peak at $\theta \approx 0.69$ one has $dN_n/d\Omega \approx 2.88q^2\sqrt{\varepsilon_1}/\hbar c$. Note that, for given values of $\beta_{1\parallel}$ and $\beta_{1\perp}$, the quantity $\varepsilon_1^{-1/2}dN_n/d\Omega$ depends on the permittivities ε_0 and ε_1 in the form of the ratio $\varepsilon_0/\varepsilon_1$. When the charge moves in the vacuum one has $\varepsilon_1 = 1$ and for the graphs presented in figure 1 $\varepsilon_0 = 3.75$. The latter corresponds to a cylinder made of quartz. In figure 2 we have presented the selected peaks from figure 1b for $n = 10$ and $n = 12$.

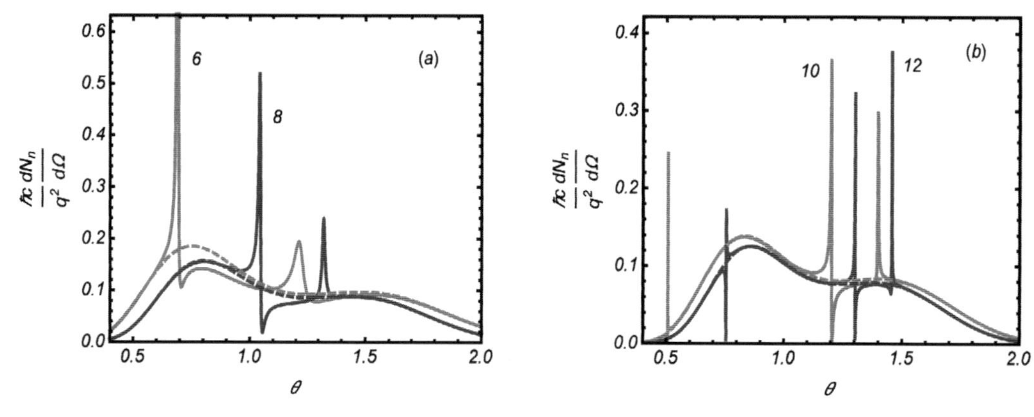

Figure 1. The angular density of the number of the quanta, radiated per period T for the motion along a circular helix on the harmonics $n = 6, 8, 10, 12$ (numbers near the curves), as a function of the angle θ. The dashed curves correspond to the radiation in a homogeneous medium with dielectric permittivity ε_0 immersed in the same medium. The other values of the parameters are $\beta_{1\parallel} = 0.4$, $\beta_{1\perp} = 0.9$, $\rho_c/\rho_0 = 0.7$.

In the case of the motion described by (1) and (2), the transversal oscillations of the charge serve as an additional tool to control the height and width of the peaks in the angular distribution of the radiation intensity. In addition, the oscillations may induce new angular peaks. Let us denote by $dN_{n,m}/d\Omega$ the contribution of the term with a given m in (5) to the angular density of the number of the radiated quanta. In table 1 we present this contribution at the angle $\theta \approx 0.69$ in the case $m = n = 6$, corresponding to the peak in the left panel of figure 1, for $n_0 = 3$ and for

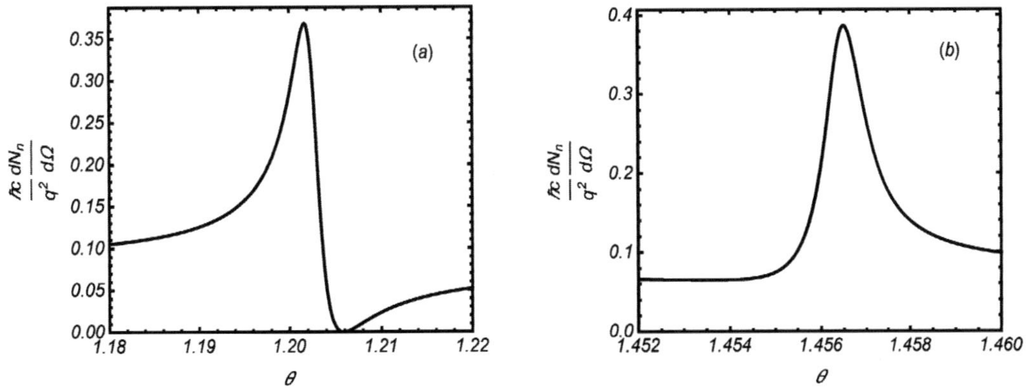

Figure 2. The angular density of the number of the quanta near the selected peaks for $n = 10$ (a) and $n = 12$ (b). The values of the parameters are the same as those for figure 1(b).

various values of the relative amplitude of the oscillations. The values of the other parameters are the same as those for figure 1. For small values of a the contribution of the term with $m = n$ dominates. For example, in the case $a = 0.1$ for the contribution $(\hbar c/q^2\sqrt{\varepsilon_1})dN_{6,m}/d\Omega$ at the peak $\theta \approx 0.69$ one has 3.6×10^{-4}, 1.1×10^{-3}, 9.4×10^{-4}, 3.4×10^{-5} for $m = 4, 5, 7, 8$, respectively.

Table 1.

a	0	0.05	0.1	0.15	0.2	0.25	0.28
$(\hbar c/q^2\sqrt{\varepsilon_1})dN_{6,6}/d\Omega$	2.88	2.90	2.93	2.78	1.97	0.42	0.016

As is seen from the table, the dependence on the amplitude of the oscillations is not monotonic. After an initial increase, it decreases with increasing a.

4. Conclusion

We have investigated the radiation intensity from a transversal oscillator rotating around a dielectric cylinder. The radial oscillations are described by (1). We have assumed that the energy of the charge is fixed. From this condition the time dependence of the azimuthal coordinate of the charge is obtained. The latter is given in the integral form (2). For general functions $\rho_e(t)$ and $\phi_e(t)$, the angular density of the radiation intensity on a given harmonic is given by the expression (5) with the coefficients defined by (6). In the case of a circular helix coaxial with the cylinder axis, in the summation over m in (5) the term $m = n$ survives only and we recover the results previously discussed in [17]. For this case, under certain conditions for the velocity of the charge and for the dielectric permittivity of the cylinder, strong narrow peaks may appear in the angular distribution of the radiation intensity. We have shown that these peaks appear also in the presence of additional transversal oscillations of the charge if the oscillations amplitude is relatively small.

Refernces

[1] Ter-Mikaelian M L 1972 *High Energy Electromagnetic Processes in Condensed Media* (New York: Wiley Interscience)

[2] Baier V N, Katkov V M and Strakhovenko V M 1998 *Electromagnetic Processes at High Energies in Oriented Single Crystals* (Singapore: World Scientific)
[3] Rullhusen R, Artru X and Dhez P 1998 *Novel Radiation Sources Using Relativistic Electrons* (Singapore: World Scientific)
[4] Potylitsyn A P 2011 *Electromagnetic Radiation of Electrons in Periodic Structures* (Berlin: Springer)
[5] Tsytovich V N 1951 *Westnik MGU* **11** 27 (in Russian)
[6] Kitao K 1960 *Prog. Theor. Phys.* **23** 759
[7] Erber T, White D and Latal H G 1976 *Acta Phys. Austriaca* **45** 29
[8] Schwinger J, Tsai W-Y and Erber T 1976 *Ann. Phys.* **96** 303
[9] Erber T, White D, Tsai W-Y and Latal H G 1976 *Ann. Phys.* **102** 405
[10] Rynne T M, Baumgartner G B and Erber T 1978 *J. Appl. Phys.* **49** 2233
[11] Villaroel D and Fuenzalida V 1987 *J. Phys. A: Math. Gen.* **20** 1387
[12] Villaroel D 1987 *Phys. Rev. A* **36** 2980
[13] Villaroel D and Milan C 1987 *Phys. Rev. D* **38** 383
[14] Grigoryan L Sh, Kotanjyan A S and Saharian A A 1995 *Sov. J. Contemp. Phys.* **30** 239
[15] Kotanjyan A S, Khachatryan H F, Petrosyan A V and Saharian A A 2000 *Sov. J. Contemp. Phys.* **35** 1
[16] Saharian A A and Kotanjyan A S 2005 *J. Phys. A: Math. Gen.* **38** 4275
[17] Saharian A A and Kotanjyan A A 2009 *J. Phys. A: Math. Theor.* **42** 135402
[18] Kotanjyan A A and Saharian A A 2013 *NIMB* **309** 177
[19] Kotanjyan A S and Saharian A A 2014 *J. Phys. Conf. Ser.* **517** 012025
[20] Hofmann A 2004 *The Physics of Sinchrotron Radiation* (Cambridge: Cambridge University Press)

Radiation from relativistic electrons in "light" and in conventional undulators. Classical and quantum approaches

A P Potylitsyn[1,2], A M Kolchuzhkin[3] and S A Strokov[1]

[1] National Research Tomsk Polytechnic University, 634050 Tomsk, Russia
[2] National Research Nuclear University "MEPhI", 115409 Moscow, Russia
[3] Moscow State Technological University, 127994 Moscow, Russia

E-mail: potylitsyn@tpu.ru

Abstract. A photon spectrum of undulator radiation (UR) is calculated in the semi-classical approach. The UR intensity spectrum is determined by an electron trajectory in the undulator neglecting by energy losses for radiation. Using the Planck's law, the UR photon spectrum can be calculated from the classical intensity spectrum both for linear and nonlinear regimes. The radiation of an electron in a field of strong electromagnetic wave (radiation in the "light" undulator) is considered in the quantum electromagnetic frame. Comparison of results obtained by both approaches has been shown that UR spectra in the whole cone coincide with high accuracy for the case $x \ll 1$. Characteristics of the collimated UR beam were simulated with taking into account the discrete process of photon emission along an electron trajectory in both kinds of undulators.

1. Introduction

One of the options of the future International Linear Collider (ILC) includes the polarized positron source based on a helical undulator with length ~100 m placed downstream the main linear accelerator. The energy of electron beam is about 150-250 GeV, the photon yield is $\sim 10^2$ photon/electron with energy ≥ 10 MeV, and polarization ~0.5 [1]. As a rule, characteristics of UR are calculated using the classical electrodynamics [2] and such an approach was used to calculate UR characteristics for the ILC positron sources [3, 4].

In our work we have compared characteristics of UR calculated using the classical electrodynamics and the quantum electrodynamics. We used the analogy between radiation of electrons in a "light" undulator and in the conventional undulator. In the former case, radiation is treated as the Compton back scattering process [5].

2. Photon spectrum from conventional magnetic undulator.

For the sake of simplicity, we consider UR from a helical undulator with a period λ_u, an undulator parameter K (nonlinearity parameter), and number of period N_u. In such an undulator, the electron

Content from this work may be used under the terms of the Creative Commons Attribution 3.0 licence. Any further distribution of this work must maintain attribution to the author(s) and the title of the work, journal citation and DOI.
Published under licence by IOP Publishing Ltd

trajectory is the helix. In the system where an electron is at rest (in average) (*R*-system) the trajectory of an electron is a circular with a radius [6]:

$$R \approx \frac{K\lambda_u}{\pi\gamma_0}, \text{ if } K \sim 1, \tag{1}$$

here γ_0 is the Lorentz-factor.

Using the well known formulas for synchrotron radiation, one can easily obtain formulas for radiation intensity after Lorentz-transformation from *R*-system to laboratory system [7]

$$\frac{dW}{d\Omega} = \frac{8\alpha\hbar\omega_0 N_u \gamma_0^4}{(1+K^2+\gamma^2\theta^2)^3} K^2 \sum_{n=1}^{\infty} n^2 \left[J_n'^2(nZ) + \left(\frac{\gamma_0\theta}{K} - \frac{1}{Z} \right)^2 J_n^2(nZ) \right], \tag{2}$$

where $\omega_0 = \frac{2\pi c}{\lambda_u}\left(1 - \frac{1+K^2}{2\gamma_0^2}\right)$ is the frequency of the fundamental harmonic, θ is outgoing photon angle, n is the number of the harmonic, $Z = \frac{2K\gamma_0\theta}{1+K^2+\gamma_0^2\theta}$.

The formula (2) is obtained in the small outgoing photon angle approximation for the long undulator ($N_u >> 1$). There is the well known relation connecting the frequency of *n*-th harmonic and outgoing angle [6]:

$$\omega^{(n)} = n \frac{2\gamma_0^2}{1+(\gamma_0\theta)^2+K^2} \frac{2\pi c}{\lambda_u}. \tag{3}$$

Using this relation, it is conveniently to transform the expression (2) into the spectral distribution ($dW/d\Omega \to dW/d\omega$) and after that to use a dimensionless spectral variable instead a frequency:

$$S^{(n)} = \frac{\omega^{(n)}}{2\gamma_0^2 2\pi/\lambda_u} = \frac{n}{1+(\gamma_0\theta)^2+K^2}, \quad 0 \le S^{(n)} \le S_{\max}^{(n)}, \quad S_{\max}^{(n)} = \frac{n}{1+K^2}. \tag{4}$$

The photon UR spectrum for *n*-th harmonic can be obtained from the UR intensity spectrum dividing by the emitted photon energy $\hbar\omega^{(n)}$:

$$\frac{dN^{(n)}}{dS^{(n)}} = 4\pi\alpha K^2 N_u \left\{ \frac{\left[n - S^{(n)} 2(1+K^2)\right]^2}{4S^{(n)}K^2\left[n - S^{(n)}(1+K^2)\right]} J_n^2(nZ) + J_n'^2(nZ) \right\}. \tag{5}$$

The number of photons emitted at the *n*-th harmonic can be calculated after integration of the spectral distribution (5):

$$N^{(n)} = \int_0^{S_{\max}^{(n)}} \frac{dN^{(n)}}{dS^{(n)}} dS^{(n)}. \tag{6}$$

The total number of emitted photons can be found by summing up the contributions from the *n*-th harmonic:

$$N_{tot} = \sum_{n=1}^{\infty} N^{(n)} \approx \sum_{n=1}^{n_{\max}} N^{(n)}. \tag{7}$$

For the typical case in UR ($K \sim 1$), the sum (7) is converges quickly and criterium for the upper harmonic number $n_{\max}$ can be chosen from the following condition [8]:

$$\frac{N^{(n_{\max}+1)}}{N_{tot}} \leq 0.005. \tag{8}$$

The dependence of the emitted photons number on the harmonic number per an electron and per an undulator period for the case $K = 1$ is shown in figure 1.

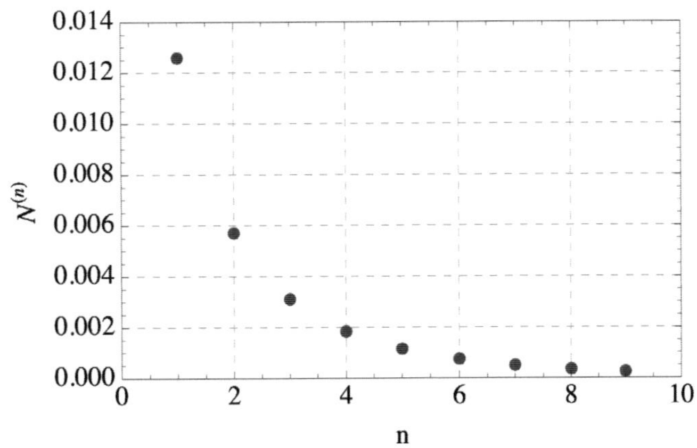

Figure 1. Dependence of the emitted photons number on the harmonic number.

The criterion (8) is fulfilled for $n_{\max} = 9$.
The spectral distribution for different harmonic numbers are presented in figure 2 and the resulting spectrum in figure 3. Both distributions were calculated per an undulator period.

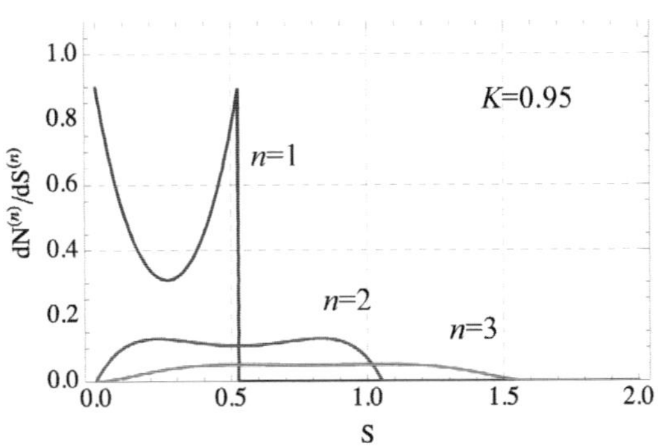

Figure 2. Spectral distribution for the first three harmonics.

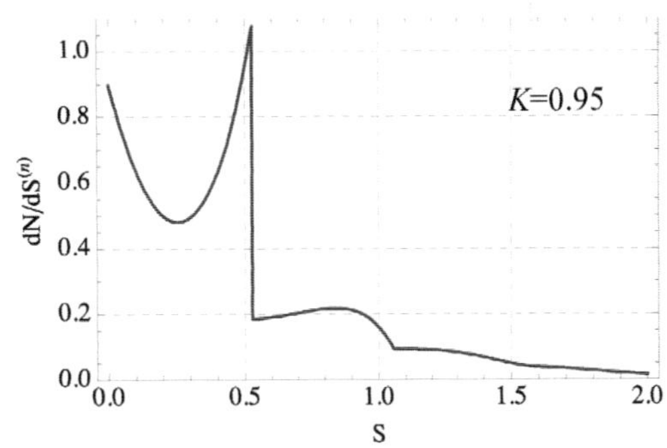

Figure 3. Resulting spectrum of the summarized harmonics.

The spectra presented above depend on the parameter K only and have an universal character allowing to transform them into real photon spectra for any values of γ_0 and K.

The circular polarization of UR depends on the photon outgoing angle or, using relation (3), can be written as the spectral dependence:

$$\xi_2(S^{(n)}) = \sum_{n=1}^{n_{\max}} 2\pi\alpha K N_u \frac{\left[2(1+K^2)S^{(n)} - n\right]}{\sqrt{S^{(n)}}\left[n - S^{(n)}(1+K^2)\right]} J_n(nZ) J_n'(nZ) \Bigg/ \sum_{n=1}^{n_{\max}} \frac{dN^{(n)}}{dS^{(n)}} . \qquad (9)$$

Such a dependence is shown in figure 4, which also can be used for calculations of energy dependence for fixed parameters γ_0 and λ_u.

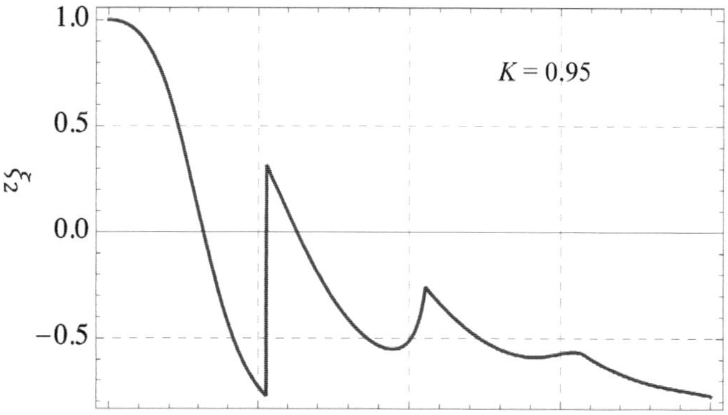

Figure 4. Degree of circular polarization.

The circular polarization of collimated UR is calculated using the same expression (9) but all harmonics are calculated in the limit:

$$S^{(n)}_{\min coll} \le S^{(n)} \le S^{(n)}_{\max}, \; S^{(n)}_{\min coll} = \frac{n}{1+(\gamma_0\theta_c)^2 + K^2}, \qquad (10)$$

where θ_c is the angular collimator aperture.

As was shown in [9, 10], the distribution of the number of emitted UR photons is described by the Poisson law with the mean value N_{tot}. As a rule $N_{tot} > 1$ and it means that a photon may be emitted from an arbitrary point along a trajectory of an electron.

The simulation of such a process in a "light" undulator (the quantum Compton backscattering process) was performed in [9, 10] with taking into account a stochastic process of radiation along a trajectory. The quantum recoil effect for the case of a "light" undulator leads to so-called "red shift" in resulting spectra, but for the case $x = \frac{4\gamma_0\hbar\omega_0}{mc^2} << 1$ ($\hbar\omega_0$ is the energy of a laser photon) such an effect is negligible but the quantum discreetness of radiation process continues to play the important role. The comparison of the calculated "semi classical" UR spectrum using formulas (5, 6) with simulated spectra based on the quantum approach is shown in figure 4a for radiation in whole cone and for collimated UR spectrum in figure 5b.

Comparison were performed for an undulator with following parameters: $E_0 = 150$ GeV (γ_0 = 293500); λ_u = 11.5 mm; K = 0.95, (see [11]) and for "light" undulator with the wavelength twice less due to difference between both kinds of undulators [5].

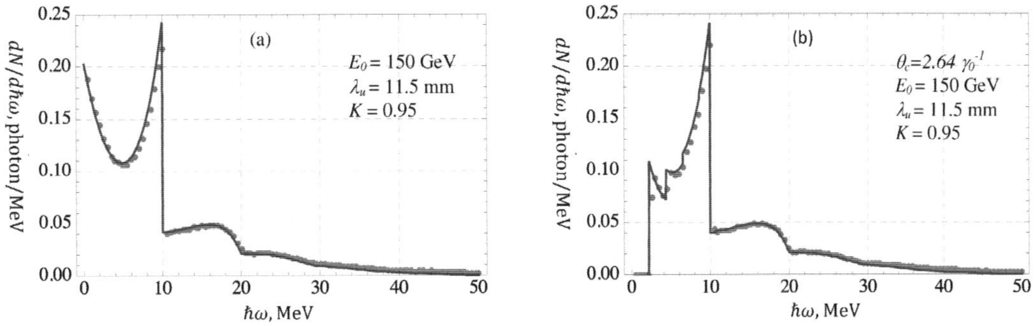

Figure 5. Comparison of the semiclassical UR spectrum (solid line) with simulated spectra based on the quantum approach in the whole cone (a) and collimated (b).

We would like to emphasize that spectra from conventional and "light" undulators were obtained in absolute units (photon per 1 MeV) and comparison showed the good agreement between them.

A tight collimation of UR generated in a long undulator which can be used for the ILC positron source design may lead to some distortion of spectra and, essentially, of resulting polarization due to different distances from an emission point and the collimator.

We simulated the effect of a "long" undulator for the following undulator parameters [12]:

$L_u = 231$ m
$\lambda_u = 11.5$ mm
$N_u = 2 \times 10^4$
$K = 0.92$
$E_e = 250$ GeV
$R_c = 0.7$ mm
Distance from the center of
the undulator to collimator = 412 m

and calculated the number of emitted photons per electron per period for each harmonic (table 1).

Table 1. Number of emitted photons per electron per period for each harmonic.

n	1	2	3	4	5	6	7	8
$N^{(n)}$	0.01167	0.00494	0.00251	0.00138	0.00079	0.00047	0.00028	0.00017

It means that the number of photons emitted from the all undulator length

$$N_0 = N_{tot} \frac{L_u}{\lambda_u} \approx 446.$$

For this case, instead of the Poisson distribution it is possible to use the Gaussian distribution

$$F_g(k, \bar{k}, \sigma) = \frac{1}{\sigma\sqrt{2\pi}} e^{-\frac{(k-\bar{k})^2}{2\sigma^2}}, \quad (11)$$

where $\bar{k} = N_0$, $\sigma = \sqrt{N_0}$.

Because the distribution F_g is very "narrow" for the sake of simplicity, we have changed simulation of the random emission points with taking into account a different number of emitted photons $\bar{k}$ by the uniform distribution along trajectory with fixed number of emitted photons N_0.

The simulation was performed using the following algorithm:

- calculation of $n_{\max}$, N_{tot} and spectral distributions for $n \le n_{\max}$;
- simulation of the harmonic number n;
- simulation of the spectral variable $S^{(n)}$ and corresponding angle θ;
- the condition of passing collimator $\theta \le \theta_c$ is applied for each point of photon creation;
- for "useful" events, creation of histograms according to the formulas (6) and (9);
- averaging by points of photons' creation along the trajectory where length L_u was divided uniformly by 200 points. Further increasing of the trajectory dividing doesn't lead to improvement of simulation results;
- total number of simulated photons was two million.

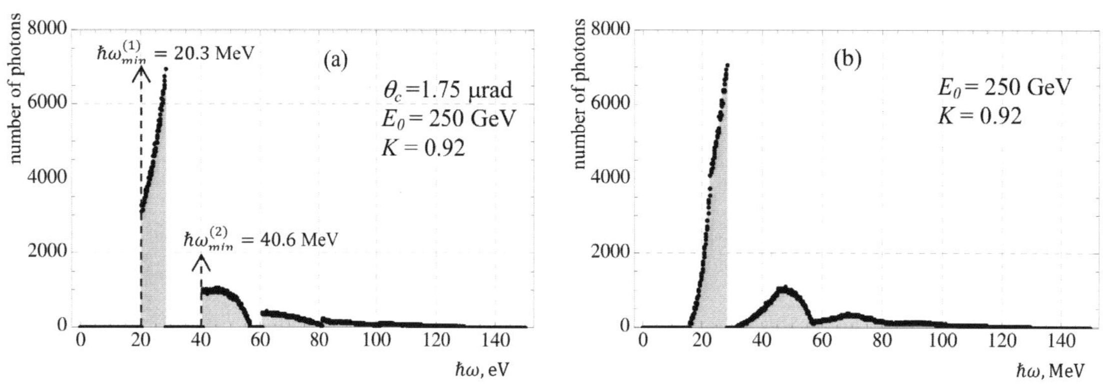

Figure 6. Spectrum of collimated UR emitted from the center of undulator (a) and along the length of undulator (b).

The simulation results are shown in figure 6. The effect of a "long" undulator leads to the "smoothing" of the spectral line. In comparison with the "ideal" case (figure 6a), such a smoothing leads to an appearance of photons with energies less than $\hbar\omega_{min}$. For the considered case, a part of such photons is about 8% for the first harmonic and about 15% for the second one.

The distortion of the circular polarization degree due to this reason will be considered elsewhere.

References

[1] *ILC Technical Design Report*, www.linearcollider.org/ILC/Publications/Technical-Design-Report

[2] Kincaid B M 1977 *J. Appl. Phys.* **48** 2684

[3] Flottmann K 1993 *Preprint DESY* 93-161

[4] Alexander G, Barley J, Batygin Y, Berridge S, Bharadwaj V, Bower G, Bugg W, Decker F-J, Dollan R, Efremenko Y, Flöttmann K, Gharibyan V, Hast C, Iverson R, Kolanoski H, Kovermann J W, Laihem K, Lohse T, McDonald K T, Mikhailichenko A A, Moortgat-Pick G A, Pahl P, Pitthan R, Pöschl R, Reinherz-Aronis E, Riemann S, Schälicke A, Schüler K P, Schweizer T, Scott D, Sheppard J C, Stahl A, Szalata Z, Walz D R, Weidemann A, 2009 *Nucl. Instrum. and Meth.* A **610** 451-487

[5] Corde S, Ta Phuoc K, Lambert G, Fitour R, Malka V, Rousse A, Beck A, and Lefebvre E 2013

Rev. Mod. Phys. **85** 1
[6] Sarachik E S and Skappert G T 1970 *Phys. Rev.* D **1** 2738
[7] Potylitsyn A P 2011 *Electromagnetic Radiation of Electrons in Periodic Structures* (Springer)
[8] Potylitsyn A P, Kolchuzhkin A M 2013 *Nucl. Instrum. and Meth.* B **309** 15
[9] Potylitsyn A P, Kolchuzhkin A M 2012 *Eur. Phys. Lett.* **100** 24006
[10] Potylitsyn A P, Kolchuzhkin A M 2014 *Physics of Particles and Nuclei* **45** 1000
[11] Alrashdi A, Bailey I, Newton D, et al 2014 *Proc. of IPAC 2014*, Dresden, Germany, 586
[12] Ushakov A 2012 *Polarization issues (presentation)* LCWS12 International Workshop on Future Linear Colliders (University of Texas at Arlington)

Electron beam diagnostics tool based on Cherenkov radiation in optical fibers

A V Vukolov, A I Novokshonov, A P Potylitsyn, S R Uglov

Department of Applied Physics, National Research Tomsk Polytechnic University, 30 Lenina Street, Tomsk, Russia

E-mail: novokshonov@tpu.ru

Abstract. The results of experimental investigations of Cherenkov radiation in optical fibers with 0.6 mm thickness which were used to scan an electron beam of 5.7 MeV energy are presented. Using such a technique for beam profile measurements it is possible to create a compact and reliable device compared to existing systems based on ionization chambers.

1. Introduction

A wide application of electron accelerators with energy ~ 10 MeV in different fields (essential for medical treatment) determines a necessity for beam parameter diagnostics in on-line mode. One of these important parameters is the transverse beam profile. The most widespread technique of transverse beam profile measurement is based on detection of bremsstrahlung generated by the electrons passing through a thin metal wire [1, 2]. Such a diagnostics device has several disadvantages such as weak sensitivity and inability of gamma beam diagnostics.

There is another technique for beam diagnostics, which is used in intraoperative therapy, based on the ionization chamber. But this device has insufficient spatial resolution exceeding 1 mm [3, 4]. One can use X-ray films for beam profile diagnostics to reach submillimeter resolution [5], but such a technique is an off-line one.

The development of the new approaches for beam profile measurements is the actual task.

In the works [6-8] the feasibility of Cherenkov radiation in glass fibers for high energy beam diagnostics was demonstrated but application of such a technique for moderate energy electron beams was lacking. In our work the glass fibers were used for a few MeV energy electron beams with submillimeter spatial resolution. This technique is based on Cherenkov radiation detection which is generated by an electron beam in a bulk of 0.6 mm thickness optical fiber. The fiber was covered by the scattering lead foil to ensure electrons with broad angular distributions overlapped the Cherenkov angle inside the fiber.

2. Experimental setup and theoretical estimations

Investigations of characteristics of the developed device were carried out at the microtron MI-6 of Tomsk Polytechnic University. The main microtron parameters are shown in table 1 and the schematic of experiment is shown in figure 1.

Content from this work may be used under the terms of the Creative Commons Attribution 3.0 licence. Any further distribution of this work must maintain attribution to the author(s) and the title of the work, journal citation and DOI.
Published under licence by IOP Publishing Ltd

Table 1. The main microtron parameters.

Parameter	Value
Energy of electrons, MeV	5.7
Monochromaticity	less than 3 %
Spill frequency, Hz	25
Duration of the macropulse, μs	0.5
Number of bunches in a macropulse	$\sim 10^4$
Bunch population	$\sim 10^7$

Cherenkov radiation was generated by the electron beam inside the fiber with 0.6 mm thickness. Construction of a fiber represents a glass optical wire with 0.6 mm thickness with a plastic outer layer. This fiber was wrapped by a lead foil with 0.2 mm thickness. So the total diameter of optical cable was ~ 2 mm. Registration of radiation was carried out by two different photomultipliers (PMT): silicon PMT Sensl MicroSB and vacuum PMT.

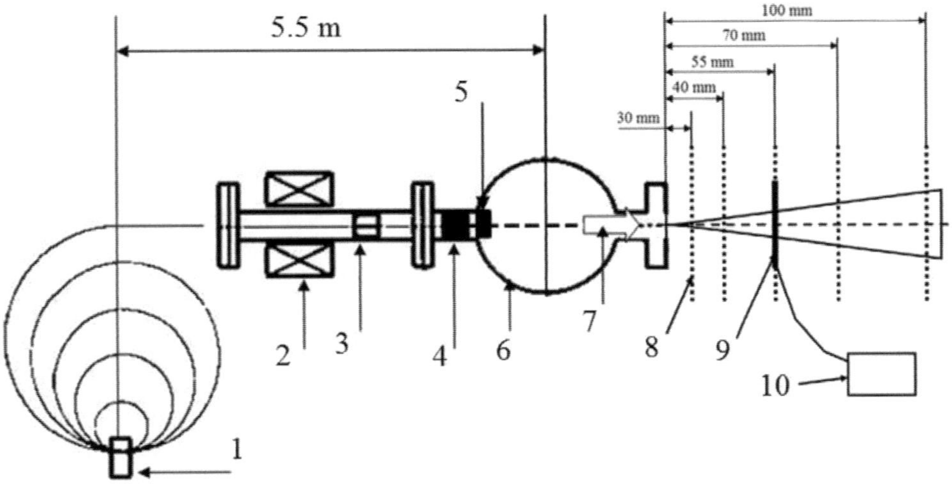

Figure 1. Scheme of the setup: 1 – microtron injector; 2 – focusing magnet; 3 – beam collimator; 4 – inductive sensor; 5 – second beam collimator; 6 – experimental chamber; 7 – electron beam; 8 – X-ray films (dotted lines); 9 – fiber; 10 – silicon/vacuum PMT.

As shown in figure 1, the electron beam passes through the focusing magnet, beam collimator, inductive sensor, second beam collimator with 1 mm diameter, experimental chamber, and then it goes from vacuum chamber into air through Al window with 100 μm thickness. There were used X-ray films on different distances from the output window. X-ray films were positioned at a range of distances from the output window. Using the same range of positions as for the X-ray films, a fiber was located and moved perpendicular to the electron beam axis in the horizontal or vertical direction to measure the beam profile. Such a technique allowed to measure beam profile and then to compare it with measurements using X-ray films.

It is well-known that Cherenkov radiation is generated in a cone with axis that lies on the electron trajectory. Opening angle of such a cone $\theta_{ch}(\lambda)$ is [5]:

$$\cos\theta_{ch}(\lambda) = \frac{1}{\beta n(\lambda)}, \tag{1}$$

where $\beta = v/c$, λ is wavelength, n is the refraction index of the medium.

Number of photons, generated by one electron per length unit, can be estimated also [9]:

$$\frac{d^2N}{dxd\lambda} = 2\pi\alpha z^2 \sin\theta_{ch}(\lambda)\frac{1}{\lambda^2}, \tag{2}$$

where x is a particle path, α is the thin structure constant, z is a charge of the particle.

Estimation of the Cherenkov photons yield in visible range (from 400 nm to 700 nm) in glass is about 30 photons per one electron and per 1 mm path. The photon outgoing angle in glass is $\theta_{ch} \approx 46^{\circ}$ (refraction index $n = 1.47$). The number of transition radiation photons is about 10^3 times less than the Cherenkov photons yield in 1 mm thickness glass.

The piece of the fiber which was located on the electron beam axis can be inclined relative to the beam axis. With the inclination the number of Cherenkov photons captured in total reflection angle is increased thereby the signal rises.

The length of the fiber was 2 m and it was connected by one of its ends to the PMT.

The main parameters of the silicon PMT are shown in table 2 [10].

Table 2. The main parameters of silicon PMT Sensl MicroSB.

Parameter	Value
Square of PMT sensitive area, mm^2	6×6
Number of pixels	18980
Spectral sensitivity, nm	from 300 to 800
Gain coefficient	3×10^6
Efficiency of photon registration (420 nm)	47%
Supply voltage, V	24.5

Additional advantages of such a PMT are its compact size, insensitivity to magnetic field, low supply voltage compared to vacuum PMT, mechanical reliability, weak reaction on ionizing radiation and possibility of operation in vacuum.

The main parameters of the vacuum PMT are shown in table 3.

Table 3. The main parameters of vacuum PMT.

Parameter	Value
Diameter of sensitive area, mm	25
Spectral sensitivity, nm	from 300 to 850
Gain coefficient	10^6
Light sensitivity of photocathode, A/lm	8×10^{-5}
Sensitivity of photocathode for 410 nm wavelength, A/W	4×10^{-2}
Supply voltage, kV	1.5

The signal which was registered by the PMT, was then treated by analog-to-digital converter (ADC) for digitizing and for program processing.

3. Results of experiments

Oscillograms of the signal from the vacuum and silicon PMTs are presented in figure 2. These oscillograms were measured with different fibers: 5 mm thickness fiber for the vacuum PMT and 0.6 mm thickness fiber for the silicon PMT. One can see that the amplitude of signal from silicon PMT even with the thinner fiber is more than two times higher than the amplitude of signal from vacuum PMT. Such a low level of signal didn`t allow us to use the vacuum PMT in our measurements.

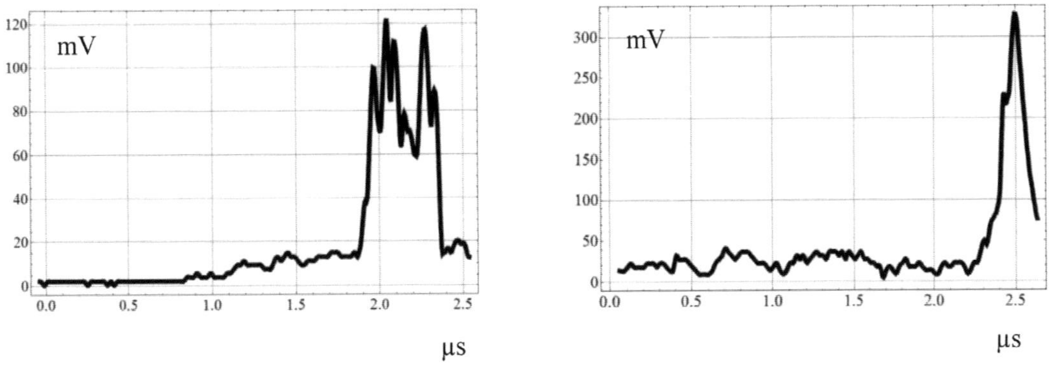

Figure 2. Oscillograms of signal from the vacuum (left one) and from the silicon (right one) PMT.

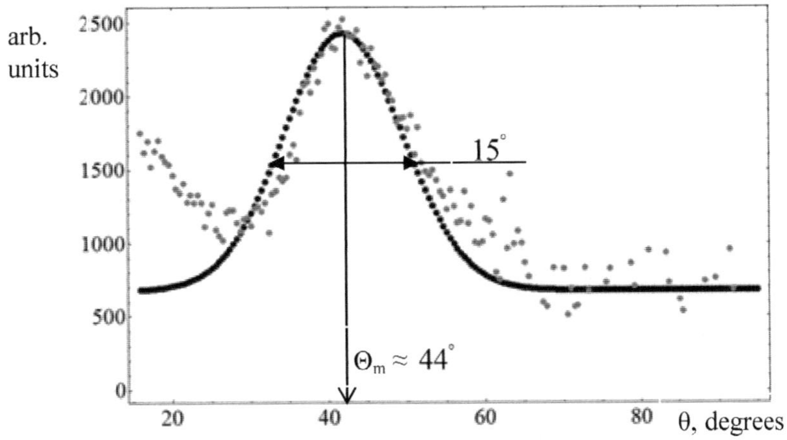

Figure 3. Orientation curve. Grey points – experimental data, black points – fitting of experimental data by normal distribution.

At the first stage of our experiment we have measured orientation curve which is represented in figure 3. The orientation curve was measured by inclination of the fiber relatively to the electron beam axis in horizontal plane. The angle in figure 3 is the angle between fiber and electron beam axis. The cause of high intensity in small angle region is that fiber was held by a metal holder and halo of electron beam interacted with the holder for small inclination angle. The exposure time at each angle was 2 seconds (50 pulses). An averaging of the PMT signal was carried out during 50 pulses.

The orientation curve gives information about the angle of fiber inclination corresponding to the maximal signal and about possibility of the electron beam diagnostics without special scatterer (lead foil) when the inclination angle of the fiber equal to 90°. From figure 3 one can see that when the fiber inclination angle is close to 90° the intensity of Cherenkov radiation is more than 10 times less than in maximum (44°). For increasing of Cherenkov radiation yield for $\theta = 90^{\circ}$ the additional scatterer (lead foil) was used.

In figure 4 experimental measurements of the transverse beam profile are shown with 0.6 mm thickness fiber without scatterer (figure 4a) and with the lead foil (figure 4b) when the distance between the fiber and output window was 70 mm. From figure 4 one can see that such a beam diagnostics is useless in the case when there is no scatterer.

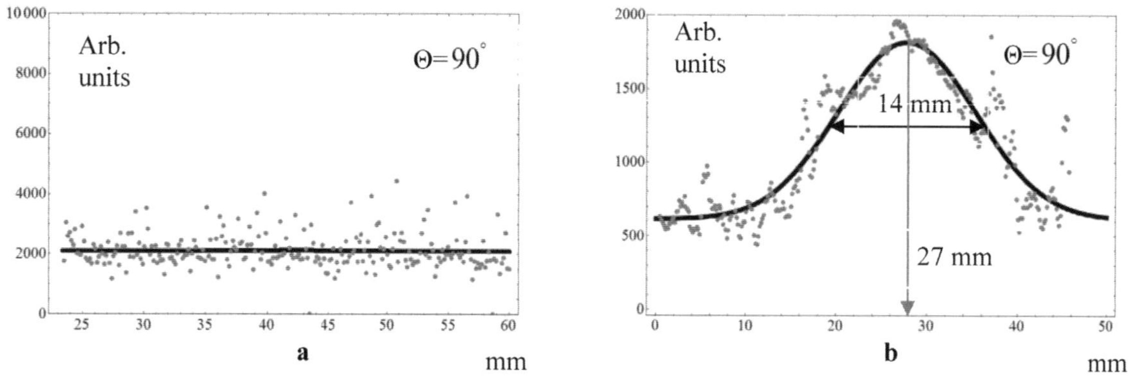

Figure 4. Transverse beam profile: **a** – without scatterer; **b** – with scatterer.

As shown in figure 1, measurements of beam profile were carried out using X-ray films Gafchromic EBT-3 with distances between the films and output window L = 30, 40, 55, 70 and 100 mm. The details of the X-ray films treatment are described in the work [11]. The results from these measurements were compared with results obtained with the fiber. One of the electron beam profiles obtained by X-ray film is shown in figure 5. The beam profile measured for 70 mm distance is shown (figure 5a), as well as its horizontal (figure 5b) and vertical (figure 5c) distributions.

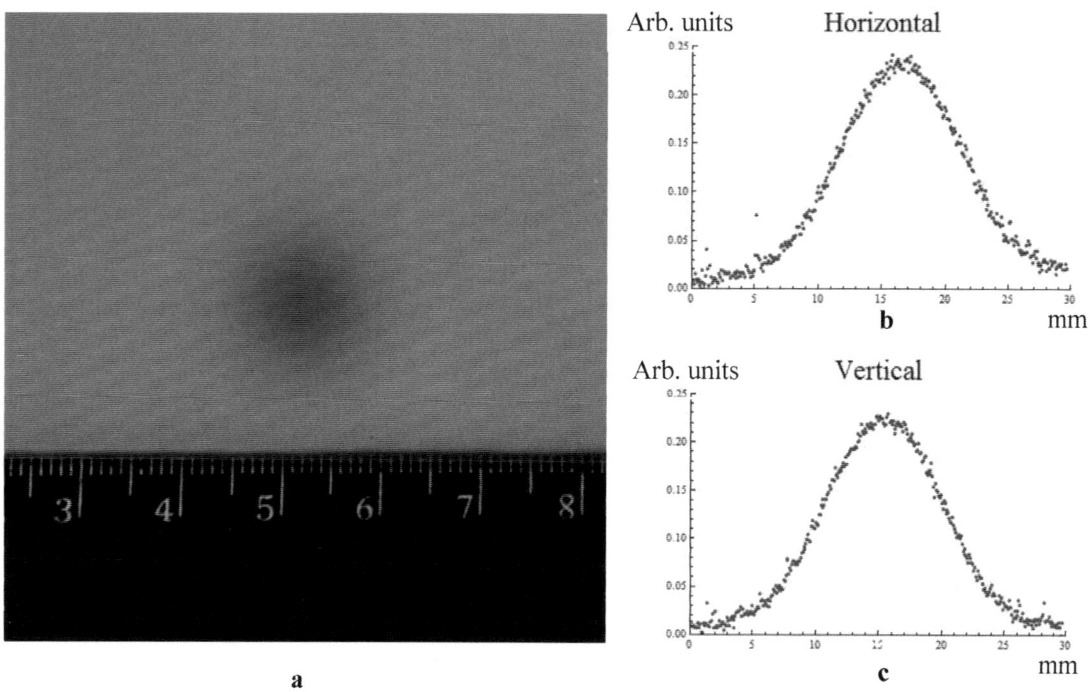

Figure 5. Electron beam profile for 70 mm distance: **a** – X-ray film; **b** – horizontal distribution; **c** – vertical distribution.

In figure 6 horizontal profiles of the electron beam for different distances from the output window measured using the X-ray films are shown.

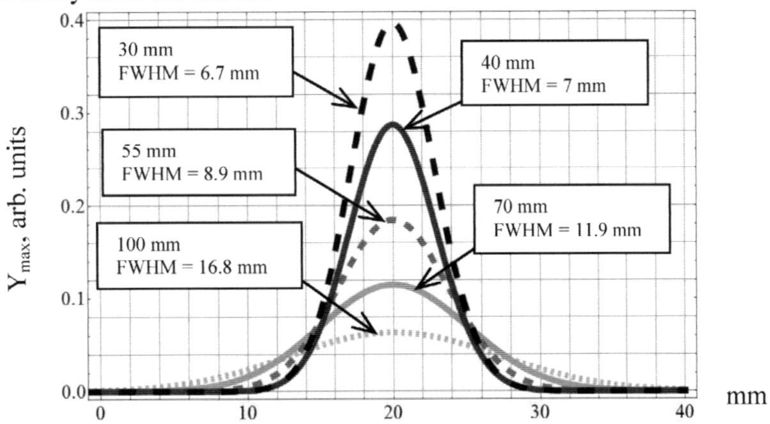

Figure 6. Normal distribution fits for horizontal profiles of electron beam measured by X-ray films.

Figure 7 demonstrates horizontal profiles of the electron beam measured by the fiber with the silicon PMT. Profiles measured by the fiber and profiles measured using the X-ray films are in good agreement. The difference between these two techniques is about 3 mm, which can be explained by the fibers diameter with taking into account a scatterer thickness.

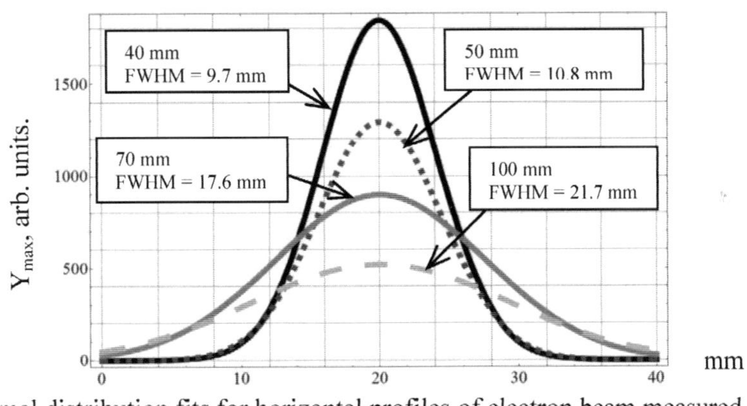

Figure 7. Normal distribution fits for horizontal profiles of electron beam measured by fiber with the silicon PMT.

Experimental dependencies of the transverse beam profile on a distance L between detector and output window measured by X-ray films and fiber, and also simulation results obtained in PCLab (manual for this program can be found in [12]) are presented in figure 8. Experimental dependencies are fitted by polynomials.

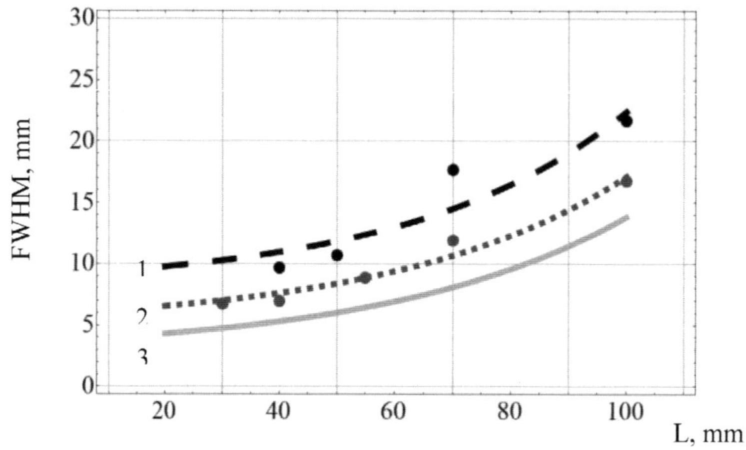

Figure 8. Comparing of experimental results obtained from X-ray films, from fiber and result of simulation curve: 1 – experiment with fiber curve; 2 – experiment with X-ray film curve; 3 – simulating curve using Monte-Carlo technique.

4. Conclusions

Results of the experiments demonstrate a feasibility of transverse electron beam profile measurements for a few MeV energy and low intensity beams using Cherenkov radiation generated in a bulk of optical fibers. The technique proposed in this work has simple realization, small sizes and good signal/noise ratio. The main advantage of this technique is the perpendicular orientation of the fiber relative to the electron beam axis. It means that there are no distortions of experimental results in comparison with the case when the fiber is inclined at the Cherenkov angle θ_m. The proposed technique allows to measure beam profiles in on-line mode.

Acknowledgements

This work was partially supported by the Russian Ministry of Education and Science within the program "Nauka" Grant № 3.709.2014/K.

Authors would like to acknowledge Dr. L. Sukhikh for helping in X-ray film data treatment.

References

[1] Rawnsley W R, Ries T C, Mackenzie G H 1987 *Proc. PAC*
[2] Steibach C, Rooij M van A, 1985 *J Trans. Nucl. Sci.* **5**
[3] Almond P R, Biggs P J, Coursey B M 1999 *J Medical Physics* **26** 1847
[4] Low D A, Moran J M, Dempsey J F et al. 2011 *J Medical Physics* **38** 1313
[5] Schiaparelli P, Zefiro D, Massone F and Tacichi G 2010 *J Medical Physics* **37** 3510
[6] Baudrenghien P and Mastoridis T 2012 *Proc. Advanced Beam Dynamics Workshop on High-Intensity and High-Brightness Hadron Beams* 565-569
[7] Wulf F and Korfer M 2009 *Proc. DIPAC2009* 411
[8] Murokh A, Agustsson R, Boucher S, Frigola P, Hodgetts T, Ovodenko A, Ruelas M and Tikhoplav R 2012 *Proc. IPAC2012* 996
[9] Landau L D and Lifshitz E M 1984 *Electrodynamics of Continuous Media.*
[10] http://sensl.com/downloads/ds/DS-MicroJseries.pdf
[11] Sukhikh E, Sukhikh L and Malikov E 2015 *Advanced Material Research* **1084** 572
[12] Bespalov V I Programm package EPHCA for statistic modeling of photons and charged particles radiation field http://portal.tpu.ru/SHARED/b/BVI/pclab

Electromagnetic field of a charge moving in a waveguide and intersecting a boundary between vacuum and a resonance dispersion medium

Tatiana Yu Alekhina[1] **and Andrey V Tyukhtin**

St. Petersburg State University, 7/9 Universitetskaya nab., St. Petersburg, 199034 Russia

E-mail: t.alehina@spbu.ru

Abstract. Electromagnetic field of a charge that uniformly moves in a circular waveguide from a vacuum into a dielectric possessing frequency dispersion of a resonant type is under consideration. The main attention is paid to the analytical and numerical investigation of the waveguide modes excited by the charge. It is shown that there is a region in the medium where partial compensation of Cherenkov radiation (CR) takes place similar to the case of a nondispersive dielectric. However, the electromagnetic field pattern is complex and the role of resonant dispersion is essential. In the vacuum area of the waveguide, the main part of the field is a response from the resonant medium with frequency which is close to the resonance frequency. This effect might be used for diagnostics of medium properties. The typical plots for the field demonstrate the process of forming the so-called 'wakefield' in the dielectric part of waveguide.

1. Introduction

In this paper, the electromagnetic field of a charged particle that moves along the axis of a circular waveguide and flights from a vacuum area into an area filled up with a medium possessing frequency dispersion of a resonant type is under investigation. Note that the case when the charge flights from the medium into vacuum has been considered in paper [1], and such an interesting effect as a single mode Cherenkov-transition radiation (CTR) has been pointed out. This phenomenon is interesting in connection with development of new methods of generation of gigahertz and terahertz radiation [2-4].

The problems with one or two transversal boundaries between a nondispersive dielectric and a vacuum in a waveguide were analyzed in a series of papers [5-9]. In the case of a charge flying from the vacuum area into a dielectric one [5, 7], the CTR (which is Cherenkov radiation (CR) generated in a dielectric and reflected off the boundary) compensates CR (or the so-called wakefield) in some area near the border. This effect is important for the wakefield acceleration technique [10, 11]. However, the medium dispersion that was not taken into account in mentioned works can radically influence the radiation. We consider here the case of a semi-infinite dielectric that possesses a resonant dispersion. The field of a charge in the unbounded resonant medium was analyzed in [12-14]. The case of a regular waveguide filled with this medium was analyzed in [15-17]. These papers show that the role of resonant dispersion is essential in such problems.

[1] To whom any correspondence should be addressed.

Content from this work may be used under the terms of the Creative Commons Attribution 3.0 licence. Any further distribution of this work must maintain attribution to the author(s) and the title of the work, journal citation and DOI.
Published under licence by IOP Publishing Ltd

2. Analytical investigation

2.1. General aspects

Let us consider a small charge particle q that uniformly moves with the velocity $\vec{V} = c\beta\vec{e}_z$ along the z - axis of a metal circular waveguide with the radius a and intersects the border ($z = 0$) between two semi-infinite media (figure 1). The medium on the left ($z < 0$) is a vacuum with $\varepsilon_1 = 1$.

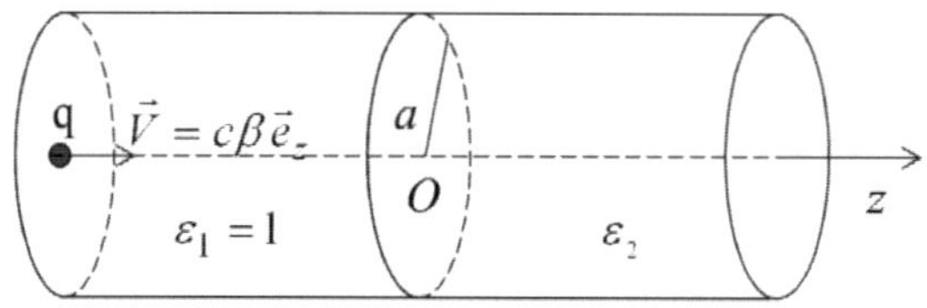

Figure 1. Geometry of the problem.

The nonmagnetic homogeneous isotropic medium on the right from the boundary ($z > 0$) is described by the following frequency-dependent electric permittivity:

$$\varepsilon_2 = \varepsilon_2(\omega) = 1 + \frac{\omega_p^2}{\omega_r^2 - \omega^2}, \tag{1}$$

where ω_p and ω_r represent a plasma frequency and a resonant frequency, respectively. Note that the permittivity at zero frequency is $\varepsilon_2 = \varepsilon_0 = 1 + \omega_p^2 / \omega_r^2$, and $\varepsilon_2 \to 1$ at $\omega \to \infty$. The charge intersects the boundary at the moment $t = 0$.

The general analytical solution to the problem for two arbitrary isotropic homogeneous media is given in the form of a decomposition in infinite series of normal modes [18]. In this paper, we write only expressions for the magnetic field:

$$\vec{H}_{1,2} = \vec{H}^q_{1,2} + \vec{H}^b_{1,2}, \tag{2}$$

where subscript 1 and subscript 2 relate to the areas $z < 0$ and $z > 0$, respectively. The first summand in (2) is the 'forced' field, which is the field of a charge in a regular waveguide (we use the terms following V.A. Ginzburg [19]). The forced field is written in the form [5]:

$$H^q_{\varphi 1,2} = \frac{2q\beta^2}{\pi a^2} \sum_{n=1}^{\infty} \frac{\omega_n J_1(\chi_n r/a)}{J_1^2(\chi_n)} \int_{-\infty}^{+\infty} \frac{\exp[-i(\omega t - z/c\beta)]}{\omega^2(1 - \beta^2\varepsilon_{1,2}) + \omega_n^2} d\omega, \tag{3}$$

where $\omega_n = \chi_n c/a$, χ_n is the nth zero of the Bessel function ($J_0(\chi_n) = 0$).

The second summand in (2) is the 'free' field, which is connected with the influence of the boundary. It is written in the form [5]

$$H^b_{\varphi 1,2} = \frac{2q\beta}{\pi c a^2} \sum_{n=1}^{\infty} \frac{\omega_n J_1(\chi_n r/a)}{J_1^2(\chi_n)} \int_{-\infty}^{+\infty} B_{n1,2} \exp[i(k_{z1,2}|z| - \omega t)] d\omega, \tag{4}$$

where

$$B_{n1,2} = \frac{\omega(\varepsilon_{1,2} - \varepsilon_{2,1})\left[\omega(\pm\varepsilon_{1,2}\beta^2 \mp 1) - c\beta\, k_{z2,1}\right]}{(c\varepsilon_2 k_{z1} + c k_{z2})\left[\omega^2(1 - \varepsilon_{1,2}\beta^2) + \beta^2\omega_n^2\right](\omega \pm c\beta\, k_{z2,1})},$$

$$k_{z1,2} = \sqrt{\omega^2\varepsilon_{1,2} - \omega_n^2} \Big/ c, \quad \operatorname{Im} k_{z1,2} > 0. \tag{5}$$

The forced field was earlier investigated in detail [1, 16-18]. Note that if the condition $\varepsilon_2(\omega)\beta^2 > 1$ is fulfilled, the forced field in the medium includes CR:

$$H_{\phi 2}^{CR} = \frac{q}{a^2}\sum_{n=1}^{\infty}\frac{\omega_n J_1(\chi_n r/a)}{J_1^2(\chi_n)}A^{CR}\sin\left[\omega_r x_{0n}\left(\frac{z}{c\beta}-t\right)\right]\theta(\beta ct - z), \tag{6}$$

where
$$A^{\mathrm{CR}} = \frac{2\sqrt{2}\beta^2}{\omega_{\mathrm{r}}\sqrt{1-\beta^2}}\frac{S+\xi_n-2(1-\beta^2)}{S\sqrt{S+\xi_n}},$$

$$x_{0n} = (S+\xi_n)^{1/2}(2-2\beta^2)^{-1/2}, \tag{7}$$

$$y_n = \omega_n/\omega_{\mathrm{r}} = \chi_n/x_{\mathrm{r}},\ x_{\mathrm{r}} = a\omega_{\mathrm{r}}/c, \tag{8}$$

$\xi_n = 1-\beta^2(\varepsilon_0+y_n^2)$, $S=\sqrt{(1+\xi_n)^2+4\beta^2(1-\beta^2)y_n^2}$, and $\theta(x)$ is the Heaviside step function. One can see that the frequency of the radiated waves lays within the interval $\omega_{\mathrm{c}}<\omega<\omega_{\mathrm{r}}$, where $\omega_{\mathrm{c}}=\omega_{\mathrm{r}}(1-\varepsilon_0\beta^2)^{1/2}(1-\beta^2)^{-1/2}$ if $\varepsilon_0\beta^2<1$ and $\omega_{\mathrm{c}}=0$ if $\varepsilon_0\beta^2>1$. Thus, CR is generated in a medium with resonance dispersion for any velocity of the charge, in contrast to the nondispersive dielectric, when the CR is excited if the charge velocity exceeds the Cherenkov threshold, that is $\varepsilon_0\beta^2>1$. Note that the considered case can be reduced to the nondispersive dielectric case with the permittivity ε_0 in the limiting process when $\omega_{\mathrm{r}}\to\infty$, $\omega_{\mathrm{p}}\to\infty$ and $\varepsilon_2\to\varepsilon_0$.

The forced field in the vacuum part of the waveguide and in the medium near the charge has a quasi-Coulomb character. It is written in the form

$$H_{\varphi 1,2}^{q} = \frac{q}{a^2}\sum_{n=1}^{\infty}\frac{J_1(\chi_n r/a)}{J_1^2(\chi_n)}A_{1,2}^{q}\exp\left(-\omega_{\mathrm{r}}x_{0n}^{(1),(2)}\left|\frac{z}{c\beta}-t\right|\right), \tag{9}$$

where
$$x_{0n}^{(1)} = \beta y_n(1-\beta^2)^{-1/2},\ x_{0n}^{(2)} = (S-\xi_n)^{1/2}(2-2\beta^2)^{-1/2}, \tag{10}$$

$$A_1^q = 2\beta(1-\beta^2)^{-1/2},\ A_2^q = A_1^q\beta y_n\left[S-\xi_n+2(1-\beta^2)\right]S^{-1}\left[2(S-\xi_n)\right]^{-1/2}.$$

In subsequent research, the main attention is focused on the free-field components (4). They are analyzed using two methods: an analytical method and a numerical one (they were developed in our papers [1, 5-9, 20] for different problems with sectionally homogeneous waveguides). In this paper, we only provide the most important results.

2.2. *The field in the vacuum area*

The analytical investigation is based on the complex variable function theory. Asymptotic expressions for the free field components of each mode can be obtained with the steepest decent technique [21], as it was made in our previous papers [5-9, 20]. The first step in such research is to study the integrand singularities in (4) in the complex plane of ω which are listed for the field in the vacuum area as follows:

– the branch points of the radical k_{z1} $\pm\omega_n - \mathrm{i}0$;

– the branch points of the radical k_{z2}

$$\pm\tilde{\omega}_n^{(1),(2)} = \pm\omega_{\mathrm{r}}\left\{\left[\varepsilon_0+y_n^2\mp\sqrt{(\varepsilon_0+y_n^2)^2-4y_n^2}\right]/2\right\}^{1/2} - \mathrm{i}0,\ \pm\omega_{\mathrm{r}}-\mathrm{i}0; \tag{11}$$

– the poles $\pm\omega_{0n}^{(1)} = \pm\mathrm{i}\omega_{\mathrm{r}}x_{0n}^{(1)}, \pm\omega_{0n}^{(22)} = \pm\mathrm{i}\omega_{\mathrm{r}}x_{0n}^{(2)}$, where $x_{0n}^{(1),(2)}$ described by equations (10). The singularities placed on the real axis are slightly shifted downward from the axis if small losses are taken into account (figure 2(a) shows the comparative disposition of an initial integration pass and the singularities). Note that, further, we tend these losses to zero.

The branch cuts are defined by the following equations: $\operatorname{Re}\sqrt{\omega^2-\omega_n^2}=0$, $\operatorname{Re}\sqrt{\omega^2\varepsilon_2-\omega_n^2}=0$, and the initial integration path goes along the upper edge of the cuts. Before using the steepest decent technique, it is convenient to make the following replacement of variables:

$$\omega=\omega_n\cosh\chi,\ \sqrt{\omega^2-\omega_n^2}=\omega_n\sinh\chi. \tag{12}$$

This replacement removes the pair of branch points $\pm\omega_n$. Disposition of singularities and the integration path in a complex plane of (ω) and (χ) are shown in figure 2. There are two saddle points on contour Γ: χ_0 and $\chi_1=\chi_0+i\pi$, where $\cosh(\chi_0)=ct/R$, $R=\sqrt{c^2t^2-z}$. The steepest descent paths (SDPs) consist of two branches (Γ_0^* and Γ_1^*). The contributions of the saddle points $\chi_{0,1}$ represent the space transition radiation and the front of TR propagates with the light velocity c. Singularities (in the case under consideration - the branch points $\tilde{\chi}_n^{(2)}$, $\tilde{\chi}_r$ which are analogues of $\pm\tilde{\omega}_n^{(2)}$, $\pm\omega_r$) can be crossed in the transformation of contour Γ into the SDP $\Gamma_{0,1}^*$ passing through the saddle points $\chi_{0,1}$, and the contributions from the corresponding singularities should be included in asymptotic expressions.

So, if $\operatorname{Re}\chi_{0,1}>\operatorname{Re}\tilde{\chi}_n^{(2)}$, then there is no intersection of the branch points and the saddle points contributions only give asymptotic expression for the free field in the vacuum part of the waveguide. However, in the case $\operatorname{Re}\chi_{0,1}<\operatorname{Re}\tilde{\chi}_n^{(2)}$, the saddle points can lie on the edges of the branch cuts, and the branch points $\tilde{\chi}_n^{(2)}$ are crossed in transformation of Γ into $\Gamma_{0,1}^*$. If $\operatorname{Re}\chi_{0,1}<\operatorname{Re}\tilde{\chi}_n^{(2)}$, then the branch cut contributions are added to the free field (figure 2(b)). The branch cut contributions are included into the free field in the area $|z|<\tilde{z}$, $\tilde{z}=ct\,\tilde{\omega}_n^{(1)}/\tilde{\omega}_n^{(2)}$ and the frequencies of this part of the field lay within the interval $[\omega_r,\tilde{\omega}_n^{(2)}]$, where $\tilde{\omega}_n^{(1)}$, $\tilde{\omega}_n^{(2)}$ described by formulas (11). The contributions from the branch points can be evaluated analytically by the Laplace method [20,21] or numerically by integration along the branch cuts.

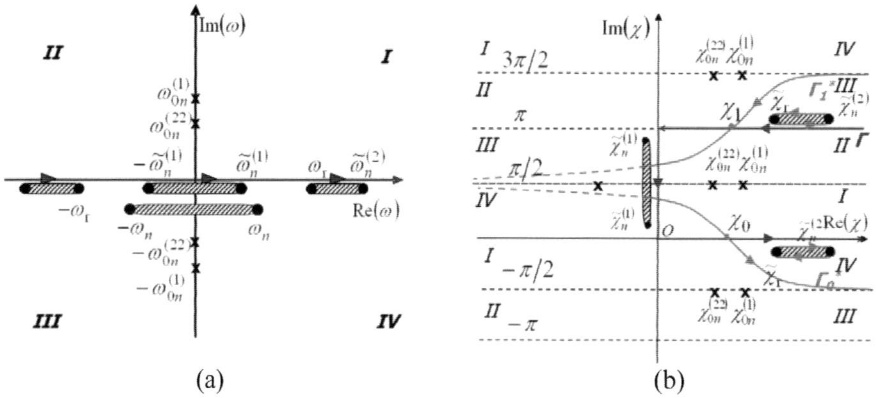

Figure 2. Disposition of singularities of integrands, branch cuts, and the integration paths in a complex plane of (a) ω and (b) χ for the field in a vacuum (the case of $\operatorname{Re}\chi_{0,1}<\operatorname{Re}\tilde{\chi}_n^{(2)}$). The dashed parts of the SDP $\Gamma_{0,1}^*$ are situated in the lower sheet of the Riemann surface.

2.3. *The field in the resonant medium*

For the free-field component in the resonant medium, analysis of the singularities of integrand in (4) in the complex plane of ω shows that, in addition to singularities described above, there is one more pair of poles $\pm\omega_{0n}^{(21)} = \pm\omega_r x_{0n}$, where x_{0n} is defined by formula (7). These poles coincide with ones in the forced field (they are connected with CR in the medium). Note that magnitude of $\omega_{0n}^{(21)}$ tends to ω_r with increasing in the mode number n and, for some n, is decreasing in increase in β. If $\beta << 1$, then $x_{0n} \approx 1-(\varepsilon_0-1)\beta^2/2$; if $\beta \approx 1$, then $x_{0n} = y_n\left(y_n^2+\varepsilon_0-1\right)^{-1/2}\left[1+O\left(\gamma^{-2}\right)\right]$, where $\gamma = \left(1-\beta^2\right)^{-1/2}$.

Disposition of singularities and the integration path in a complex plane of (ω) is presented in figure 3. In this situation, it is not convenient to make any change of variables (like (12)), and we apply the steepest decent technique directly in the complex plane of (ω).

The saddle point analysis shows a complex wave picture because there are three pairs of saddle points with different frequencies $\pm\Omega_n^{(1)}$, $\pm\Omega_n^{(2)}$, and $\pm\Omega_n^{(3)}$ in some domain near the boundary at $z < z_0$. These frequencies can be approximately described by following expressions:

$$\Omega_n^{(1)} \approx \tilde{\omega}_n^{(1)} ct/R,\ \Omega_n^{(2)} \approx \tilde{\omega}_n^{(2)} ct/R \text{ and } \Omega_n^{(3)} \approx \omega_r (ct-z)/R, \tag{13}$$

where $R = \sqrt{c^2t^2 - z}$. However, there is only one pair of saddle points $\Omega_n^{(2)} \approx \tilde{\omega}_n^{(2)} ct/R$ in the area $z_0 < z < ct$. In figure 4, dependences of the saddle point frequencies on distance are shown. Magnitude z_0 can be evaluated numerically as analytical approximation for $\Omega_n^{(1)}$ is not quite correct in this domain.

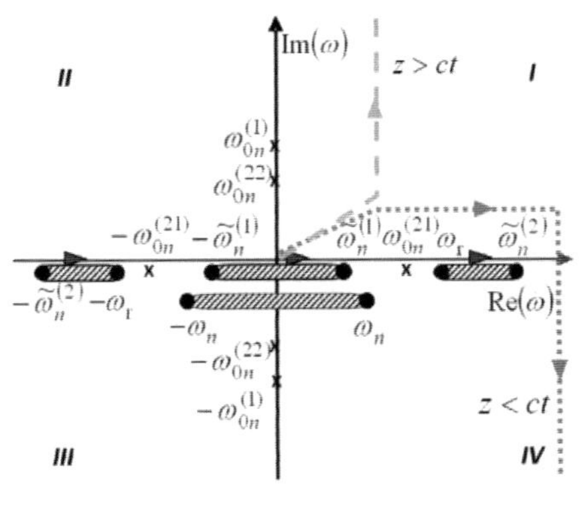

Figure 3. Singularities, branch cuts and integration paths in the complex plane of ω for the field components in the resonant dielectric.

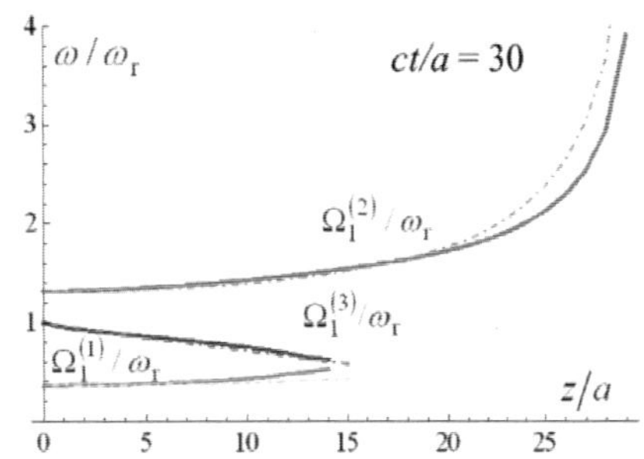

Figure 4. Frequencies of saddle points at different distances in the medium (solid lines are for numerical results, dot-dashed lines are for analytical approximations): $n=1$; $x_r = 5$; $\omega_p = 0.8\omega_r$.

In addition to the saddle point contributions and the branch cut contributions, the poles $\pm\omega_{0n}^{(21)}$ can contribute into the free field under some condition. These contributions (calculated with the residue theorem) give the CTR, which is described by formula (6) but taken with an opposite sign. So, there is a compensation for the forced field with a part of free one in some domain from the boundary at $z < z_2$. However, it is difficult to estimate the dimension of this zone in general. We managed to find it analytically for rather greater velocities of the charge motion only ($\beta \approx 1$). In this situation, the disposition of the poles $\pm\omega_{0n}^{(21)}$ in close proximity to just one of pair of saddle points $\pm\Omega_n^{(1)}$ is only important (figure 4). Therefore, for $\beta \approx 1$, we have a generalization of the case of nondispersive dielectric [5]:

$$z_2 \approx ct\varepsilon_2^{-1}\Big|_{\omega=\omega_{0n}^{(21)}}, \tag{14}$$

where the dimension of this zone z_2 is determined with the group velocity $v_{g2} \approx c\varepsilon_2^{-1}$ in a regular waveguide.

3. Numerical investigation, results and discussion

In the numerical approach, the exact integral representations (4) are used to calculate the free-field components. The numerical algorithm is based on a certain transformation of the initial integration path in the complex plane of ω as it was earlier used for the computation of the field in different dispersive unbounded [14] or semi-bounded [22] media including different problems with waveguides [1, 4-9, 20]. First, we rewrite the exact formulas into the integrals between the limits 0 and $+\infty$. Second, we transform the integration path into a new contour in the upper-half plane for $|z|>ct$ (before the 'wave front') and into another contour in the lower-half plane for $|z|<ct$ (behind the 'wave front'). These new contours are presented in figure 3 with green dashed line and red dotted line. They bypass all of the singularities and subsequently go parallel to the SDP. Note that calculations can be optimized with matching of parameters of new contours.

The results obtained are presented in figure 5. Ammonia [23] is considered to be a typical medium; it has a resonance of 23.87 GHz and nondimensional value $x_r = 5$ (refer to expression (8)) at $a \approx 1$ cm. Figure 5 shows the component H_φ of the first mode of the total field in the vacuum area and in the resonant medium for different values of velocity β at different time moments. The forced field (6) and (9) in the medium and the branch cuts contributions (evaluated numerically) in a vacuum are also presented.

In the vacuum part of the waveguide ($z<0$) the free field consist of TR and the branch cuts contributions. As can be seen, the branch cuts contributions are the main part of the field in a vacuum especially for relatively greater velocities (figure 5(d)-(i)). Therefore, the branch cuts contributions give a response from the resonant medium with frequencies being about resonance frequency. The maximum amplitude of this response is comparable with the amplitude of CR in the medium. The front of this wavepacket propagates in a vacuum with the velocity $\tilde{v} = c\tilde{\omega}_n^{(1)}/\tilde{\omega}_n^{(2)}$, where $\tilde{\omega}_n^{(1)}$, $\tilde{\omega}_n^{(2)}$ defined by formulas (11).

In the resonant medium ($z>0$), the wave picture is rather complex. For relatively low velocity of the charge motion (figure 5(a)-(c)), the amplitude of TR is comparable to the amplitudes of CR and the CTR which considerably increase with an increase in velocity β (figure 5(d)-(i)). As opposed to the case of nondispersive dielectric [5,7], dispersion in the medium leads to the complex wave behaviour of TR connected with essential contributions of three different saddle points with frequencies (13). As the amplitudes of these different waves are comparable, this results in complex effects pronounced in figure 5(b), (c). For relatively greater velocities β (figure 5(d)-(i)), TR is not very essential and the CTR effect is dominant. The forced field, which consists of CR (wakefield) and a quasi-Coulomb field, form practically the total field in certain area after the charge $z_2 < z < c\beta t$. However, there is also domain $z < z_2$ where the influence of the boundary is principal. In this domain, the part of the free field (the CTR) compensates the wakefield (CR) as well as in the case of nondispersive dielectric. However, for relatively low velocity of the charge motion (figure 5(a)-(c)), the compensation is not quite visible because of the effect of TR. Note that the dimension of the compensation area is smaller if ε_2 takes on a larger value (14). It might be achieved, for example, for greater mode number when the CR frequency tends to the resonant frequency, and $\varepsilon_2\big|_{\omega_{0n}^{(21)}}$ is considerably increasing.

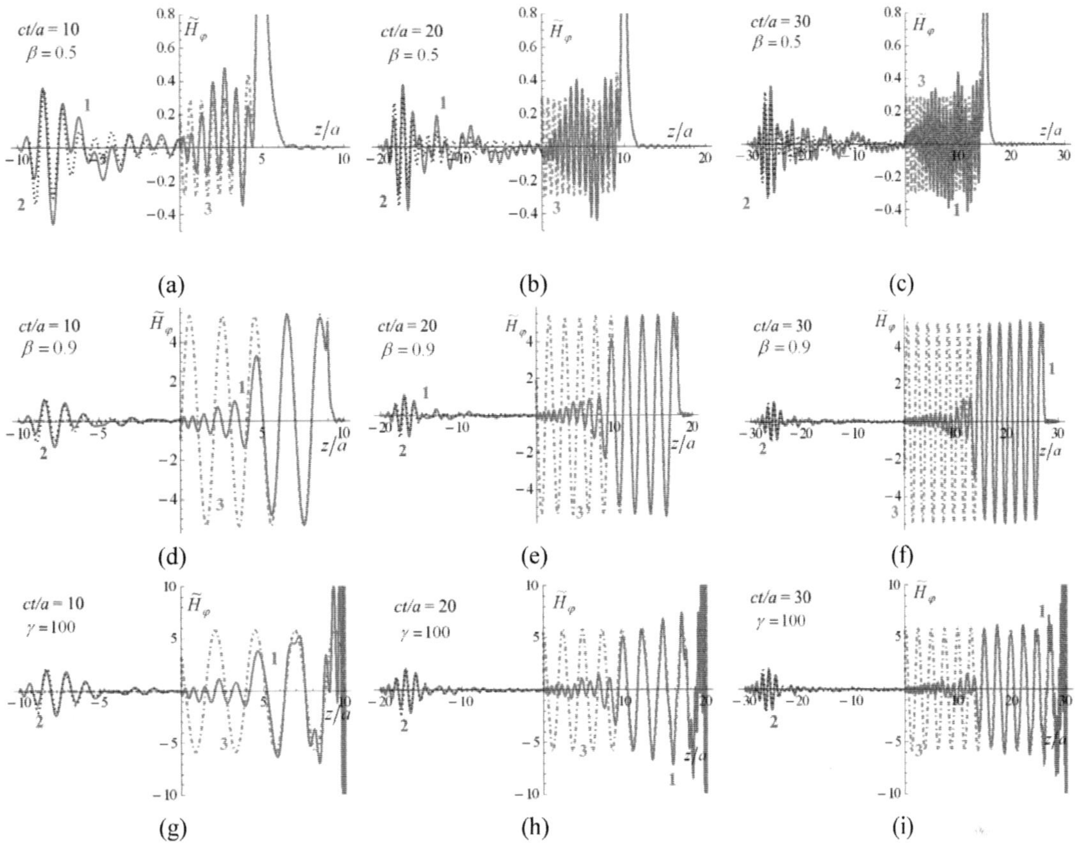

Figure 5. Dependence of normalized component $\tilde{H}_\varphi = H_\varphi a^2 q^{-1}$ of the first mode of the total field (red solid line 1) on distance z/a for different dimensionless times ct/a and velocities β (or $\gamma = \left(1-\beta^2\right)^{-1/2}$). The branch point contributions (blue dashed line 2) and the forced field (green dot-dashed line 3) are presented as well; n = 1; $x_r = 5$; $\omega_p = 0.8\omega_r$; $a = 1\,\mathrm{cm}$. Sighting point is $r = 0.5a$.

Note that, for ultrarelativistic particle (figure 5(g)-(i)), the TR increases and concentrates near the front which propagates with the light velocity c [20].

4. Conclusion

We have analyzed the electromagnetic field of a small charged bunch that moves in a circular waveguide and intersects the boundary between a vacuum and a resonant medium. The analysis of the field components has been performed both analytically and numerically. The analytical investigation has been based on the steepest descent method. The algorithm for the computation has been also presented.

It has been shown that, in any cross-section in the vacuum area, the radiation field consists of two parts. The first one is the single-frequency field (the contribution of the saddle points), and the second one is the field having certain limited continuous spectrum lying near the resonance frequency (the contribution of the branch cuts). It is interesting that, as a rule, the second part is predominant. This result might be used for diagnostics of material properties.

In the resonant medium, there is the area after the bunch where the wave field practically coincides with the wakefield in the regular waveguide (the 'wakefield area') similar to the case of nondispersive dielectric. There is also the domain where the boundary influence is principal, the total field is relatively small, and wave picture is rather complex (the 'compensation area'). Note that the compensation area is decreasing and, correspondingly, the wakefield area is increasing with increase in the mode number and the velocity of the bunch motion. The results obtained can be useful for wakefield acceleration technique.

Acknowledgments
This research was supported by the Russian Foundation for Basic Research (grant 15-02-03913a).

References

[1] Alekhina T Yu and Tyukhtin A V 2015 *Phys. Rev. ST AB* **18** 091302
[2] Cook A M, Tikhoplav R, Tochitsky S Y, Travish G, Williams O B and Rosenzweig J B 2009 *Phys. Rev. Lett.* **103** 095003
[3] Andonian G, Williams O, Wei X, Niknejadi P, Hemsing E, Rosenzweig J B, Muggli P, Babzien M, Fedurin M, Kusche K, Malone R and Yakimenko V 2011 *Appl. Phys. Lett.* **98** 202901
[4] Antipov S, Jing C, Kanareykin A, Butler J E, Yakimenko V, Fedurin M, Kusche K and Gai W 2012 *Appl. Phys. Lett.* **100** 132910
[5] Alekhina T Yu and Tyukhtin A V 2012 *J. Phys.: Conf. Series* **357** 012010
[6] Alekhina T Yu and Tyukhtin A V 2012 *Phys. Rev. ST AB* **15** 091302
[7] Alekhina T Yu and Tyukhtin A V 2013 *Phys. Rev. ST AB* **16** 081301
[8] Alekhina T Yu and Tyukhtin A V 2014 *Phys. Rev. ST AB* **17** 071302
[9] Alekhina T Yu and Tyukhtin A V 2014*J. Phys.: Conf. Series* **517** 012012
[10] Power J G, Conde M E, Gai W, Kanareykin A, Konecny R and Schoessow P 2000 *Phys. Rev. ST-AB* **3** 101302-1
[11] Kanareykin A 2010 *J. Phys.: Conf. Series* **236** 012032
[12] Afanasiev G N, Kartavenko V G and Magar E N 1999 *Physika B* **269** 95-113
[13] Afanasiev G N and Kartavenko V G 1998 *J. Phys D.: Appl.Phys.* **31** 2760-2778
[14] Tyukhtin A V and Galyamin S N 2008 *Phys. Rev. E* **77** 066606
[15] Bolotovskii B M 1962 *Sov. Phys. Usp.* **4** 781
[16] Tyukhtin A V 2004 *Tech. Phys. Lett.* **30** 7, 605-608
[17] Tyukhtin A V 2005 *Tech. Phys. Lett.* **31** 2, 150-153
[18] Barsukov K A 1960 *Sov. Phys. JETP* **37** 787
[19] Ginzburg V L and Tsytovich V N 1990 *Transition Radiation and Transition Scattering* (London: Hilger) p 445
[20] Alekhina T Yu and Tyukhtin A V 2011 *Physical Review E* **83** 066401
[21] Felsen B and Marcuvitz N 2003 *Radiation and Scaterring of Waves* (New Jersey: Wiley Interscience)
[22] Galyamin S N and Tyukhtin A V 2011 *Phys. Rev. E* **84** 056608
[23] Schachter L 2000 *Physical Review E* **62** 1, 1252-1257

Project on the superposition of beamlines for parametric X-ray radiation and coherent transition radiation in the THz region at LEBRA

Y Hayakawa[1], K Hayakawa[1], M Inagaki[1], T Kaneda[2], K Nakao[1], K Nogami[1], T Sakae[2], T Sakai[1], N Sei[3], Y Takahashi[4] and T Tanaka[1]

[1] Laboratory for Electron Beam Research and Application (LEBRA), Institute of Quantum Science, Nihon University, Narashinodai 7-24-1, Funabashi 274-8501, Japan

[2] School of Dentistry at Matsudo, Nihon University, Sakaecho-Nishi 2-870-1, Matsudo 271-8587, Japan

[3] Research Institute for Measurement and Analytical Instrumentation, National Institute of Advanced Industrial Science and Technology (AIST), Tsukuba 305-8568, Japan

[4] Institute of Materials Structure Science, High Energy Accelerator Research Organization (KEK-PF), 1-1 Oho, Tsukuba, Ibaraki 305-0801, Japan

E-mail: yahayak@lebra.nihon-u.ac.jp

Abstract. A new project to develop a terahertz (THz)-wave light source is in progress at the parametric X-ray (PXR) beamline of the Laboratory for Electron Beam Research and Application (LEBRA) at Nihon University. The THz-wave source is based on coherent transition radiation (CTR) emitted from a metal foil inserted downstream from a crystal target that is the PXR radiator. Beryllium or titanium foil is the most promising candidate for a THz-wave radiator. Since the electron linac of LEBRA was developed for a free electron laser (FEL), electron beam with bunch length of 1 ps (rms) can be provided by magnetic bunching at the bending magnet section. Thus, very intense coherent transition radiation (CTR) can be obtained in the frequency region around 1 THz. The results of preliminary experiments for CTR production suggested that sufficiently intense THz-CTR can be obtained using the LEBRA linac. In order to realize a THz-wave source for practical application studies, we have a plan to add the extraction feature for THz waves to the PXR beamline.

1. Introduction

The Laboratory for Electron Beam Research and Application (LEBRA) has been running as a light source facility by Nihon University. The light sources have been conducted using the electron beam from a conventional electron linac. The first light source developed at LEBRA was an infrared free electron laser (FEL) with a wavelength range from 1.3 to 6 μm [1]. The second one was a monochromatic X-ray source based on parametric X-ray radiation (PXR) [2, 3]. Both of these sources were constructed on the dedicated beamline and have selectivity for the photon energy (wavelength) as a feature of the light sources. Usually, the FEL beamline and the PXR beamline are used alternately by the week. Various user studies employing these monochromatic light sources have been conducted since 2004 at LEBRA.

In addition to the light sources, a project to develop new sources in the wavelength region of THz has been started at LEBRA. Terahertz (THz)-wave sources are based on coherent radiation

Content from this work may be used under the terms of the Creative Commons Attribution 3.0 licence. Any further distribution of this work must maintain attribution to the author(s) and the title of the work, journal citation and DOI.
Published under licence by IOP Publishing Ltd

because of the electron bunch length, estimated to be 1 ps [4]. First, coherent synchrotron radiation (CSR) from a bending magnet (BM) has been applied to the THz-wave source at the FEL beamline [5]. The second 45°-BM of the 90° bending section upstream of the FEL undulator is used as a CSR radiator. Since the magnetic compression of the electron bunches is halfway achieved at the source point, the available THz wave mainly consists of components with frequencies below 0.3 THz. To extract a THz wave to an experimental hall where a person can work, the THz-CSR beam is superposed onto the FEL transport line using a reflector of $LiTaO_3$ crystal. The THz-wave power in the experimental hall will be significantly reduced due to the diffraction loss in the long vacuum duct. Actually, the intensity measured by a D-band wave detector was roughly 100 nJ per macropulse when the macropulse duration of the electron beam was 20 μs.

In addition to the THz-CSR source at the FEL beamline, we are planning to develop THz-wave sources at the PXR beamline. Originally, the design concept of the PXR beamline assumed a target would be inserted against the electron beam for the utilization of the beamline. Thus, coherent transition radiation (CTR) is a promising candidate for a THz radiator at the PXR beamline.

2. LEBRA PXR beamline

The PXR beamline was developed as a dedicated beamline for PXR generation. Figure 1 shows the geometric configuration of the PXR beamline. The PXR generating system is connected to the linac through the 90° bending section. The PXR generator consists of a double crystal system, where the first crystal is the X-ray radiator and the second one is the X-ray reflector for transport, in a large vacuum chamber. The beamline has a transport line penetrating a 2-m-thick shield wall to extract X-rays to the experimental hall. In front of the shield wall, a multipurpose vacuum chamber is connected to both the electron beamline and the X-ray line.

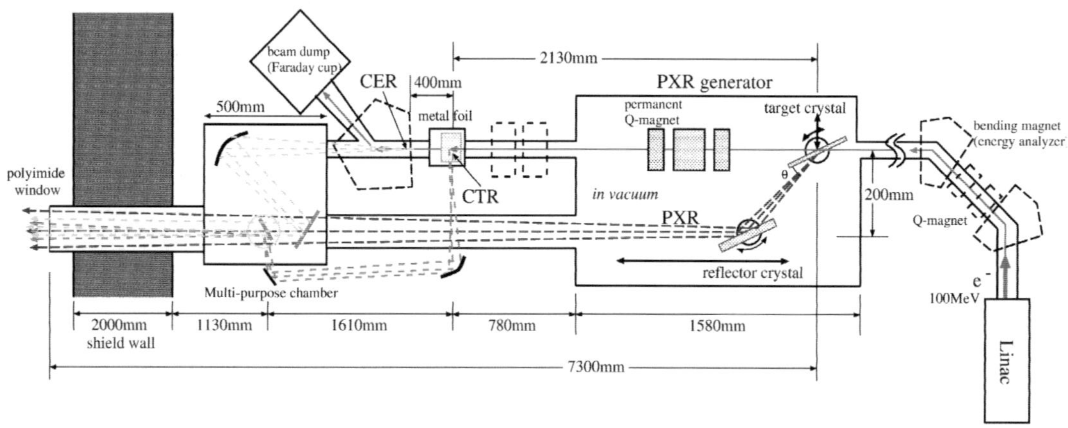

Figure 1. Schematic top view of the LEBRA-PXR beamline. The promising THz-wave sources based on CTR or CER are also indicated downstream from the PXR radiator on the electron beamline.

The LEBRA linac usually provides a 100-MeV electron beam for PXR production. The typical macropulse of the electron beam has an intensity from 100 to 135 mA, depending on the condition of the electron gun. Since the average beam current is limited to 5 μA because of legal regulations for radiation safety, the linac is ordinarily operated in the macropulse condition of 5-μs duration and 5-pps repetition rate.

The PXR source has provided X-rays from 4 to 34 keV employing a Si(111) or Si(220) plane as an X-ray radiator. Typical parameters of the LEBRA-PXR beamline are listed in table 1. The main application of the LEBRA-PXR source is X-ray imaging effectively using energy-selective monochromaticity and/or the spatial coherence of PXR. In particular, diffraction-enhanced imaging (DEI) that can give a phase contrast or contrast based on small-angle scattering is the most characteristic application of PXR.

Table 1. Typical parameters of the linac and the PXR source of LEBRA.

Electron energy	100 MeV (typ.)
Accelerating frequency	2856 MHz (S-band)
Macropulse beam current	70 – 135 mA (typ.)
Macropulse duration	4 – 5 μs (typ.)
Macropulse repetition rate	5 pps
Average beam current	1 – 3.5 μA
electron bunch length	1 – 3 ps (rms)
bunch charge	25 – 47 pC
Beam size on the target	0.5 – 2 mm in dia. (FWHM)
PXR energy range	
Si(111) target:	4.0 – 20.5 keV
Si(220) target:	6.5 – 34 keV
Irradiation field at the X-ray exit	100 mm in diameter
Total X-ray photon rate	$\sim 10^7$ /s @17.5 keV

3. THz-wave sources at the PXR beamline

The PXR beamline is equipped with two target-insertion devices to monitor the beam profile. One is installed between the 90° bending section and the PXR generator. The other one is installed downstream from the PXR chamber. Since the magnetic compression for electron bunches is fully achieved at these points, a metal foil inserted using the insertion device could be a THz-wave source based on transition radiation having frequency components higher than those of the existing wave source in the FEL beamline. The most downstream bending magnet of the PXR beamline is also a candidate for a THz-wave radiator.

3.1. Coherent transition radiation

In principle, it is possible to adopt the upstream metal foil target and the crystal target for PXR as a CTR radiator. These targets, however, are more than ~4.8 m away from the entrance of the through-hole penetrating the shield wall. Because of the diffraction effect, it is difficult to guide the THz-wave beam to the PXR transport line without a considerable intensity reduction. Thus, the downstream target is the most preferable CTR source on the PXR beamline. Since the backward CTR beam can be extracted from the view window just above the target, it is possible to effectively transport the THz-wave beam in the atmosphere without serious diffraction loss by using a focusing mirror system.

In the case of forward CTR, the light beam reaching the multipurpose chamber will be treated by using mirrors in the vacuum. However, the duct aperture between the target and the multipurpose chamber is not sufficient. Therefore, the diffraction loss cannot be ignored, especially for the lower-frequency components.

3.2. Coherent edge radiation

When a relativistic electron crosses the boundary of a magnetic field, electromagnetic radiation, referred to as edge radiation, is emitted. In the THz region, coherent edge radiation (CER) from an electron bunch has substantial intensity, and its characteristics are very similar to those of transition radiation [6]. On the LEBRA-PXR beamline, the entrance of the last BM can be a THz-CER source in addition to THz-CSR from the bending section. Since the CER source is pointlike, it will be easier to prepare a beam-shaping optical system than in the case of CSR.

The distance between the CTR target and the entrance of the BM is estimated to be 400 mm. This value suggests that the shadow effect of the CTR target will considerably suppress CER having longer wavelengths than 1 mm when the electron beam energy is 100 MeV [7, 8]. Thus, CER without the CTR target is more practical for a THz-wave source emitted in the forward direction from the electron beam.

4. Observation of CTR at the PXR beamline

The intensity of transition radiation depends on the plasma frequency ω_p of the target material and the incident electron energy E. The spectral intensity $I_{TR}(\omega)$ in the bandwidth $\Delta\omega$ at the frequency ω within the radiation cone from a single electron is written as [9],

$$I_{TR}(\omega) = \frac{\alpha}{\pi}\left[\left(1+\frac{2\omega^2}{\omega_p^2}\frac{m^2c^4}{E^2}\right)\ln\left(1+\frac{\omega_p^2}{\omega^2}\frac{E^2}{m^2c^4}\right)-2\right]\frac{\Delta\omega}{\omega}, \tag{1}$$

where c, α, and m are the speed of light, the fine-structure constant, and the electron rest mass, respectively. When N electrons longitudinally form a bunch, the CTR intensity becomes

$$I_{CTR}(\omega) = N\,\left[1+(N-1)\,f(\omega)\right]\,I_{TR}(\omega). \tag{2}$$

Here, $f(\omega)$ is the longitudinal bunch form factor defined by

$$f(\omega) = \left|\int F(z)\,e^{i\frac{\omega}{c}z}\,\mathrm{d}z\right|^2, \tag{3}$$

where $F(z)$ is the normalized distribution of the electron bunch along the longitudinal direction [10, 11]. According to eq. (1), materials having higher plasma frequencies are suitable for CTR generation in the THz region when the electron energy is higher than several tens of MeV. With respect to electron beam-loss and background radiation, however, it is desirable to use a thin target of light materials. Taking heat resistance into account, light metals such as beryllium and titanium are expected to serve as materials for the CTR target.

To investigate the characteristics of THz-CTR, a preliminary observation for backward CTR extracted from a view window was carried out in the accelerator room using a 100-MeV electron beam from the linac. The experimental setup is explained in figure 2. A 50-μm-thick Ti foil was used for the CTR target mounted on the insertion device, and the Si crystal for PXR production was retracted from the electron beam axis. According to eq. (2, 3), the CTR intensity strongly depends on the electron bunch length and is almost proportional to N^2 when the bunch length is comparable to the wavelength [12]. Using magnetic bunch compression at the 90° bending section, the rms bunch length shorter than 1 ps can be achieved at the PXR beamline as same as the FEL beamline. Figure 3 shows the CTR pulse-energy measured by a pyroelectric detector (Gentec-EO THZ5I-MT-BNC) as a function of the electron-macropulse charge. Here, the bunch length might change depending on the macropulse charge because of the variation of the beam loading at acceleration tube. The CTR intensity, however, seems to obey N^2 law and this behaviour suggests that the radiation is enhanced by the coherent effect.

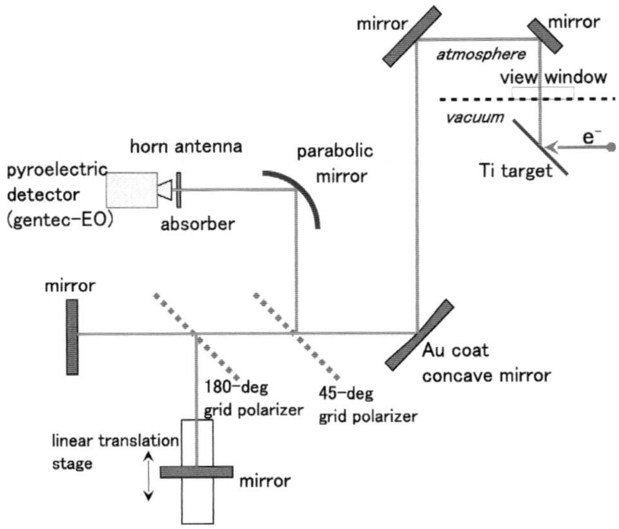

Figure 2. Schematic explanation of the setup for the THz-CTR observation in the accelerator room. A Martin-Puplett interferometer was temporarily assembled just at the side of the PXR beamline. THz-wave spectra can be obtained by scanning the mirror mounted on the linear translation stage. The distance between the CTR target and the pyroelectric detector is roughly estimated to be 2 m.

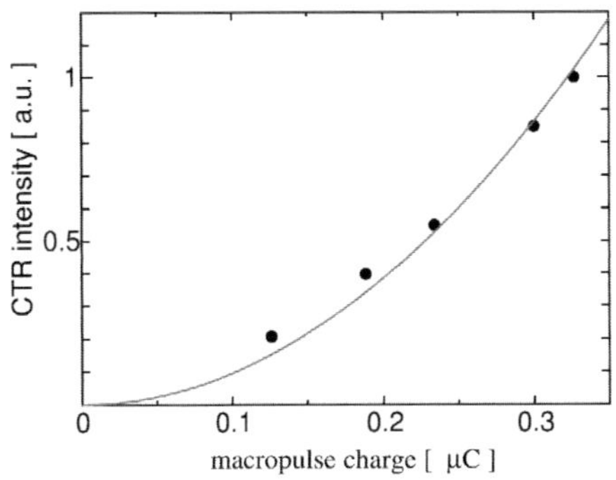

Figure 3. The CTR intensity as a function of the electron-macropulse charge. A solid line is a quadratic curve fitting the data.

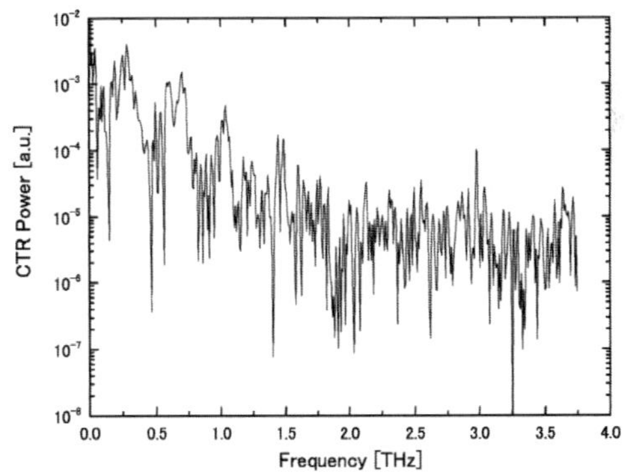

Figure 4. The spectrum of CTR measured using a Martin-Puplett interferometer.

The spectrum of THz-CTR was obtained using a Martin-Puplett interferometer and the typical result is shown in figure 4 [13]. Since the THz-wave was too intense for the pyroelectric detector, black polyethylene films were used as an absorber just in front of the detector. Although the structure corresponding to the interference due to the reflected waves from the absorber is found in the spectrum, it can be seen that it has frequency components from 0.1 to 2 THz higher than those of CSR extracted from the FEL beamline.

The measurement using a calibrated thermal sensor (Ophir 3A-P-THz) was carried out to evaluate the absolute power obtained from the THz-CTR source. Since the electron bunch length is highly sensitive to the RF phase fluctuation of the linac, the pulse-to-pulse stability of the CTR output is relatively poor. The sensor used in the experiment, however, has a long response time, 2.5 s. Thus, the output signal was averaged as shown in figure 5. When the current, the duration, and the repetition rate of the electron macropulse are approximately 70 mA, 5 μs and 5 pps, respectively, the average power of the THz wave was approximately 4.6 mW, corresponding to 0.9 mJ per macropulse. This result suggests that the LEBRA THz-CTR source could be one of the most intense THz-wave sources based on accelerators, except for THz-FELs. Further research on CTR and CER at the PXR beamline will be reported in our next publication to be prepare.

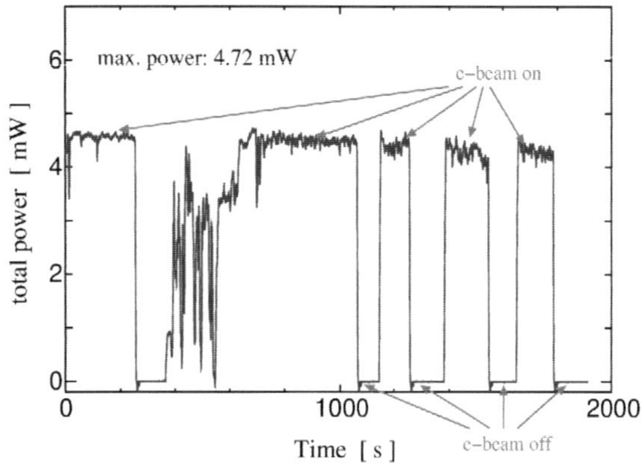

Figure 5. The result of absolute power measurement for THz-CTR at the PXR beamline. The average power was continually monitored using a calibrated thermal sensor having a response time of 2.5 s. The output value means the average over several tens of macropulses.

5. Superposing the THz-CTR beam on the PXR beam

The source points of CTR and CER are not visible directly from the experimental hall through the PXR transport duct and are at least 4.5 m away from the exit window. An optical mirror system is therefore necessary to extract the THz-wave to the experimental hall for application studies. Since these THz-wave sources are more than 2 m distant from the PXR radiator, it is supposed that the insertion of the PXR target would not seriously affect THz-wave generation with respect to the shadow effect. After passing through the PXR target, the electron beam divergence should grow up to the order of 10 mrad resulted from the multiple scattering in the target crystal. The permanent quadrupole (Q)-magnets in the PXR generator chamber, however, can suppress the extreme increase in the transverse beam size [2]. In addition, the longitudinal structure of the electron bunch should be almost preserved across the thin crystal target. This means that the insertion of the PXR target does not disturb the scheme of coherent radiation. Thus, it is desirable to introduce a system that enables the simultaneous use of the PXR beam and the THz-wave beam. For this purpose, we are planning to utilize the multipurpose chamber downstream from the last BM of the PXR beamline, as shown in figure 1.

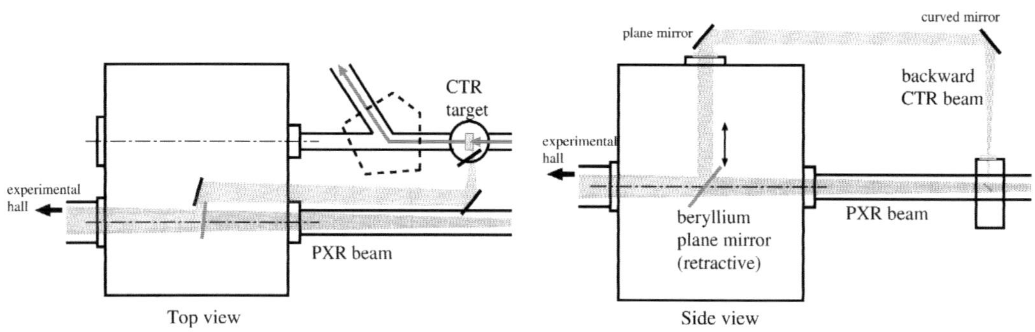

Figure 6. Schematic drawing of the system superposing the backward CTR beam onto the PXR beam. The CTR beam is temporarily extracted to the atmosphere and reenters the vacuum via the chamber roof.

Figure 6 schematically shows the transport of the backward CTR beam from the target inserted into the electron beamline. The THz-CTR beam will be extracted to the atmosphere through the optical view window and will be collimated and/or shaped using a curved mirror. The THz-CTR beam will then be transported toward the roof of the chamber through the air and reenter the vacuum through the other view window. A thin plane mirror made of beryllium

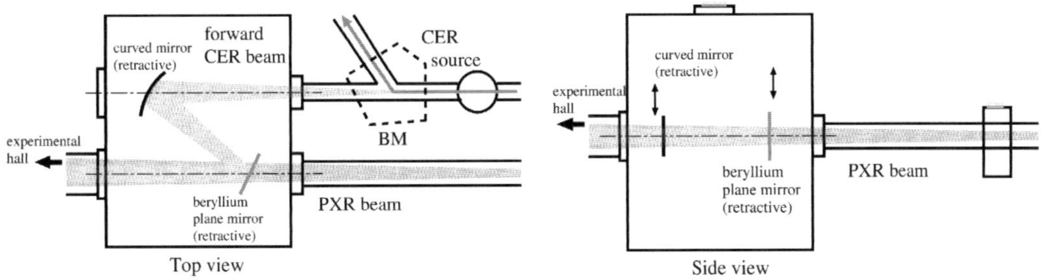

Figure 7. Schematic drawing of the system superposing the forward CER beam onto the PXR beam. The CER beam is shaped with a coupled mirror system in the vacuum chamber.

will be used to superpose the THz-wave beam on the X-ray beam because of high transmittance of X-rays. The beryllium mirror has to be retractive for the usual mode of the PXR application.

In the case of the forward direction, CER from the BM entrance will be utilized as the THz-wave source. The CTR target upstream of the BM should be retracted from the electron beam axis to avoid the reduction in CER intensity due to the shadow effect. The CER beam, upon reaching the multipurpose chamber, will be collimated by a parabolic mirror focused on the CER source point. Another beryllium mirror will be used for the superposition of the THz wave and of X-rays, as in the case of backward CTR. The scheme of the coupled mirrors in the chamber is shown in figure 7.

The superposing system in the multipurpose chamber should alternate between the backward CTR and the forward CER. A switchable retracting system for the mirrors is currently in the design stage.

6. Application of the superposing beam

X-ray imaging, including advanced techniques such as DEI, has been the main application of the LEBRA-PXR source. Similarly, non destructive imaging is one of the most popular applications of THz waves. Therefore, simultaneous imaging of THz waves and X-rays seems to be a prospective application of the superposing beam extracted to the experimental hall. The THz-wave beam can be easily separated from the X-ray beam using a thin beryllium mirror or Al-coated-polyimide mirror, as shown in figure 8. THz waves are sensitive to the molecular state in a sample, in contrast with monochromatic X-rays that can detect the kinds of elements [14]. Simultaneous imaging may be a powerful tool for material science and biology.

Using a short pulse of THz-CTR corresponding to a one-cycle or a half-cycle wave, pump-probe spectroscopy is another interesting application of the superposing beam. The timing of the THz-wave pulse to the X-ray pulse is adjustable by optical path length. Collaboration with

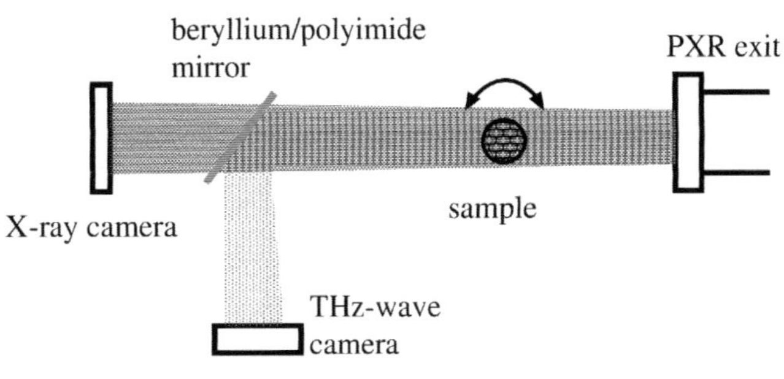

Figure 8. Schematic of a typical imaging application simultaneously using THz waves and X-rays in the experimental hall. Not only simple projection imaging but also three-dimensional tomography may be possible.

XAFS (X-ray absorption fine structure) analysis, which is another successful application of PXR, is attractive as a unique application [15].

7. Summary

The development of a THz-wave source at the LEBRA-PXR beamline is in progress, and a preliminary observation of THz-CTR from the Ti foil target was carried out using a 100-MeV electron beam of the PXR beamline. The obtained spectrum of THz-CTR had frequency components from 0.1 to 2 THz, and the available power was close to 5 mW. This result suggests that the CTR source has sufficient performance to serve as a practical THz-wave source.

To provide the THz-wave beam to application studies, it is planned that the THz-CTR beam or the THz-CER beam will be transported through the X-ray beamline by the superposition of the THz-wave beam onto the PXR beam. The new system is expected to allow users to simultaneously utilize THz-wave and X-rays in the experiment hall where radiation safety is guaranteed.

Acknowledgments

The author's work was supported in part by KAKENHI grants in aid (25286087 and 24560069) from the Japanese Ministry of Education, Culture, Sports, Science, and Technology.

References

[1] Hayakawa Y, Sato I, Hayakawa K, Tanaka T, Nakazawa H, Yokoyama K, Kanno K, Sakai T, Ishiwata K, Enomoto A, Fukuda S, Ohsawa S, Tsuchiya K and Kato M 2002 *Nucl. Instrum. and Meth.* A **483** 29
[2] Hayakawa Y, Sato I, Hayakawa K and Tanaka T 2005 *Nucl. Instrum. and Meth.* B **227** 32
[3] Hayakawa Y, Sato I, Hayakawa K, Tanaka T, Mori A, Kuwada T, Sakai T, Nogami K, Nakao K and Sakae T 2006 *Nucl. Instrum. and Meth.* B **252** 102
[4] Sei N, Ogawa H, Hayakawa K, Tanaka T, Hayakawa Y, Nakao K, Sakai T, Nogami K and Inagaki M 2013 *J. Phys. D: Appl. Phys.* **46** 045104
[5] Sei N, Ogawa H, Hayakawa K, Tanaka T, Hayakawa Y, Nakao K, Sakai T, Nogami K and Inagaki M 2014 *J. Opt. Soc. Am.* B **31** 2150
[6] Andonian G, Cook A, Dunning M, Hemsing E, Marcus G, Murokh A, Reiche S, Schiller D, Rosenzweig J B, Babzien M, Kusche K and Yakimenko V 2009 *Phys. Rev. ST Accel. Beams* **12** 030701
[7] Artru X and Ray C 2008 *Nucl. Instrum. and Meth.* B **266** 3725
[8] Naumenko G, Artru X, Potylitsyn A, Popov Y, Sukhikh L and Shevelev M 2010 *J. Phys.: Conf. Series* **236** 012004
[9] Baryshevsky V G and Feranchuk I D 1985 *Nucl. Instrum. and Meth.* **228** 490
[10] Hirschmugl C J, Sagurton M and Williams G P 1991 *Phys. Rev.* A **44** 1316
[11] Yim C, Ko J, Jung S, Han D, Kim C, Park J, Kang H S and Ko I S 2012 *Phys. Rev. ST Accel. Beams* **15** 030706
[12] Nakazato T, Oyamada M, Niimura N, Urasawa S, Konno O, Kagaya A, Kato R, Kamiyama T, Torizuka Y, Nanba T, Kondo Y, Shibata Y, Ishi K, Ohsaka T and Ikezawa M 1989 *Phys. Rev. Lett.* **63** 1245
[13] Martin D and Puplett E 1970 *Infrared Physics* **10** 105
[14] Hayakawa Y, Hayakawa K, Inagaki M, Kaneda T, Nakao K, Nogami K, Sakae T, Sakai T, Sato I, Takahashi Y and Tanaka T 2015 *Nucl. Instrum. and Meth.* B **355** 251
[15] Inagaki M, Hayakawa Y, Nogami K, Tanaka T, Hayakawa K, Sakai T, Nakao K and Sato I 2008 *Japan. J. Appl. Phys.* **47** 8081

Comparison of DTR spectral-angular characteristics of divergent beam of relativistic electrons in scattering geometry of Laue and Bragg

S V Blazhevich, T V Koskova, A Z Ligidov, A V Noskov
Belgorod State University, Belgorod, Russia

E-mail: noskovbupk@mail.ru

Abstract. Diffracted transition radiation (DTR) generated by a divergent beam of relativistic electrons crossing a single-crystal plate in different (Laue, Bragg) scattering geometry has been considered for the general case of asymmetric reflection of the electron coulomb field relative to the entrance target surface. The expressions for spectral-angular density of DTR and parametric X-ray Radiation (PXR) has been derived. Then DTR and PXR has been considered in case of a thin target, when multiple scattering of electron is negligibly small, which is important for divergence measurement in real time regime. Numerical calculation of spectral-angular density of DTR by a beam of relativistic electrons has been made using averaging over the bivariate Gauss distribution as angular distribution of relativistic electrons in the beam. It has been shown that in Bragg scattering geometry the angular density of DTR is bigger, than in Laue geometry, which can be explained by the existence of the frequency range, in which the incident wave propagation vector takes complex value even under absence of absorption. In this range, all of photons are reflected in Bragg direction. It means that the range of total reflection defines the width of DTR spectrum.

1. Introduction

In the physics of interaction of relativistic electrons with matter, is important for the experimental data interpretation to know spatial and angular distributions of particles in the incident beam. That is why working out express methods of obtaining information about the characteristics of the beam used in the experiment is actual problem. One of the approaches is to use different types of radiation excited by relativistic charged particles in matter. Recently the possibility of use of parametric X-radiation (PXR) for the diagnostics of relativistic electron beams was experimentally studied in [1, 2]. In [3] it was suggested using PXR generated in a thin crystal to get operative information on spatial distribution of the relativistic electron beam. The applicability of transition radiation (TR) of vacuum ultraviolet range to measure the electron beam cross dimensions was demonstrated in [4]. The authors of [5] suggest the use of X-ray Cherenkov radiation by ultrarelativistic charged particles in the photon energy range, which includes K-absorption edges for some of the materials, to reveal the cross-dimensions of the beam.

In all the above-listed works, the beam parameters estimation was carried out in the framework of kinematic PXR theory, therefore studying the influence of dynamic effects on the characteristics of coherent radiation by relativistic electron beams remains an important task.

Content from this work may be used under the terms of the Creative Commons Attribution 3.0 licence. Any further distribution of this work must maintain attribution to the author(s) and the title of the work, journal citation and DOI.
Published under licence by IOP Publishing Ltd

As we know PXR appears due to the scattering of a relativistic electron Coulomb field on a system of parallel atomic planes in the crystal [6–8]. When a charged particle crosses the crystal plate surface, the transition radiation (TR) takes place [9, 10]. TR appearing on the border diffracts then on a system of parallel atomic planes of the crystal which forms DTR in a narrow spectral range [11–14]. The DTR photons move near the Bragg scattering direction.

The process of coherent X-ray radiation by a single relativistic electron in a crystal is described in the framework of the dynamical theory of x-rays diffraction in [15–18]. In these papers, the dynamic theory of coherent X-ray radiation generated by a relativistic electron in a crystal are built for general case of asymmetric relative to the crystal (target) surface reflection of the electron Coulomb field. In this case, the system of the parallel reflecting atomic planes in the target can be located at arbitrarily given angle to the target surface. Under these conditions, coherent X-radiation in the direction of the Bragg scattering appears because of two coherent radiation mechanisms, namely PXR and DTR.

In the present work, a dynamical theory of coherent X-ray radiation generated in a single-crystal target by a divergent beam of relativistic electrons has been developed both in Laue and Bragg scattering geometries.

2. Radiation process geometries

Let us consider a beam of relativistic electrons crossing a monocrystalline plate in of Laue (figure 1) and Bragg (figure 2) scattering geometries.

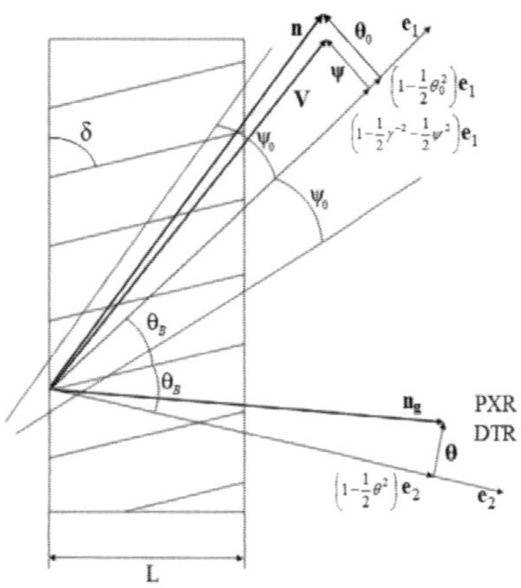

Figure 1. Laue scattering geometry.

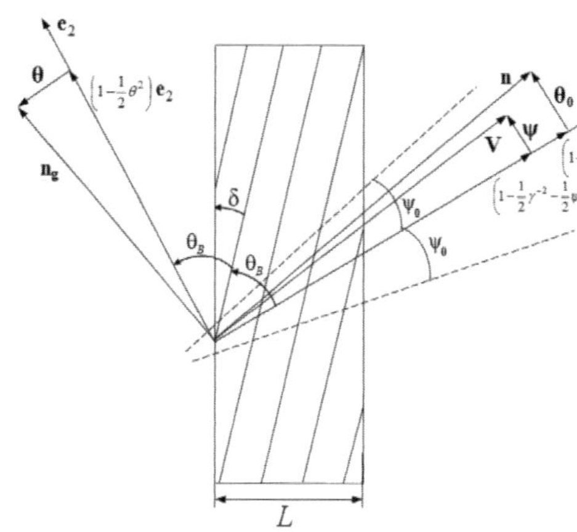

Figure 2. Bragg scattering geometry.

Let us involve the angular variables $\boldsymbol{\psi}$, $\boldsymbol{\theta}$ and $\boldsymbol{\theta}_0$ in accordance with the definition of relativistic electron velocity $\mathbf{V}$ and unit vectors in direction of momentum of the photon radiated in the direction near electron velocity vector $\mathbf{n}$ and in the Bragg scattering direction $\mathbf{n}_g$:

$$\mathbf{V}=\left(1-\frac{1}{2}\gamma^{-2}-\frac{1}{2}\psi^2\right)\mathbf{e}_1+\boldsymbol{\psi},\ \mathbf{e}_1\boldsymbol{\psi}=0,\ \mathbf{n}=\left(1-\frac{1}{2}\theta_0^2\right)\mathbf{e}_1+\boldsymbol{\theta}_0,\quad \mathbf{e}_1\boldsymbol{\theta}_0=0,\quad \mathbf{e}_1\mathbf{e}_2=\cos 2\theta_B$$

$$\mathbf{n}_g=\left(1-\frac{1}{2}\theta^2\right)\mathbf{e}_2+\boldsymbol{\theta},\ \mathbf{e}_2\boldsymbol{\theta}=0, \tag{1}$$

where $\boldsymbol{\theta}$ is the radiation angle, counted from direction of axis of radiation detector $\mathbf{e}_2$, $\boldsymbol{\psi}$ is the incidence angle of an electron in the beam counted from the electron beam axis $\mathbf{e}_1$, $\boldsymbol{\theta}_0$ is the angle between the movement direction of incident photon and axis $\mathbf{e}_1$, $\gamma=1/\sqrt{1-V^2}$ is Lorentz-factor of the particle. The angular variables are decomposed into the components parallel and perpendicular to the figure plane: $\boldsymbol{\theta} = \boldsymbol{\theta}_{\parallel} + \boldsymbol{\theta}_{\perp}$, $\boldsymbol{\theta}_0 = \boldsymbol{\theta}_{0\parallel} + \boldsymbol{\theta}_{0\perp}$, $\boldsymbol{\psi} = \boldsymbol{\psi}_{\parallel} + \boldsymbol{\psi}_{\perp}$, ψ_0 is the divergence parameter of the beam of radiating electrons.

Let us consider the electromagnetic processes in the crystalline medium characterized by the complex permittivity

$$\varepsilon(\omega, \mathbf{r}) = 1 + \chi(\omega, \mathbf{r}), \tag{2}$$

where $\chi(\omega,\mathbf{r}) = \chi_0(\omega) + \sum_{\mathbf{g}} \chi_{\mathbf{g}}(\omega)\exp(i\mathbf{gr})$, $\chi(\omega,\mathbf{r})$ is the dielectric susceptibility, $\chi_{\mathbf{g}}(\omega) = \chi'_{\mathbf{g}}(\omega) + i\chi''_{\mathbf{g}}(\omega)$ is the Fourier coefficient of the expansion of the dielectric susceptibility of the crystal in reciprocal lattice vectors $\mathbf{g}$, and $\chi_0(\omega)$ is the average dielectric susceptibility.

In this work the two-wave approach of dynamic diffraction theory are used, in which both the incident and diffracted waves are considered as equitable in process of self-repumping one into another in crystalline target. While solving the problem, let us consider an equation for a Fourier image of an electromagnetic field $\mathbf{E}(\mathbf{k},\omega) = \int dt\, d^3\mathbf{r}\, \mathbf{E}(\mathbf{r},t)\exp(i\omega t - i\mathbf{kr})$. The strengths of electromagnetic fields, excited by electron in the crystal are

$$\begin{aligned} \mathbf{E}(k,\omega) &= E_0^{(1)}(\mathbf{k},\omega)\mathbf{e}_0^{(1)} + E_0^{(2)}(\mathbf{k},\omega)\mathbf{e}_0^{(2)}, \\ \mathbf{E}(\mathbf{k}+\mathbf{g},\omega) &= E_{\mathbf{g}}^{(1)}(\mathbf{k},\omega)\mathbf{e}_1^{(1)} + E_{\mathbf{g}}^{(2)}(\mathbf{k},\omega)\mathbf{e}_1^{(2)}, \end{aligned} \tag{3}$$

$\mathbf{e}_0^{(1)} \perp \mathbf{k}$, $\mathbf{e}_0^{(2)} \perp \mathbf{k}$, $\mathbf{e}_1^{(1)} \perp \mathbf{k}_{\mathbf{g}}$ и $\mathbf{e}_1^{(2)} \perp \mathbf{k}_{\mathbf{g}}$, $\mathbf{k}_{\mathbf{g}} = \mathbf{k} + \mathbf{g}$. Vectors $\mathbf{e}_0^{(2)}$, $\mathbf{e}_1^{(2)}$ are situated on the plane of vectors $\mathbf{k}$ и $\mathbf{k}_{\mathbf{g}}$ (π -polarization) and $\mathbf{e}_0^{(1)}, \mathbf{e}_1^{(1)}$ are perpendicular to this plane (σ -polarization);

The system of equations for the Fourier transform images of the electromagnetic field in a two wave approximation of dynamic theory of diffraction has the following form [19].

For the Laue scattering geometry the system of equations has a view

$$\begin{cases} (\omega^2(1+\chi_0) - k^2)E_0^{(s)} + \omega^2\chi_{-\mathbf{g}}C^{(s)}E_{\mathbf{g}}^{(s)} = 8\pi^2 ie\omega\mathbf{e}_0^{(s)}\mathbf{V}\delta(\omega - \mathbf{kV}), \\ \omega^2\chi_{\mathbf{g}}C^{(s)}E_0^{(s)} + (\omega^2(1+\chi_0) - k_{\mathbf{g}}^2)E_{\mathbf{g}}^{(s)} = 0, \end{cases} \tag{4}$$

where $C^{(s)} = \mathbf{e}_0^{(s)}\mathbf{e}_1^{(s)}$, $C^{(1)} = 1$, $C^{(2)} = \cos 2\theta_B$. Equations set (4) under $s = 1$ describes the fields of σ -polarization, and under s = 2 the fields of π -polarization.

For the Bragg scattering geometry the system of equations has a view

$$\begin{cases} (\omega^2(1+\chi_0) - k^2)E_0^{(s)} + \omega^2\chi_{-\mathbf{g}}C^{(s,\tau)}E_{\mathbf{g}}^{(s)} = 8\pi^2 ie\omega\mathbf{e}_0^{(s)}\mathbf{V}\delta(\omega - \mathbf{kV}), \\ \omega^2\chi_{\mathbf{g}}C^{(s,\tau)}E_0^{(s)} + (\omega^2(1+\chi_0) - k_{\mathbf{g}}^2)E_{\mathbf{g}}^{(s)} = 0, \end{cases} \tag{5}$$

where $C^{(s,\tau)} = \mathbf{e}_0^{(s)}\mathbf{e}_1^{(s)} = (-1)^{\tau}C^{(s)}$, $C^{(1)} = 1$, $C^{(2)} = |\cos 2\theta_B|$.

The system of equations (5) for $s = 1$ and $\tau = 2$ describes the σ - polarized fields. For $s = 2$, the system of equations (5) describes π - polarized fields; note that if $2\theta_B < \frac{\pi}{2}$, then $\tau = 2$, otherwise $\tau = 1$. In the system of equations (4) и (5) following table of symbols is denoted

$$P^{(1)} = \sin\varphi,\ P^{(2)} = \cos\varphi,\ \mathbf{e}_0^{(1)}\mathbf{V} = (\theta - \psi)P^{(1)} = \theta_{\perp} - \psi_{\perp},\ \mathbf{e}_0^{(2)}\mathbf{V} = (\theta + \psi)P^{(2)} = \theta_{//} + \psi_{//},$$

$$\chi'_{\mathbf{g}} = \chi'_0 (F(g)/Z)(S(\mathbf{g})/N_0)\exp(-g^2u_{\tau}^2/2),\ \chi''_{\mathbf{g}} = \chi''_0\exp\left(-\frac{1}{2}g^2u_{\tau}^2\right), \tag{6}$$

where $\chi_0 = \chi'_0 + i\chi''_0$ – is the average dielectric susceptibility, F(g) – is the form factor of atom containing Z electrons, $S(\mathbf{g})$ is the structural factor of a unit cell containing N_0 atoms, u_{τ} is the r.m.s. amplitude of thermal vibrations of crystal atoms. The work addresses the X-ray frequency range, where $\chi'_{\mathbf{g}} < 0, \chi'_0 < 0$. We will consider a crystal with the following symmetry $(\chi_g = \chi_{-g})$, φ is the azimuthal radiation angle measured from the plane formed by the vectors $\mathbf{V}$ и $\mathbf{g}$, the value of the reciprocal lattice vector is determined by the $g = 2\omega_B \sin\theta_B / V$, where ω_B is the Bragg frequency.

3. Spectral-angular density of PXR and DTR in a thin crystal

Let us consider the relativistic electron coherent X-radiation in Laue scattering geometry (see figure 1). If we perform the analytical procedures similar to those used in [20, 21] we will obtain the expressions for the spectral-angular density of PXR and DTR for the propagation direction of the emitted photon $\mathbf{k_g} = k_\mathbf{g}\mathbf{n_g}$ (see. figure 1) taking into account the direction deviation of the electron velocity $\mathbf{V}$ relative to the electron beam axis $\mathbf{e}_1$

$$\omega\frac{d^2N^{(s)}_{\mathrm{PXR}}}{d\omega d\Omega} = \frac{e^2}{\pi^2}\frac{\Omega^{(s)2}}{(\gamma^{-2}+(\theta_\perp-\psi_\perp)^2+(\theta_{//}+\psi_{//})^2-\chi'_0)^2}R^{(s)}_{\mathrm{PXR}}, \tag{7a}$$

$$R^{(s)}_{\mathrm{PXR}} = \left(1-\frac{\xi^{(s)}}{\sqrt{\xi^{(s)2}+\varepsilon}}\right)^2 \sin^2\left(\frac{b^{(s)}}{2}\left(\sigma^{(s)}+\frac{\xi^{(s)}-\sqrt{\xi^{(s)2}+\varepsilon}}{\varepsilon}\right)\right)\left(\sigma^{(s)}+\frac{\xi^{(s)}-\sqrt{\xi^{(s)2}+\varepsilon}}{\varepsilon}\right)^{-2}, \tag{7b}$$

$$\omega\frac{d^2N^{(s)}_{\mathrm{DTR}}}{d\omega d\Omega} = \frac{e^2}{\pi^2}\Omega^{(s)2}\left(\frac{1}{\gamma^{-2}+(\theta_\perp-\psi_\perp)^2+(\theta_{//}+\psi_{//})^2}-\frac{1}{\gamma^{-2}+(\theta_\perp-\psi_\perp)^2+(\theta_{//}+\psi_{//})^2-\chi'_0}\right)^2 R^{(s)}_{\mathrm{DTR}}, \tag{8a}$$

$$R^{(s)}_{DTR} = \frac{\varepsilon^2}{\xi^{(s)2}+\varepsilon}\sin^2\left(b^{(s)}\frac{\sqrt{\xi^{(s)2}+\varepsilon}}{\varepsilon}\right), \tag{8b}$$

where

$$\Omega^{(1)} = \theta_\perp - \psi_\perp,\ \Omega^{(2)} = \theta_{//}+\psi_{//}, \sigma^{(s)} = \frac{1}{|\chi'_\mathbf{g}|C^{(s)}}\left(\gamma^{-2}+(\theta_\perp-\psi_\perp)^2+(\theta_{//}+\psi_{//})^2-\chi'_0\right),$$

$$\varepsilon = \frac{\sin(\delta+\theta_B)}{\sin(\delta-\theta_B)},\ b^{(s)} = \frac{1}{2\sin(\delta-\theta_B)}\frac{L}{L^{(s)}_{ext}},\ \xi^{(s)}(\omega) = \eta^{(s)}(\omega)+\frac{1-\varepsilon}{2\nu^{(s)}},\ L^{(s)}_{ext} = 1/\omega|\chi'_\mathbf{g}|C^{(s)},$$

$$\eta^{(s)}(\omega) = \frac{2\sin^2\theta_B}{V^2|\chi'_\mathbf{g}|C^{(s)}}\left(1-\frac{\omega(1-\theta_{//}\cot\theta_B)}{\omega_B}\right),\ \nu^{(s)} = \frac{\chi'_\mathbf{g}C^{(s)}}{\chi'_0}. \tag{9}$$

The parameter ε is an important parameter in (2) - (4); it determines the degree of asymmetry of the reflection of the field in a crystal plate with respect to the target surface. Note that the angle of electron incidence on the target surface $\delta-\theta_B$ increases when the ε parameter decreases, and vice versa.

Parameter $b^{(s)}$ characterizing the thickness of the crystal plate is the ratio of half of the path of the electron in the target $L_e = L/\sin(\delta-\theta_B)$ to the extinction length $L^{(s)}_{ext}$. Parameter $\nu^{(s)}$ can take the values in the interval $0 \le \nu^{(s)} \le 1$ and determines the degree of reflection of the waves from the crystal, which is caused by the nature of the interference of the waves reflected from different planes (constructive ($\nu^{(s)} \approx 1$) or destructive ($\nu^{(s)} \approx 0$)). The spectral function $\eta^{(s)}(\omega)$ rapidly changes with the frequency of the radiation therefore this function is convenient for use as an argument in the diagrams demonstrating the spectra of PXR and DTR .

Let us consider the relativistic electron coherent X-radiation in Bragg scattering geometry (see figure 2). If we perform the analytical procedures similar to those used in [22] we will obtain the expressions for the spectral-angular density of PXR and DTR for the propagation direction of the emitted photon $\mathbf{k_g} = k_\mathbf{g}\mathbf{n_g}$ (see. figure 2) taking into account the direction deviation of the electron velocity $\mathbf{V}$ relative to the electron beam axis $\mathbf{e}_1$

$$\omega\frac{d^2N^{(s)}_{\mathrm{PXR}}}{d\omega d\Omega}=\frac{e^2}{\pi^2}\frac{\Omega^{(s)2}}{(\gamma^{-2}+(\theta_\perp-\psi_\perp)^2+(\theta_{//}+\psi_{//})^2-\chi'_0)^2}R^{(s)}_{\mathrm{PXR}},\tag{10a}$$

$$R^{(s)}_{\mathrm{PXR}}=\frac{\left(\xi^{(s)}+\sqrt{\xi^{(s)2}-\varepsilon}\right)^2}{\xi^{(s)2}-\varepsilon+\varepsilon\sin^2\left(\frac{b^{(s)}\sqrt{\xi^{(s)2}-\varepsilon}}{\varepsilon}\right)}\frac{\sin^2\left(\frac{b^{(s)}}{2}\left(\frac{\xi^{(s)}+\sqrt{\xi^{(s)2}-\varepsilon}}{\varepsilon}-\sigma^{(s)}\right)\right)}{\left(\frac{\xi^{(s)}+\sqrt{\xi^{(s)2}-\varepsilon}}{\varepsilon}-\sigma^{(s)}\right)^2},\tag{10b}$$

$$\omega\frac{d^2N^{(s)}_{\mathrm{DTR}}}{d\omega d\Omega}=\frac{e^2}{\pi^2}\Omega^{(s)2}\left(\frac{1}{\gamma^{-2}+(\theta_\perp-\psi_\perp)^2+(\theta_{//}+\psi_{//})^2}-\frac{1}{\gamma^{-2}+(\theta_\perp-\psi_\perp)^2+(\theta_{//}+\psi_{//})^2-\chi'_0}\right)^2R^{(s)}_{\mathrm{DTR}},\tag{11a}$$

$$R^{(s)}_{\mathrm{DTR}}=\frac{\varepsilon^2}{\xi^{(s)^2}-\left(\xi^{(s)^2}-\varepsilon\right)\coth^2\left(\frac{b^{(s)}\sqrt{\varepsilon-\xi^{(s)^2}}}{\varepsilon}\right)},\tag{11b}$$

where the notations analogous to (4) are used:

$$C^{(2)}=\left|\cos 2\theta_B\right|,\ \varepsilon=\frac{\sin(\theta_B-\delta)}{\sin(\theta_B+\delta)},\ b^{(s)}=\frac{1}{2\sin(\theta_B+\delta)}\frac{L}{L^{(s)}_{ext}},\ \xi^{(s)}(\omega)=\eta^{(s)}(\omega)+\frac{1+\varepsilon}{2\nu^{(s)}}.\tag{12}$$

Under a fixed value of θ_B the value ε defines the orientation of crystal plate in relation to the system of diffracting atomic planes. When the angle of electron incidence on the target surface $\theta_B+\delta$ decreases the value of δ parameter can become negative and then will increase in magnitude (in extreme case $\delta\to-\theta_B$) that leads to increase of ε. On the contrary, when the angle of electron incidence decreases the value of ε decrease (in extreme case $\delta\to\theta_B$). The expressions (7), (10) and (8), (11) describe the spectral-angular density of PXR and DTR of the relativistic electron crossing a crystal plate at an angle $\boldsymbol{\psi}(\psi_\perp,\psi_{//})$ relative to the axis of the electron beam $\mathbf{e}_1$. The expressions are obtained in the framework of two-wave approximation of dynamic diffraction theory taking into account the angle between the reflecting system of parallel atomic planes of the crystal and the target surface (angle δ).

In Bragg scattering geometry, the frequency range of total external absorption (extinction) of pseudo photons of coulomb field of relativistic electron in single crystal exists that was known for real X-ray photons. In this region, the incident wave vector takes on a complex value even in absence of absorption and as the result all the photons reflect. If absorption is absent the of wave vector lengths have such a view:

$$k^{(1,2)}=\omega\sqrt{1+\chi'_0}+\frac{\omega\left|\chi'_g C^{(s)}\right|}{2\varepsilon}\left(\xi^{(s)}(\omega)\pm\sqrt{\xi^{(s)}(\omega)^2-\varepsilon}\right).\tag{13}$$

The region of total reflection defines by following inequality:

$$-\sqrt{\varepsilon}<\xi^{(s)}(\omega)<\sqrt{\varepsilon},\ -\sqrt{\varepsilon}-\frac{1+\varepsilon}{2\nu^{(s)}}<\eta^{(s)}(\omega)<\sqrt{\varepsilon}-\frac{1+\varepsilon}{2\nu^{(s)}},\tag{14}$$

from which one can see that the width of this range is defined by value of $2\sqrt{\varepsilon}$. It can be shown from (11b) that the width of region of total reflection practically coincides with the of DTR spectrum width.

The magnitude and the width of angular density of PXR produced by relativistic electron with energy exceeding several hundred MeV practically do not depend on the electron energy. In this connection angular density of PXR turn out insensitive to divergence of electron beam therefore we will consider the contribution in coherent radiation of relativistic electron by DTR only.

As the result, we will obtain from (8) and (11) the expression describing the angular density of DTR in scattering geometries Laue (16) and Bragg (15):

$$\frac{dN_{\mathrm{DTR}}^{(s)}}{d\Omega}=\frac{e^2\chi_0'^2}{2\pi^2\sin^2\theta_B|\chi_g'|C^{(s)}}\frac{\Omega^{(s)2}}{\sigma^{(s)2}\left(|\chi_g'|C^{(s)}\sigma^{(s)}+\chi_0'\right)^2}\int_{-\infty}^{\infty}\frac{\varepsilon^2}{\xi^{(s)2}+\varepsilon}\sin^2\left(b^{(s)}\frac{\sqrt{\xi^{(s)2}+\varepsilon}}{\varepsilon}\right)d\xi^{(s)}(\omega), \quad (15)$$

$$\frac{dN_{\mathrm{DTR}}^{(s)}}{d\Omega}=\frac{e^2\chi_0'^2}{2\pi^2\sin^2\theta_B|\chi_g'|C^{(s)}}\frac{\Omega^{(s)2}}{\sigma^{(s)2}\left(|\chi_g'|C^{(s)}\sigma^{(s)}+\chi_0'\right)^2}\varepsilon\sqrt{\varepsilon}\pi\cdot\tanh\left(\frac{b^{(s)}}{\sqrt{\varepsilon}}\right). \quad (16)$$

4. The averaging of coherent radiation generated by a relativistic electron of the divergent beam in a crystal over angles of electron incidence.

Let us consider the effect of the electron beam divergence on the spectral and angular characteristics of the radiation. For that, let us average radiation of an electron over all of its possible straight trajectories in the beam. As an example, we will carry out the averaging of the spectral-angular density of DTR for the electron beam with the Gauss angular distribution:

$$f(\psi)=\frac{1}{\pi\psi_0^2}\cdot\exp\left\{-\frac{\psi^2}{\psi_0^2}\right\}, \quad (17)$$

where ψ_0 is the divergence of the beam of radiating electrons (see figure 1-figure 2).

The magnitude and the width of angular density of PXR produced by relativistic electron with energy exceeding several hundred MeV practically do not depend on the electron energy. In this connection angular density of PXR turn out insensitive to divergence of electron beam therefore we will consider the contribution in coherent radiation of relativistic electron by DTR only. By averaging the expressions (7), (8) and (15), (16) we obtain

$$\left\langle\omega\frac{d^2N_{\mathrm{DTR}}^{(s)}}{d\omega d\Omega}\right\rangle=\frac{1}{\pi\psi_0^2}\iint d\psi_\perp d\psi_{//}\exp\left\{-\frac{\psi_\perp^2+\psi_{//}^2}{\psi_0^2}\right\}\omega\frac{d^2N_{\mathrm{DTR}}^{(s)}}{d\omega d\Omega}, \quad (18a)$$

$$\left\langle\frac{dN_{\mathrm{DTR}}^{(s)}}{d\Omega}\right\rangle=\frac{1}{\pi\psi_0^2}\iint d\psi_\perp d\psi_{//}\exp\left\{-\frac{\psi_\perp^2+\psi_{//}^2}{\psi_0^2}\right\}\frac{dN_{\mathrm{DTR}}^{(s)}}{d\Omega}. \quad (18b)$$

The calculation have been made for W(110) single crystal plate of thickness $L = 1\ \mu\mathrm{m}$, Bragg angle $\theta_B \approx 20.5^o$, electron energy $E=10$ GeV for photons with σ - polarization. In figure 3–figure 4 the curves are plotted by formulas (7), (8) and (18a) demonstrate the dependence of DTR spectral-angular densities on initial divergence of electron beam in Bragg and Laue scattering geometries under the fixed observation angle $\mathbf{\theta}$. The curves are plotted for the case of symmetric in relation to the target surface reflection of the radiation ($\varepsilon=1$). One can see in figure 3 that in Bragg scattering geometry the spectrum is wider than in Laue geometry. The DTR spectrum width practically coincides with the width of spectral range of total external reflection of radiation and does not depend on the initial beam divergence ψ_0 as it is seen in figure 3. Because of these facts the angular density of DTR in Bragg geometry turns out bigger than in Laue geometry, as it is seen in figure 5 and figure 6. If the asymmetry parameter ε increases, then the range of total absorption, as well as the width of DTR spectrum increase (figure 7) too and as the result an increase will be observed in angular density of DTR (figure 8). So, DTR of the beam of relativistic electrons in Bragg scattering geometry can be more intensive than in Laue geometry. It is necessary to note that the values of spectral-angular and angular density of DTR strongly depend on initial divergence of the electron beam.

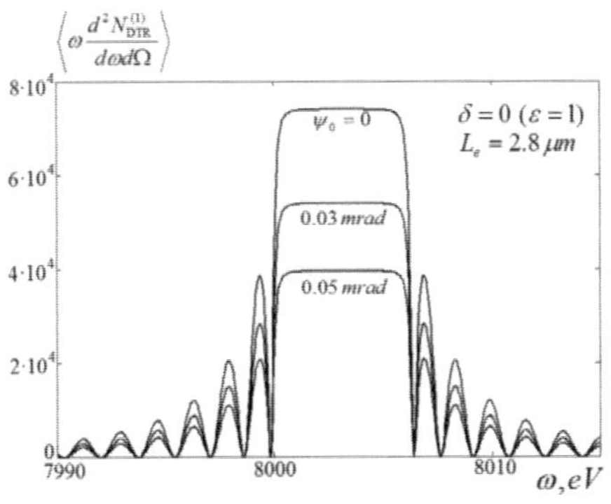

Figure 3. DTR spectra in Bragg scattering geometry in the range of angular density maximum under different values of electron beam divergence ψ_0. Symmetric reflection case ($\delta = 0$), $\theta_\perp = 0.05$ mrad, $\theta_{//} = 0$.

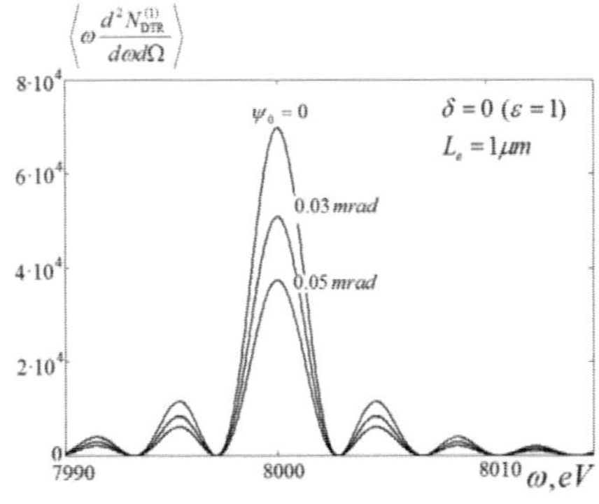

Figure 4. DTR spectra in Laue scattering geometry in the range of angular density maximum for different values of electron beam divergence ψ_0. Symmetric reflection case ($\delta = 90^o$), $\theta_\perp = 0.05$ mrad, $\theta_{//} = 0$.

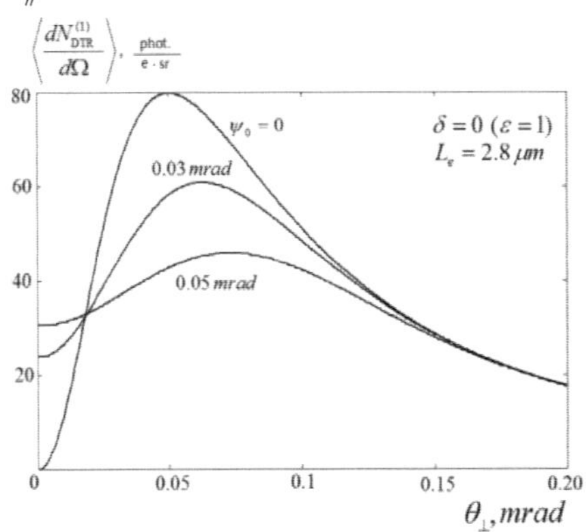

Figure 5. DTR angular density in Bragg scattering geometry under different values of electron beam divergence.

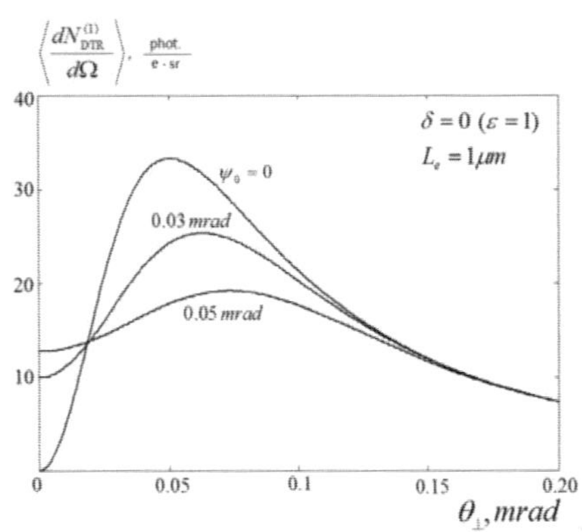

Figure 6. DTR spectra in Laue scattering geometry in the range of angular density maximum for different values of electron beam divergence ψ_0.

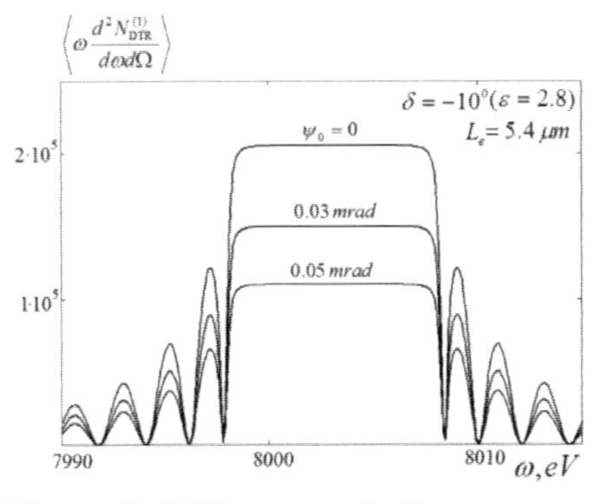

Figure 7. DTR spectra in Bragg scattering geometry in range of angular density maximum for different values of electron beam divergence ψ_0. Asymmetric reflection case ($\delta \neq 0$).

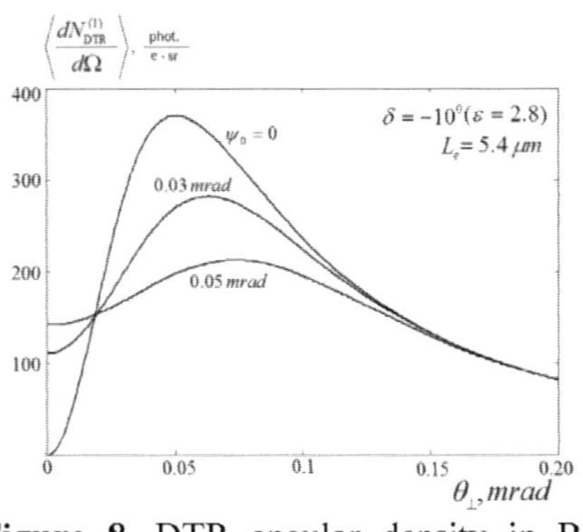

Figure 8. DTR angular density in Bragg scattering geometry in range of angular density maximum for different values of electron beam divergence ψ_0. Asymmetric reflection case ($\delta \neq 0$).

5. Conclusion

In the present work, the dynamic theory of coherent X-ray radiation of divergent beam of the relativistic electrons crossing a single crystal plate is considered. We have obtained the expressions for spectral-angular density parametric X-ray radiation and diffracted transition radiation of relativistic electron taking into account the deviation of electron velocity direction relative to the electron beam axis. In the framework of two-wave approximation of diffraction theory the expression for PXR and DTR are obtained for general case of asymmetric reflection of pseudo photons of the electron coulomb field. The expressions have been derived both for Laue and for Bragg scattering geometries. Based on the obtained expression for spectral-angular and angular density of the radiation of an electron in the beam using averaging over the distribution of the electron incidence angles the expressions for radiation of the electron beam as a whole have been obtained. The effect of the electron beam divergence on the angular characteristics of PXR and DTR was investigated for both cases of scattering geometry.

As an example, in the present work we carried out the averaging of the spectral-angular density of PXR and DTR for the electron beam having the Gauss angular distribution function. There was shown considerable depend of spectral-angular characteristic of DTR on divergence of electron beam. It was shown that in Bragg geometry the generation of DTR is more effective than in Laue Geometry and in condition of asymmetric reflection, the intensity of DTR can be increased additionally.

Acknowledgements

The Russian Science Foundation (project № 15-12-10019) supported this work.

References

[1] Takabayashi Y 2012 *Phys. Lett.* A **376** 2408
[2] Takabayashi Y, Sumitani K 2013 *Phys. Lett.* A **377** 2577
[3] Gogolev A, Potylitsyn A, Kube *G* 2011 *J. Phys. Conference Series* **357** 012018
[4] Sukhikh L G, Gogolev S Yu, and Potylitsyn A P 2010 *J. Phys. Conference Series* **236** 012011
[5] Konkov A S, Karataev P V, Potylitsyn A P and Gogolev A S 2014 *J. Phys. Conference Series* **517** 012003
[6] Ter-Mikaelian M 1972 *High-Energy Electromagnetic Process in Condensed Media* (New York: Wiley)
[7] Garibian G M, Yang C 1971 *J. Exp. Theor. Phys.* **61** 930
[8] Baryshevsky V G, Feranchuk I D 1971 *J. Exp. Theor. Phys.* **61** 944
[9] Ginzburg V L and Frank I M 1946 *J. Exp. Theor. Phys.* **15** 15
[10] Ginzburg V L and Tsytovich V N 1984 *Transition Radiation and Transition Scattering* (Nauka, Moscow, 1984; Adam Hilger, Bristol, United Kingdom, 1990)
[11] Caticha A 1989 *Phys. Rev.* A **40** 4322
[12] Baryshevsky V 1997 *Nucl. Instr. and Meth.* A **122** 13
[13] Artru X, Rullhusen P 1998 Nucl. *Instr. and Meth.* B **145** 1
[14] Nasonov N 1998 *Phys. Lett.* A **246** 148
[15] Blazhevich S V, Noskov A V 2008 *Nucl. Instr. and Meth.* B **266** 3770
[16] Blazhevich S, Noskov A 2009 *J. Exp. Theor. Phys.* **109** 901
[17] Blazhevich S V, Boltenko Yu A, Koskova T V, Mazilov A.A., Noskov A V 2015 *Problems of Atomic Science and Technology* **№5 (99)** 3
[18] Blazhevich S V, Koskova T V, Noskov A V 2015, *Russian Physics Journal* **58 No. 5** 585
[19] Bazylev V, Zhevago N 1987 *Emission From Fast Particles Moving in a Medium and External Fields* Moscow Nauka
[20] Blazhevich S, Noskov A 2008 *Nucl. Instr. Meth. In Phys. Res.* B **266** 3770
[21] Blazhevich S, Noskov A 2010 *J. Phys.: Conf. Ser.* **236** 012013
[22] Blazhevich S, Noskov A 2010 *J. Technical Physics* **55** 317–325

Theoretical Analysis of Orientation Distribution Function Reconstruction of Textured Polycrystal by Parametric X-rays

I Lobach and A Benediktovitch

Physics Department, Belarusian State University, 4, Nezavisimosti Av., 220030, Minsk, Belarus

E-mail: lobachihor@mail.ru

Abstract. The possibility of quantitative texture analysis by means of parametric x-ray radiation (PXR) from relativistic electrons with Lorentz factor $\gamma > 50$MeV in a polycrystal is considered theoretically. In the case of rather smooth orientation distribution function (ODF) and large detector ($\theta_D \gg 1/\gamma$) the universal relation between ODF and intensity distribution is presented. It is shown that if ODF is independent on one from Euler angles, then the texture is fully determined by angular intensity distribution. Application of the method to the simulated data shows the stability of the proposed algorithm.

1. Introduction

The orientation distribution function (ODF) of a polycrystalline material provides quantitative description of crystallograhic texture and enables to calculate the anisotropic physical properties of a polycrystalline material [1, 2]. There are several methods of quantitative texture analysis that provide the ODF. The most common is based on measurement of Bragg x-ray diffraction peak as a function of polycrystalline sample orientation, the so called pole figures. Provided with pole figures measured for several reciprocal lattice vectors there are several approaches to reconstruct the ODF [3–6]. The same methodology is used for processing the neutron diffraction data, the neutron technique being advantageous for large samples with coarse grains, e.g. geological samples [7]. Another widespread technique for quantitative texture analysis is electron backscattering diffraction (EBSD) [8]. This technique directly provides orientation of individual grains at the investigated cross section, unlike the x-ray and neutron diffraction techniques that provide the texture information in indirect way, but from large statistical ensemble of grains from the gauge volume.

Parametric x-ray radiation (PXR) is radiation in x-ray frequency range from relativistic electrons interacting with crystal target, the radiation parameters being directly related to crystallographical properties of the target [9]. Theoretical description [10, 11] and experiments [12–15] on PXR from a polycrystal show that there is a connection between texture of a polycrystal and PXR spectrum and intensity distribution. One of the obstacles in applications of PXR is its low total photon yield, however, the experiments [14, 15] revealed that the intensity of PXR from a polycrystalline Mo can be comparable with the intensity of its characteristic K_α-line. In the present contribution we investigate theoretically if it is feasible to perform quantitative texture analysis based on PXR data and to reconstruct the ODF.

Content from this work may be used under the terms of the Creative Commons Attribution 3.0 licence. Any further distribution of this work must maintain attribution to the author(s) and the title of the work, journal citation and DOI.
Published under licence by IOP Publishing Ltd

Journal of Physics: Conference Series **732** (2016) 012015 doi:10.1088/1742-6596/732/1/012015

2. Calculation of PXR from polycrystals

2.1. Basic relations for PXR in polycrystals

Orientation distribution function defines the volume part of crystallines having certain orientation:

$$\frac{dV}{V} = f(\varphi, \psi, \theta) \sin\theta d\varphi d\psi d\theta, \tag{1}$$

here φ, ψ, θ are the 3 Euler angles, see figure 1a; the blue coordinate system corresponds to the sample, xy-plane being the sample's surface, the red coordinate system is attached to the crystallographic unit cell of the material.

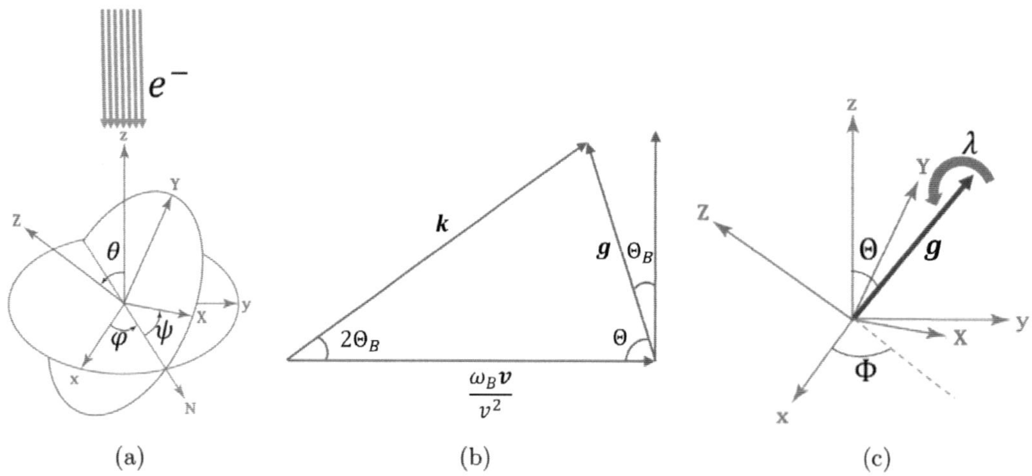

Figure 1. (a) Geometry of the problem, (b) Geometry of PXR, (c) The angles Φ, λ, Θ.

If the ODF is known, the distribution of PXR intensity resulting from interacting of relativistic electrons with polycrystal can be found. If one can use kinematical theory of PXR, which is justified for most cases of polycrystaline materials [15], the distribution of PXR intensity is given by (see [10] for more details):

$$I(\Theta, \Phi) = \frac{dN}{d\Phi d\Theta \sin\Theta} = T(|\boldsymbol{g}|, \Theta_B) \int_0^{2\pi} f(\varphi, \psi, \theta) d\lambda, \tag{2}$$

$$T(|\boldsymbol{g}|, \Theta_B) = \alpha L \frac{J}{e} \frac{|\cos 2\Theta_B|}{|\boldsymbol{g}|^3} |\kappa_g|^2 ((1 + \cos^2 2\Theta_B) \ln(\xi_{Deff} + \sqrt{\xi_{Deff}^2 + 1}) - \cos^2 2\Theta_B),$$

here N is a number of emitted photons, the introduced angles are shown in figure 1, φ, ψ, θ are expressed through Φ, λ, Θ, see [10] and figures 1a, 1c; $\xi_{Deff} = \gamma\theta_D$ and θ_D is the angular size of the detector (in radians); $\gamma = E/mc^2$ is the Lorentz factor of an electron with energy E; $\boldsymbol{g}$ is the reciprocal lattice vector, $\alpha = 1/137$, J is the electron beam current, e is the electron charge, L is the thickness of the polycrystal film, $\kappa_g = \chi_g(\omega)\omega^2/c^2$, κ_g does not depend on w with an appropriate accuracy, see [10]. $\omega, \chi_g(\omega)$ are the frequency and the X-ray susceptibility, respectively. Here for the convenience of further calculations we consider effective intensity I, which is a function of angles Θ, Φ. It is connected with observable intensity $I_0(2\Theta_B, \Phi)$ as

$$I = 4 \sin\Theta_B I_0. \tag{3}$$

3. Inverse Problem

In this section we describe the way one can reconstruct ODF through measured angular intensity distribution. One can obtain experimentally measured $I_0(2\Theta_B, \Phi)$ from a group of reciprocal lattice vectors $\boldsymbol{g}$ with the same $|\boldsymbol{g}|$ by the method suggested in [10], employing a detector with narrow range of detectable energies. However, the last procedure should be elaborated as it was shown in [16–19] that at relatively low electron energies (conventionally $\gamma < 50$) the contributions of "side" reflexes [16] in the resulting intensity distribution are considerable. Thus, the intensity, related to the vectors $\boldsymbol{g}$ with the same $|\boldsymbol{g}|$, cannot be distinguished with appropriate accuracy. It is caused by widening of PXR peaks under low γ. Therefore we have to restrict the realm of application of the presented approach to electron energies, corresponding to $\gamma > 50$. Suppose one has experimentally measured $I_0(2\Theta_B, \Phi)$, then $I_1(\Theta, \Phi) = dN/d\Phi d\Theta \sin\Theta$ is calculated in a straightforward way by equation (3), $\Theta \in [0, \pi/2]$.

Below we make an assumption that ODF does not depend on ψ, that is $f(\varphi, \psi, \theta) \to f(\varphi, \theta)$. It corresponds to quite large class of fibre textures that are observed in thin film samples [20]. Let us rewrite equation (2) in detail taking into account the summation over $\boldsymbol{g}$ with the same $|\boldsymbol{g}| \equiv g$.

$$\frac{I(\Theta, \Phi)}{T(|\boldsymbol{g}|, \Theta_B)} = \sum_{\boldsymbol{g}:|\boldsymbol{g}|=g} \int_0^{2\pi} f(\varphi(\Theta, \Phi, \lambda, \boldsymbol{g}), \theta(\Theta, \Phi, \lambda, \boldsymbol{g}))d\lambda. \quad (4)$$

To proceed further let us use spherical function decomposition [20] and note a useful property of spherical function integration over the rotation around reciprocal lattice vector $\boldsymbol{g}$:

$$\int_0^{2\pi} Y_{lm}(\varphi(\Theta, \Phi, \lambda, \boldsymbol{g}), \theta(\Theta, \Phi, \lambda, \boldsymbol{g}))d\lambda = c_{lm}(z)Y_{lm}(\Theta, \Phi), \quad (5)$$

here $c_{lm}(z) = 2\pi i^m P_l(z)$ and $Y_{lm}(\theta, \varphi)$ is a spherical harmonic, $P_l(z)$ is a Legendre polynomial. $(x, y, z) = \boldsymbol{g}/|\boldsymbol{g}|$ in the red coordinate system. Thus, if we regard substitution $\varphi \to \varphi(\Theta, \Phi, \lambda, \boldsymbol{g}), \theta \to \theta(\Theta, \Phi, \lambda, \boldsymbol{g})$ and following integration over λ as an operator, say $\widehat{R}_g$ then Y_{lm} is its eigenfunction and c_{lm} is the corresponding eigenvalue. By virtue of equation (5) it follows from equation (4) that

$$(\frac{I}{T})_{lm} = \sum_{\boldsymbol{g}:|\boldsymbol{g}|=g} c_{lm}(z) f_{lm}. \quad (6)$$

Some of the coefficients in equation (6) vanish due to symmetry considerations, in particular if l is odd then $(\frac{I}{T})_{lm} = 0$. It happens because in this case $c_{lm}(z)$ is an odd function and in the summation in equation (6) for every $\boldsymbol{g}$ there is always $-\boldsymbol{g}$ as well. Therefore $\frac{I}{T}$ should be defined as

$$\frac{I}{T} = \begin{cases} \dfrac{I_1(\Theta, \Phi)}{T(\Theta)} & , \Theta \leq \dfrac{\pi}{2} \\ \dfrac{I_1(\pi - \Theta, \pi + \Phi)}{T(\pi - \Theta)} & , \Theta \geq \dfrac{\pi}{2} \end{cases}$$

Further simplifications can be obtained if one considers crystallographic symmetry of the material under study. As an example we will consider diamond-like crystal lattice, in this case it appears that $\sum_{\boldsymbol{g}:|\boldsymbol{g}|=g} c_{2m}(z) = 0$ for any $|\boldsymbol{g}|$. Hence we can determine only harmonics of $f(\varphi, \theta)$ with even (except for 2) l. Similar situation takes place for quantitative texture analysis based on x-ray diffraction pole figure measurement [20].

Journal of Physics: Conference Series **732** (2016) 012015 doi:10.1088/1742-6596/732/1/012015

3.1. Nanodiamond film example

The ODF in equation (1) does not take into consideration lattice symmetries. In fact, there are orientations of the unit cell, say (φ, ψ, θ) and $(\varphi_1, \psi_1, \theta_1)$, that cannot be distinguished. Therefore, if one wants to find volume part of crystallites having orientation (φ, ψ, θ), then he has to add contributions corresponding to indistinguishable orientations as well. That is

$$\begin{aligned}\frac{dV}{V} &= f(\varphi, \psi, \theta)\sin\theta d\varphi d\psi d\theta + f(\varphi_1, \psi_1, \theta_1)\sin\theta_1 d\varphi_1 d\psi_1 d\theta_1 + \\ &... + f(\varphi_n, \psi_n, \theta_n)\sin\theta_n d\varphi_n d\psi_n d\theta_n \equiv F(\varphi, \psi, \theta)\sin\theta d\varphi d\psi d\theta,\end{aligned} \tag{7}$$

Below we will seek for the function $F(\varphi, \psi, \theta)$, which under considered assumption turns into $F(\varphi, \theta)$ and use F as ODF. Consider two diamond lattice symmetries, that will let us find F from f_{lm}:

1) The rotation of the unit cell in figure 2a about X-axis by π and then about Z-axis by π results in equivalent orientation. Hence, one has to take the following combination instead of $f(\varphi, \theta)$

$$F_0(\varphi, \theta) = f(\varphi, \theta) + f(\varphi + \pi, \pi - \theta),$$

It appears that F_0 does not contain f_{lm} with odd l.

2) Another symmetry can be found if one considers elementary unit cell, see figure 2b. Green line represents the diagonal of cubic unit cell. There is one atom on it. The facets of the parallelepiped are identical rhombuses. Thereby, if one rotates it around its diagonal by $\pm 2\pi/3$ he gets initial unit cell and, consequently, lattice. Thus, there is a triad axis.

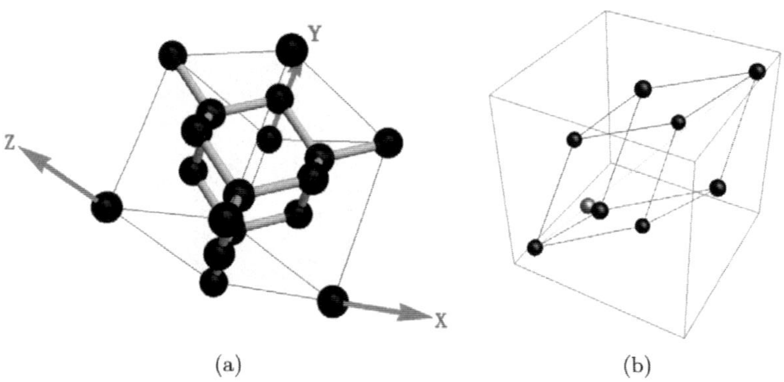

Figure 2. (a) Unit cell and attached frame of reference, (b) Elementary unit cell of the diamond lattice.

Mathematically, this triples the number of summands

$$F(\varphi, \theta) = F_0(\varphi, \theta) + F_0(\varphi^*, |\frac{\pi}{2} - \theta|) + F_0(\varphi + \frac{\pi}{2}, \frac{\pi}{2}),$$

where

$$\phi^* = \begin{cases} \varphi + \pi & , \theta \leq \frac{\pi}{2} \\ \varphi & , \theta \geq \frac{\pi}{2} \end{cases}$$

Eventually, it appears that F does not contain f_{2m}. That allows one to find F through even (except $l = 2$) harmonics of f that are known from equation (6). It is F that yields $dV(d\varphi, d\psi, d\theta)$ by virtue of equation (7). This formally allows to find ODF in terms of series in f_{lm}. It appears that for quite smooth ODFs several spherical harmonics (up to $l = 5$) are enough to get rather accurate solution.

The stability of the solution was tested. We modeled experimental intensity distributions by applying Poisson distribution to analytical solutions from Subsection 3.1 of [10]. We took the parameters of the polycrystalline diamond film from [21] and the electron beam from ASTA, Fermilab [22]. Then we found $F(\phi, \theta)$ by the above method. On the other hand, we had $F(\phi, \theta)$ expressed through initial $f_1(\varphi, \theta)$ (see [10]). See figure 3 for comparison of the results. In the plots z-axis corresponds to $F(\phi, \theta)$, in xy-plane we use polar coordinates. That is, the distance from the reference point ($x = 0, y = 0$) equals θ(rad), the angle ϕ is defined in a conventional way. It is seen in figure 3b that even if exposure time is as small as 1 second, one may obtain relatively good reconstruction of ODF. It results from considerable number of emitted photons ($\sim 10^6$), which is an advantage. In figure 3c $F(\phi, \theta)$ is strongly distorted due to high deviations of number of emitted photons from its mean value at low exposure time (or number of incident electrons). Other parameters of the case: angular size of the detector $\theta_D = 0.033$ rad, maximal l (index in spherical harmonic) used in calculations of reconstruction $l_{\max} = 3$, parameters of initial ODF $\Theta_{pref} = 1.0$ rad, $S = 6.0$ (see [10]). It is crucial to note that F can be determined by

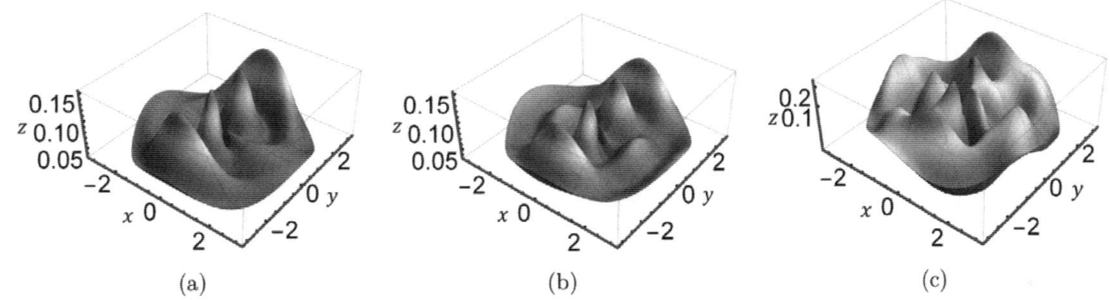

Figure 3. $z \equiv F(\phi, \theta)$, $x = \theta \cos\phi$, $y = \theta \sin\phi$. (a) Exact ODF, (b) Reconstructed, 6×10^{13} electrons (exposure time 1 sec), (c) Reconstructed, 6×10^8 electrons.

intensity distribution from all $\boldsymbol{g}$ that is easier to obtain on experiment (by direct measurement of angular intensity distribution without sorting out a contribution from a group of $\boldsymbol{g}$ with the same $|\boldsymbol{g}|$). Rewriting equation (2), but now summing over all $\boldsymbol{g}$ and taking into account structure of $T(|\boldsymbol{g}|, \Theta_B)$ one obtains

$$\begin{aligned} &\frac{I(\Theta, \Phi)}{\alpha L \frac{J}{e} |\cos 2\Theta_B| ((1 + \cos^2 2\Theta_B) \ln(\xi_{Deff} + \sqrt{\xi_{Deff}^2 + 1}) - \cos^2 2\Theta_B)} = \\ &= \sum_{\boldsymbol{g}} \frac{|\kappa_g|^2}{|\boldsymbol{g}|^3} \int_0^{2\pi} f(\varphi(\Theta, \Phi, \lambda, \boldsymbol{g}), \theta(\Theta, \Phi, \lambda, \boldsymbol{g})) d\lambda. \end{aligned} \tag{8}$$

Equation (8) is analogous to equation (4). Further derivation is essentially the same, though it is a little more cumbersome.

Journal of Physics: Conference Series **732** (2016) 012015 doi:10.1088/1742-6596/732/1/012015

4. Discussion

To reconstruct the ODF based on the approach presented above one has to measure the angular distribution of x-ray yield. In contrast to x-ray diffraction pole figure measurement, no sample rotations are needed, and, provided with large area detector with moderate energy resolution, one can collect all the necessary information in one shot. Supposing ODF independent on ψ that would be enough for the ODF reconstruction. In case of electron beam with large spot size and realistically high current (like at [22]) the 1 sec exposure time is enough to obtain satisfactory statistics, so one potentially can apply this technique for in-situ investigations, e.g. texture changes due to temperature changes or stress.

Further investigation of the method will be the subject of subsequent publications. The generalization to the case of an arbitrary ODF will be the next step. Introducing another variable, such as the angle between the film's plane and the electrons' velocity (say β) might solve this problem.

5. Conclusions

Based on relation between ODF and the intensity distribution (equation (2)) the algorithm of determining ODF (independent on ψ) through the intensity distribution is described. It is based on expansion in series of spherical harmonics, several components (up to $l = 5$ for instance) being enough to preserve a good accuracy in case of rather smooth ODF. The solution proves to be stable (with respect to Poisson distribution of number of emitted quanta) at a large number of emitted photons ($\sim 10^6$ per sec) reachable at e.g. ASTA facility [22]. Described method does not need sample tilting as it requires only angular intensity distribution (if ODF does not depend on ψ), hence, in-situ texture changes may be potentially observed. Introducing another variable (measurement of intensity distribution at different angles β) may allow one to generalize the method to arbitrary ODF.

Appendix

Below we will provide some useful relations:

We'll make use of the rotation matrix

$$\begin{pmatrix} \cos\varphi\cos\psi - \cos\theta\sin\varphi\sin\psi & -\cos\theta\cos\psi\sin\varphi - \cos\varphi\sin\psi & \sin\theta\sin\varphi \\ \cos\psi\sin\varphi + \cos\theta\cos\varphi\sin\psi & \cos\theta\cos\varphi\cos\psi - \sin\varphi\sin\psi & -\sin\theta\cos\varphi \\ \sin\theta\sin\psi & \cos\psi\sin\theta & \cos\theta \end{pmatrix} \begin{pmatrix} X \\ Y \\ Z \end{pmatrix} = \begin{pmatrix} x \\ y \\ z \end{pmatrix} = $$

$$= \begin{pmatrix} \sin\Theta\cos\Phi \\ \sin\Theta\sin\Phi \\ \cos\Theta \end{pmatrix}, \tag{A.1}$$

(X, Y, Z) and (x, y, z) correspond to red and blue coordinate systems respectively.

Further,

$$Y_{lm}(\theta, \varphi) \sim \exp(im\varphi)P_l^m(\cos\theta) = \exp(im\varphi)\sin^m\theta\frac{d^m P_l(\cos\theta)}{(d\cos\theta)^m},$$

where P_l^m is an associated Legendre polynomial. Let us define

$$M_{lm}(t) = \frac{d^m P_l(t)}{dt^m}.$$

Add up first two lines in equation (A.1), the second being multiplied by i. This way, one immediately gets $\exp(i\Phi)$ on the right-hand side. Using connections between φ, ψ, θ and Φ, λ, Θ (see Appendix of [10]) one can find $\exp(i\varphi)$ on the other side as well

$$\exp(i\varphi)\sin\theta = \exp(i\Phi)(\sqrt{1 - z^2}\sin\lambda + i(z\sin\Theta - \sqrt{1 - z^2}\cos\Theta\cos\lambda)).$$

Again, from geometrical connections between angles one has

$$\cos\theta = z\cos\Theta + \sqrt{1-z^2}\sin\Theta\cos\lambda.$$

Thus, we have parts of $Y_{lm}(\theta,\phi)$ in a convenient form. An analogue of equation (5) takes the form

$$\int_0^{2\pi}(\sqrt{1-z^2}\sin\lambda + i(z\sin\Theta - \sqrt{1-z^2}\cos\Theta\cos\lambda))^m M_{lm}(z\cos\Theta + \sqrt{1-z^2}\sin\Theta\cos\lambda)d\lambda = \\ = c_{lm}(z)P_l^m(\cos\Theta).$$

We considered $m \geq 0$. The case $m \leq 0$ is essentially the same.

References

[1] Bunge H J 1982 *Texture Analysis in Materials Science* ed Bunge H J (Butterworth-Heinemann) p 294
[2] Wenk H and Van Houtte P 2004 *Reports on Progress in Phys.* **67** 1367
[3] Nikolayev D, Savyolova T and Feldmann K 1992 *Textures and Microstruct.* **19** 9
[4] Helming K 1998 *Materials science forum* vol 273 (Trans Tech Publ) p 125
[5] Matthies S, Wenk H R and Vinel G 1988 *J. Appl. Cryst.* **21** 285
[6] Chateigner D 2013 *Combined analysis* (John Wiley & Sons)
[7] Wenk H R 2006 *Rev. in Mineral. and Geochem.* **63** 399
[8] Mason J and Schuh C 2009 *Electron Backscatter Diffraction in Material Science* (Springer US) p 35
[9] Baryshevsky V G, Feranchuk I D and Ulyanenkov A P 2005 *Parametric X-ray Radiation in Crystals. Theory, Experiment and Applications* (Springer-Verlag)
[10] Lobach I, Benediktovitch A, Feranchuk I *et al.* 2015 *Nucl. Instrum. Meth.* B **360** 75
[11] Nasonov N N 1998 *Nucl. Instrum. Meth.* B **145** 19
[12] Blazhevich S, Chepurnov A, Grishin *et al.* 1999 *Phys. Lett.* A **254** 230
[13] Chouffani K, Andreyashkin M, Endo I *et al.* 2001 *Nucl. Instrum. Meth.* B **173** 241
[14] Nawang S, Endo I, Iinuma M *et al.* 2006 *J. Phys. Society of Japan* **75**
[15] Takabayashi Y, Endo I, Ueda K *et al.* 2006 *Nucl. Instrum. Meth.* B **243** 453
[16] Gogolev A S and Potylitsyn A P 2008 *Technical Phys.* **53** 1451
[17] Shchagin A, Pristupa V and Khizhnyak N 1990 *Phys. Lett.* A **148** 485
[18] Shchagin A and Khizhnyak N 1996 *Nucl. Instrum. Meth.* B **119** 115
[19] Feranchuk I D, Ulyanenkov A, Harada J *et al.* 2000 *Phys. Rev.* E **62** 4225
[20] Bunge H J 2013 *Texture analysis in materials science: mathematical methods* (Elsevier)
[21] Lobko A, Golubeva E, Kuzhir P *et al.* 2015 *Nucl. Instrum. Meth.* B **355** 261
[22] Sen T and Seiss T 2015 *arXiv preprint arXiv:1506.06719*

Coherent X-ray radiation by relativistic electron in a structure "amorphous layer-vacuum-periodic layered medium"

S V Blazhevich, J P Gladkikh, S N Nemtsev, R A Zagorodnyuk, A V Noskov
Belgorod State University, Belgorod, Russia

E-mail: noskovbupk@mail.ru

Abstract. The dynamic theory of coherent X-ray radiation by relativistic electron crossing a three-layer structure consisting of an amorphous substance layer, a layer of vacuum and a layer with artificial periodic structure has been developed. The process of radiation and propagation of X-ray waves in an artificial periodic structure have been considered based on two-wave approximation of dynamic diffraction theory in Laue scattering geometry.

1. Introduction

When a relativistic electron crosses an amorphous plate, the transition radiation (TR) arises near the boundaries of the plate and then the TR photons propagate at a small angle to direction of electron velocity vector [1]. In the case of a monocrystalline plate, the TR will undergo the dynamical diffraction on a system of atomic planes in the crystal and will be reflected in the Bragg direction, forming the diffracted transition radiation (DTR) [2–5]. By analogy with DTR in the crystal, an electron crossing periodic layered medium generates TR whose photons diffract on system of layers of periodical layered medium in the plate forming the DTR in the direction near Bragg direction [6]. Together with DTR in layered media the parametric X-Ray radiation (PXR) [7–9] occurs as a result of diffraction of pseudo-photons of coulomb field of relativistic electron on the system of parallel atomic plane in the crystal or on the system of layers in the multilayered target.

A theory of coherent X-ray radiation of relativistic in the crystal were developed in the network of two wave approximation of dynamical theory of X-ray waves diffraction in the works [10–14]. In the works [10–11] the coherent X-ray radiation was treated in special case of symmetric reflection, when the reflecting system of atomic planes of the crystal is situated parallel to the target surface (in the case of Bragg scattering geometry) or perpendicular (in the case of Laue scattering geometry). In the works [12–14] the dynamic theory of coherent X-ray radiation of relativistic electron in crystal was developed for the general case of asymmetric relatively to the crystal surface reflection of the electron coulomb field, when a system of parallel reflecting atomic planes in the target can be situated at arbitrary angle to the target surface.

Traditionally the radiation of relativistic electron was considered in a separated amorphous, crystalline or multilayer target. A theoretical description of coherent radiation of relativistic electron in composite targets was not consider previously. The experimental research of generation of coherent X-ray radiation [15–19] in composite structures had shown the possibilities of considerable increase of

Content from this work may be used under the terms of the Creative Commons Attribution 3.0 licence. Any further distribution of this work must maintain attribution to the author(s) and the title of the work, journal citation and DOI.
Published under licence by IOP Publishing Ltd

intensity of DTR yield because of increase of target boundaries number. Recently in [20–22] the theory of coherent X-ray radiation of relativistic electron crossing composite structure "amorphous layer–crystalline layer" and "amorphous layer-vacuum-crystalline layer" was developed in framework of dynamic theory of diffraction. The expressions describing DTR and PXR of relativistic electron in such structures were derived and investigated. The possibility of considerable increase of spectral-angular density of DTR because of constructive interference of TR on the target boundaries.

The present work is devoted to investigation of coherent X-ray radiation of relativistic electron crossing the three-layer structure "amorphous - vacuum – periodic layered medium". In the framework of two-wave approximation of dynamic diffraction theory of X-ray waves in the single crystal the expressions describing the spectral-angular characteristics of the radiation in such structures are derived. The expressions describing the DTR and PXR spectral-angular densities and their interference in the considered structure have been obtained for general case of asymmetric reflection of the electron coulomb field from the layer with artificial periodic structure. At that, under constructive interference of TR waves from different boundaries of amorphous layer and constructive interference of TR waves from amorphous layer and entrance boundary of artificial periodic structure the spectral-angular density of DTR can be increased by orders of magnitude greater than the value of the spectral-angular density from artificial structure only. The possibility to increase the angular DTR density with increasing the substance density of amorphous layer has been shown.

2. Radiation amplitude

Let us consider the radiation of relativistic electron rectilinearly with a velocity $\mathbf{V}$ crosses a three-layers structure consisted of two amorphous layers and a layer with artificial periodic structure (see figure 1) with thicknesses of c, a and b, respectively. The dielectric susceptibility of the amorphous layers we will designate as χ_c and χ_a respectively. The dielectric susceptibility of layers in periodical layered structure are χ_1 and χ_2, and their thicknesses are l_1 and l_2, the period is $T = l_1 + l_2$.

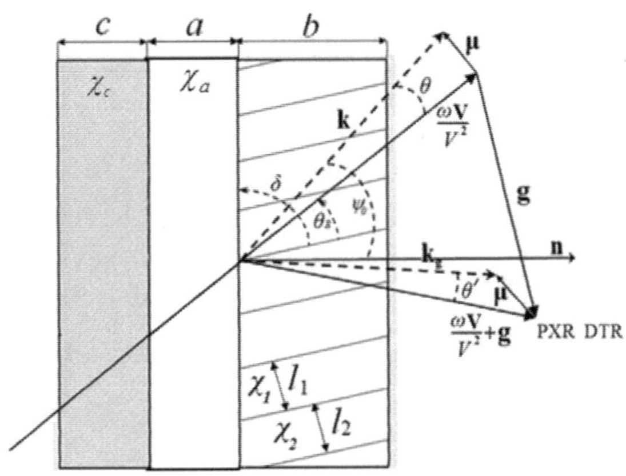

Figure 1. Geometry of the radiation process and the system of the using parameters notations, θ and θ' are the radiation angles, θ_B is Bragg angle, $\mathbf{k}$ and $\mathbf{k}_g$ are wave vectors of incident and diffracted photons.

While solving the problem, let us consider an equation for a Fourier image of an electromagnetic field

$$\mathbf{E}(\mathbf{k},\omega) = \int dt\, d^3\mathbf{r}\, \mathbf{E}(\mathbf{r},t)\exp(i\omega t - i\mathbf{k}\mathbf{r}) \; . \tag{1}$$

Since the field of a relativistic particle could, to a good accuracy, be taken as being transverse, the incident $\mathbf{E}_0(\mathbf{k},\omega)$ and diffracted $\mathbf{E}_\mathbf{g}(\mathbf{k},\omega)$ electromagnetic waves are determined by two amplitudes with different values of transverse polarization:

$$\mathbf{E}_0(\mathbf{k},\omega) = E_0^{(1)}(\mathbf{k},\omega)\mathbf{e}_0^{(1)} + E_0^{(2)}(\mathbf{k},\omega)\mathbf{e}_0^{(2)},$$
$$\mathbf{E}_\mathbf{g}(\mathbf{k},\omega) = E_\mathbf{g}^{(1)}(\mathbf{k},\omega)\mathbf{e}_1^{(1)} + E_\mathbf{g}^{(2)}(\mathbf{k},\omega)\mathbf{e}_1^{(2)}, \quad (2)$$

where the unit vectors of polarization $\mathbf{e}_0^{(1)}$ and $\mathbf{e}_0^{(2)}$ are perpendicular to vector $\mathbf{k}$, and vectors $\mathbf{e}_1^{(1)}$ and $\mathbf{e}_1^{(2)}$ are perpendicular to vector $\mathbf{k}_g = \mathbf{k}+\mathbf{g}$. Vectors $\mathbf{e}_0^{(2)}$, $\mathbf{e}_1^{(2)}$ are situated on the plane of vectors $\mathbf{k}$ and $\mathbf{k}_g$ (π-polarization) and $\mathbf{e}_0^{(1)}, \mathbf{e}_1^{(1)}$ are perpendicular to this plane (σ-polarization); $\mathbf{g}$ is similar to the reciprocal lattice vector in a single crystal medium – it is perpendicular to the layers of the structure and its magnitude is $g = \frac{2\pi}{T}n$, $n = 0,\pm1,\pm2,...$

The system of equation for the Fourier transform images of electromagnetic field in two-wave approximation of dynamic theory of diffraction has the following view [23]:

$$\begin{cases}(\omega^2(1+\chi_0)-k^2)E_0^{(s)} + \omega^2\chi_{-\mathbf{g}}C^{(s)}E_\mathbf{g}^{(s)} = 8\pi^2 ie\omega\theta VP^{(s)}\delta(\omega-\mathbf{kV}),\\ \omega^2\chi_\mathbf{g}C^{(s)}E_0^{(s)} + (\omega^2(1+\chi_0)-k_\mathbf{g}^2)E_\mathbf{g}^{(s)} = 0,\end{cases} \quad (3)$$

where $\chi_0 = \chi_0' + i\chi_0''$ is the average dielectric susceptibility, $\chi_\mathbf{g}$ and $\chi_{-\mathbf{g}}$ are the coefficients of the Fourier expansion of the dielectric susceptibility of the artificial periodic structure over the vectors $\mathbf{g}$:

$$\chi(\omega,\mathbf{r}) = \sum_\mathbf{g}\chi_\mathbf{g}(\omega)\exp(i\mathbf{gr}) = \sum_\mathbf{g}\left(\chi_\mathbf{g}'(\omega) + i\chi_\mathbf{g}''(\omega)\right)\exp(i\mathbf{gr}). \quad (4)$$

The values $C^{(s)}$ and $P^{(s)}$ are defined in the system (3) as

$$C^{(s)} = \mathbf{e}_0^{(s)}\mathbf{e}_1^{(s)},\ C^{(1)} = 1,\ C^{(2)} = \cos 2\theta_B,$$
$$P^{(s)} = \mathbf{e}_0^{(s)}(\boldsymbol{\mu}/\mu),\ P^{(1)} = \sin\varphi,\ P^{(2)} = \cos\varphi, \quad (5)$$

where $\boldsymbol{\mu} = \mathbf{k} - \omega\mathbf{V}/V^2$ is the virtual photon momentum vector component perpendicular to the particle velocity vector $\mathbf{V}$ ($\mu = \omega\theta/V$, where $\theta \ll 1$ is the angle between $\mathbf{k}$ and $\mathbf{V}$), θ_B is the Bragg angle, φ is the azimuthal angle of incidence of radiation measured from the plane formed by electron velocity vector $\mathbf{V}$ and vector $\mathbf{g}$, the value of the vector $\mathbf{g}$ is shown by expression $g = 2\omega_B \sin\theta_B / V$, ω_B is Bragg's frequency. The angle between vector $\frac{\omega\mathbf{V}}{V^2} + \mathbf{g}$ and diffracted wave vector $\mathbf{k}_\mathbf{g}$ is defined as θ'. The equation system (3) under $s = 1$, describes the fields of σ-polarization, and under $s = 2$ the fields of π-polarization

The values χ_0 and $\chi_\mathbf{g}$ for considered periodic structure have the following view:

$$\chi_0(\omega) = \frac{l_1\chi_1 + l_2\chi_2}{T}, \qquad \chi_\mathbf{g}(\omega) = \frac{\exp(-igl_1)-1}{igT}(\chi_2 - \chi_1). \quad (6)$$

If we perform the analytical procedures similar to those used in [20, 21] we will obtain the expressions for the radiation amplitude. As the result the expression for the radiation amplitude $E_\mathbf{g}^{(s)Rad} = E_{DTR}^{(s)} + E_{PXR}^{(s)}$ contained the contributions of PXR and DTR radiation mechanisms was derived:

$$E_\mathbf{g}^{(s)Rad} = E_{\mathrm{DTR}}^{(s)} + E_{\mathrm{PXR}}^{(s)}, \quad (7a)$$

$$E_{\text{DTR}}^{(s)} = \frac{8\pi^2 ieV\theta P^{(s)}}{\omega} \frac{\omega^2 \chi_{\mathbf{g}} C^{(s)}}{2\omega \frac{\gamma_0}{\gamma_{\mathbf{g}}} \left(\lambda_{\mathbf{g}}^{(1)} - \lambda_{\mathbf{g}}^{(2)}\right)} \left(\exp\left(i \frac{\lambda_{\mathbf{g}}^{(1)} - \lambda_{\mathbf{g}}^*}{\gamma_{\mathbf{g}}} b \right) - \exp\left(i \frac{\lambda_{\mathbf{g}}^{(2)} - \lambda_{\mathbf{g}}^*}{\gamma_{\mathbf{g}}} b \right) \right) \times$$

$$\left[\left(\frac{1}{\theta^2 + \gamma^{-2} - \chi_c} - \frac{1}{\theta^2 + \gamma^{-2}} \right) \exp\left(-i \frac{\omega c}{2\gamma_0} \left(\gamma^{-2} + \theta^2 - \chi_c\right) - i \frac{\omega a}{2\gamma_0} \left(\gamma^{-2} + \theta^2 - \chi_a\right) \right) + \right.$$

$$+ \left(\frac{1}{\theta^2 + \gamma^{-2} - \chi_a} - \frac{1}{\theta^2 + \gamma^{-2} - \chi_c} \right) \exp\left(-i \frac{\omega a}{2\gamma_0} \left(\gamma^{-2} + \theta^2 - \chi_a\right) \right) +$$

$$\left. + \frac{1}{\theta^2 + \gamma^{-2} - \chi_0} - \frac{1}{\theta^2 + \gamma^{-2} - \chi_a} \right] \exp\left(i \left(\frac{\omega \chi_0}{2} + \lambda_g^* \right) \frac{(c + a + b)}{\gamma_g} \right), \tag{7b}$$

$$E_{\text{PXR}}^{(s)} = \frac{8\pi^2 ieV\theta P^{(s)}}{\omega} \frac{\omega^2 \chi_{\mathbf{g}} C^{(s)}}{2\omega \frac{\gamma_0}{\gamma_{\mathbf{g}}} \left(\lambda_{\mathbf{g}}^{(1)} - \lambda_{\mathbf{g}}^{(2)}\right)} \left[\left(\frac{1}{\chi_0 - \theta^2 - \gamma^{-2}} + \frac{\omega}{2 \frac{\gamma_0}{\gamma_{\mathbf{g}}} \left(\lambda_{\mathbf{g}}^* - \lambda_{\mathbf{g}}^{(1)}\right)} \right) \times \right.$$

$$\times \left(\exp\left(i \frac{\lambda_{\mathbf{g}}^{(1)} - \lambda_{\mathbf{g}}^*}{\gamma_{\mathbf{g}}} b \right) - 1 \right) - \left. \left(\frac{1}{\chi_0 - \theta^2 - \gamma^{-2}} + \frac{\omega}{2 \frac{\gamma_0}{\gamma_{\mathbf{g}}} \left(\lambda_{\mathbf{g}}^* - \lambda_{\mathbf{g}}^{(2)}\right)} \right) \left(\exp\left(i \frac{\lambda_{\mathbf{g}}^{(2)} - \lambda_{\mathbf{g}}^*}{\gamma_{\mathbf{g}}} b \right) - 1 \right) \right] \times$$

$$\exp\left(i \left(\frac{\omega \chi_0}{2} + \lambda_g^* \right) \frac{(c + a + b)}{\gamma_g} \right), \tag{7c}$$

where

$$\lambda_{\mathbf{g}}^{(1,2)} = \frac{\omega}{4} \left(\beta \pm \sqrt{\beta^2 + 4\chi_{\mathbf{g}}\chi_{-\mathbf{g}} C^{(s)^2} \frac{\gamma_{\mathbf{g}}}{\gamma_0}} \right), \beta = \frac{1}{\omega^2}(k_{\mathbf{g}}^2 - k^2) - \chi_0 \left(1 - \frac{\gamma_{\mathbf{g}}}{\gamma_0} \right), \lambda_{\mathbf{g}}^* = \frac{\omega\beta}{2} + \frac{\gamma_{\mathbf{g}}}{\gamma_0} \lambda_0^*, \ \gamma_0 = \cos\psi_0,$$

$$\gamma_{\mathbf{g}} = \cos\psi_{\mathbf{g}}, \ \lambda_0^* = \omega \left(\frac{\gamma^{-2} + \theta^2 - \chi_0}{2} \right), \tag{8}$$

ψ_0 is the angle between incident wave vector $\mathbf{k}$ and vector normal to the plate surface $\mathbf{n}$, $\psi_{\mathbf{g}}$ is the angle between wave vector $\mathbf{k}_g$ and the vector $\mathbf{n}$ (see figure 1). As the dynamical corrections $|\lambda_0| << \omega$ and $|\lambda_{\mathbf{g}}| << \omega$, we can show that $\theta \approx \theta'$ (see in figure 1), and hereinafter will use θ in all the occasions.

3. Spectral-angular density of radiation

Let us consider the radiation of a relativistic electron in the case when the second layer is a vacuum ($\chi_a = 0$).To clarify and analyze the effects that are not associated with absorption, we consider a simple case of a thin nonabsorbing target ($\chi_0'' = \chi_c'' = 0$). From (7) we have obtained the expressions for the spectral-angular density of PXR and DTR in the propagation direction of the emitted photon $\mathbf{k}_{\mathbf{g}} = k_{\mathbf{g}}\mathbf{n}_{\mathbf{g}}$ (see figure 1):

$$\omega \frac{d^2 N_{\text{PXR}}^{(s)}}{d\omega d\Omega} = \frac{e^2}{4\pi^2} \frac{P^{(s)^2}}{|\chi_0'|} \cdot \frac{\Omega^2}{\left(\Omega_0^2 + 1\right)^2} R_{\text{PXR}}^{(s)}, \tag{9a}$$

where the expression $R_{\text{PXR}}^{(s)}$ describing PXR spectrum has a view

$$R_{\text{PXR}}^{(s)} = 4\left(1-\frac{\xi}{\sqrt{\xi^2+\varepsilon}}\right)^2 \frac{\sin^2\left(\frac{B^{(s)}}{2}\left(\sigma^{(s)}+\frac{\xi-\sqrt{\xi^2+\varepsilon}}{\varepsilon}\right)\right)}{\left(\sigma^{(s)}+\frac{\xi-\sqrt{\xi^2+\varepsilon}}{\varepsilon}\right)^2}, \tag{9b}$$

$$\omega\frac{d^2N_{\text{DTR}}^{(s)}}{d\omega d\Omega} = T_{DTR}^{(s)} = T_1^{(s)} + T_2^{(s)} + T_{\text{int}}^{(s)}, \tag{10a}$$

where

$$T_1^{(s)} = \frac{e^2}{4\pi^2}\frac{P^{(s)^2}}{|\chi_0'|}\cdot 4\Omega^2\left(\frac{1}{\Omega_0^2+\frac{\chi_c'}{\chi_0'}}-\frac{1}{\Omega_0^2}\right)^2 \sin^2\left(\frac{B^{(s)}}{2}\cdot\frac{c}{b}\cdot\frac{1}{\nu^{(s)}}\left(\Omega_0^2+\frac{\chi_c'}{\chi_0'}\right)\right)R_{\text{DTR}}^{(s)}, \tag{10b}$$

$$T_2^{(s)} = \frac{e^2}{4\pi^2}\frac{P^{(s)^2}}{|\chi_0'|}\cdot\Omega^2\left(\frac{1}{\Omega_0^2+1}-\frac{1}{\Omega_0^2}\right)^2 R_{\text{DTR}}^{(s)}, \tag{10c}$$

$$T_{\text{int}}^{(s)} = \frac{e^2}{4\pi^2}\frac{P^{(s)^2}}{|\chi_0'|}\cdot 2\Omega^2\left(\frac{1}{\Omega_0^2+1}-\frac{1}{\Omega_0^2}\right)\left(\frac{1}{\Omega_0^2}-\frac{1}{\Omega_0^2+\frac{\chi_c'}{\chi_0'}}\right)\times$$

$$\times\left[\cos\left(B^{(s)}\cdot\frac{a}{b}\cdot\frac{1}{\nu^{(s)}}\Omega_0^2\right)-\cos\left(B^{(s)}\cdot\frac{a}{b}\cdot\frac{1}{\nu^{(s)}}\Omega_0^2+B^{(s)}\cdot\frac{c}{b}\cdot\frac{1}{\nu^{(s)}}\left(\Omega_0^2+\frac{\chi_c'}{\chi_0'}\right)\right)\right]R_{\text{DTR}}^{(s)}, \tag{10d}$$

where the expression describing DTR spectrum has the following view

$$R_{\text{DTR}}^{(s)} = \frac{4\varepsilon^2}{\xi^{(s)}(\omega)^2+\varepsilon}\sin^2\left(\frac{B^{(s)}\sqrt{\xi^{(s)}(\omega)^2+\varepsilon}}{\varepsilon}\right). \tag{10e}$$

The functions $T_1^{(s)}$ and $T_2^{(s)}$ describes the spectral-angular densities of DTR corresponding to waves of the radiation generated in amorphous layer and on the entrance boundary of periodical layered mediumю The function $T_{\text{int}}^{(s)}$ describes the interference of these waves.

In expressions (9) – (10) the following notations are accepted.

$$\Omega_0^2 = \Omega^2+\Gamma^2,\ \Omega = \frac{\theta}{\sqrt{|\chi_0'|}},\ \Gamma = \frac{1}{\gamma\sqrt{|\chi_0'|}},\ \chi_0' = \frac{\chi_1'+r\chi_2'}{1+r},\ \nu^{(s)} = \frac{C^{(s)}\left|\sin\left(\frac{\pi n}{1+r}\right)\right|}{\frac{\pi n}{1+r}}\left|\frac{\chi_2'-\chi_1'}{\chi_1'+r\chi_2'}\right|,$$

$$\varepsilon = \frac{\sin(\delta+\theta_B)}{\sin(\delta-\theta_B)},\ B^{(s)} = \frac{1}{2\sin(\delta-\theta_B)}\frac{b}{L_{ext}^{(s)}},\ L_{ext}^{(s)} = \frac{1}{C^{(s)}\omega}\frac{\pi n}{\left|\sin\left(\frac{\pi n}{1+r}\right)\right||\chi_2'-\chi_1'|},\ \varepsilon = \frac{\sin(\delta+\theta_B)}{\sin(\delta-\theta_B)},$$

$$\sigma^{(s)}=\frac{\pi n}{C^{(s)}\left|\chi_2'-\chi_1'\right|\left|\sin\left(\frac{\pi n}{1+r}\right)\right|}\left(\theta^2+\gamma^{-2}-\chi_0'\right),\ \xi^{(s)}(\omega)=\frac{2\pi^2 n^2}{T^2\omega_B}L_{ext}^{(s)}\left(1-\frac{\omega}{\omega_B}\left(1-\theta_{//}\sqrt{\frac{T^2\omega_B^2}{\pi^2 n^2}-1}\right)\right)+\frac{1-\varepsilon}{2\nu^{(s)}},$$

$$r=\frac{l_2}{l_1}. \tag{11}$$

The constructive interference of the waves from different boundaries leads to a considerable increase in the spectral-angular density of DTR under the condition derived from (10b):

$$\frac{B^{(s)}}{2}\cdot\frac{c}{b}\cdot\frac{1}{\nu^{(s)}}\left(\Omega_0^2+\frac{\chi_c'}{\chi_0'}\right)=\frac{\omega c}{4\sin\left(\delta-\theta_B\right)}\left(\theta^2+\gamma^{-2}-\chi_c'\right)=(2n+1)\frac{\pi}{2},\ (n=0,1,2.....). \tag{12}$$

Additionally, DTR spectral-angular density can be increased due to the constructive interference of the waves of TR from the amorphous layer and the entrance boundary of periodical layered target under the condition following from (10d):

$$B^{(s)}\cdot\frac{a}{b}\cdot\frac{1}{\nu^{(s)}}\Omega_0^2=\frac{\omega a}{2\sin\left(\delta-\theta_B\right)}\left(\theta^2+\gamma^{-2}\right)=(2m+1)\pi\ ,(m=0,1,2.....). \tag{13}$$

When the condition $\left|\frac{\chi_1'+r\chi_2'}{1+r}\right|>\left|\chi_c'\right|$ is fulfilled, the interference term $T_{\mathrm{int}}^{(s)}$ can exceed the contribution of each term of TR in the resulting yield of DTR. In special case when averaged dielectric susceptibility of periodical medium and dielectric susceptibility of amorphous medium are equal

$$\frac{\chi_1'+r\chi_2'}{1+r}=\chi_c', \tag{14}$$

from (10) under conditions (12), (13) and (14) we can obtain the expression

$$\omega\frac{d^2N_{\mathrm{DTR}}^{(s)}}{d\omega d\Omega}=9\frac{e^2}{4\pi^2}P^{(s)^2}\theta^2\left(\frac{1}{\theta^2+\gamma^{-2}-\frac{\chi_1'+r\chi_2'}{1+r}}-\frac{1}{\theta^2+\gamma^{-2}}\right)^2 R_{\mathrm{DTR}}^{(s)}, \tag{15}$$

which shows that under these conditions the DTR yield from considered three-layered target will exceed the DTR yield from periodic layered medium 9 times.

The angular density of DTR can be calculated by formula

$$\frac{dN_{\mathrm{DTR}}^{(s)}}{d\Omega}=F_{\mathrm{DTR}}^{(s)}(\theta)=F_1^{(s)}(\theta)+F_2^{(s)}(\theta)+F_{\mathrm{int}}^{(s)}(\theta), \tag{16a}$$

where

$$F_{1,2,\mathrm{int}}^{(s)}(\theta)=\frac{T^2\omega_B}{2\pi^2 n^2 L_{ext}^{(s)}}\int_{-\infty}^{+\infty}T_{1,2,\mathrm{int}}^{(s)}d\xi^{(s)}(\omega). \tag{16b}$$

$$\frac{dN_{\mathrm{PXR}}^{(s)}}{d\Omega}=F_{\mathrm{PXR}}^{(s)}(\theta)=\frac{T^2\omega_B}{2\pi^2 n^2 L_{ext}^{(s)}}\frac{e^2}{4\pi^2}\frac{P^{(s)^2}}{\left|\chi_0'\right|}\cdot\frac{\Omega^2}{\left(\Omega_0^2+1\right)^2}\int_{-\infty}^{+\infty}R_{\mathrm{PXR}}^{(s)}d\xi^{(s)}(\omega). \tag{16c}$$

4. Radiation amplitude Spectral-angular density of radiation

The calculations were fulfilled under conditions (12), (13) and (14) i.e. under condition of nine fold increase in spectral-angular density of DTR. Parameters of calculation: amorphous medium Germanium (Ge), thickness of amorphous medium is $c=1.2$ μm, the vacuum layer thickness is $a=9.6$ μm; Periodic layered structure is carbon (C) – tungsten (W), thickness of layered structure is $b=1$ μm, thicknesses of the layers are $l_1=10.64\cdot10^{-4}$ μm and $l_2=9.36\cdot10^{-4}$ μm, period $T=l_1+l_2$, the angle between reflecting layers and target surface is $\delta=4.6^\circ$, energy of relativistic electron is $E=500\,\mathrm{MeV}$, the angle between the direction of the electron velocity and reflecting layers is

$\theta_B = 2.3°$, Bragg frequency $\omega_B = 8\,\text{keV}$. The calculations were fulfilled for σ-polarized waves ($s = 1$) for the first harmonic $n = 1$. In figure 2 the curves bilt by formulas (10) describing the spectral-angular density of DTR in maximum ($\theta_\perp = 1$ mrad) are represented. One of the curves is plotted for such thicknesses of amorphous and vacuum layers when the conditions of constructive interference (12) and (13) are satisfied. Other curve is plotted for the case of absence of amorphous layer. One can see that the amorphous layer (plate) is able to considerably increase the spectral-angular density of DTR. In figure 3 the curves describing the angular densities of DTR and PXR are plotted by formulas (16). It is seen that the contribution of PXR under condition of figure 3 is immaterial because the thickness of layer made of periodical multilayer medium is small ($b = 1$ μm).

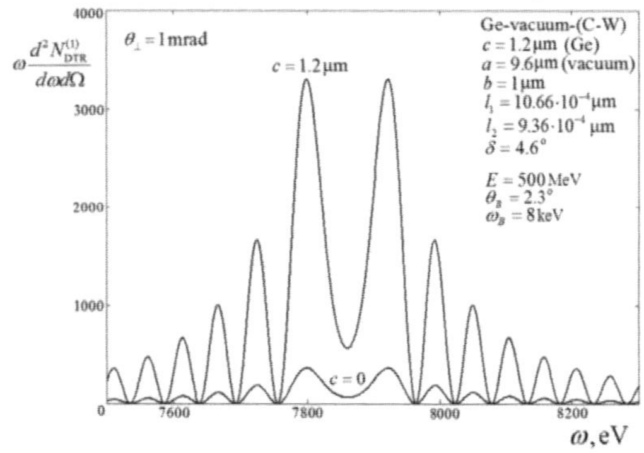

Figure 2. Spectral-angular density of DTR in the three-layer structure ($c = 1.2$ μm) and in the case of absence of amorphous layer ($c = 0$).

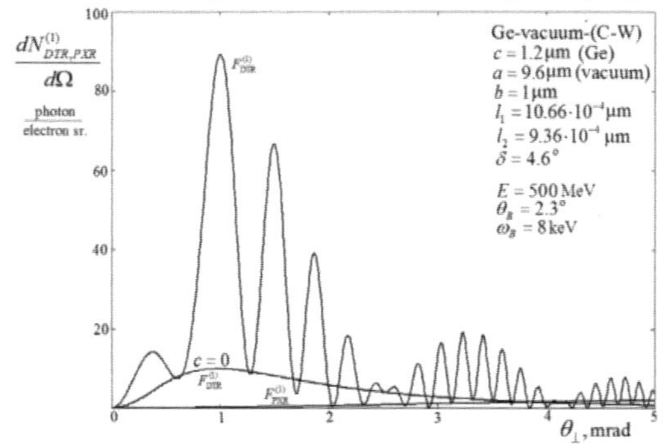

Figure 3. The angular density of DTR by a relativistic electron from three-layered structure and the same in case of absence ($c = 0$) of the amorphous layer. The angular density of PXR under such energy of electrons is considerably lower and independent on thickness of amorphous layer.

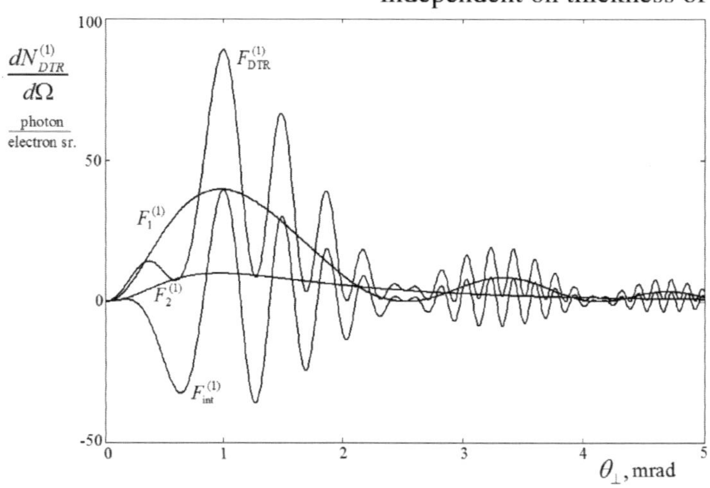

Figure 4. The contribution of TR from the amorphous layer $F_1^{(1)}$, from the entrance boundary of the multilayered medium $F_2^{(1)}$ and their interference term $F_{int}^{(1)}$ into the angular density of DTR.

5. Conclusion

A theory of coherent X-ray radiation of relativistic electrons crossing the three-layer structure "amorphous layer-vacuum-periodic layered medium" have been developed within the framework of dynamical diffraction theory. The expressions describing PXR and DTR by relativistic electron have been derived. It is shown that the spectral-angular density of the DTR can be substantially increased due to the constructive interference of TR waves from different boundaries of amorphous layer, as well as due to the constructive interference of waves of transition radiation from the amorphous layer and the entrance boundary of the periodic layered target.

Acknowledgements

This work was supported by the Ministry of Education and Science of the Russian Federation (project of the state task № 3.500.2014 / K in the field of science and state task №2014 / 420).

References

[1] Ginzburg V L and Tsytovich V N 1984 *Transition Radiation and Transition Scattering* (Nauka, Moscow, 1984;Adam Hilger, Bristol, United Kingdom, 1990)
[2] Caticha A 1989 *Phys. Rev.* A **40** 4322
[3] Baryshevsky V. 1997 *Nucl. Instr. and Meth.* A **122** 13
[4] Artru X, Rullhusen P 1998 Nucl. *Instr. and Meth.* B **145** 1
[5] Nasonov N 1998 *Phys. Lett.* A **246** 148
[6] Blazhevich S V, Kolosova I V, Noskov A V 2012 *J. Exp. Theor. Phys.* **114** 547
[7] Ter-Mikaelian M 1972 *High-Energy Electromagnetic Process in Condensed Media* (New York: Wiley)
[8] Garibian G M , Yang C 1971 *Sov. Phys. JETP* **34** 495
[9] Baryshevskii V G, Feranchuk I. D. 1997 *Sov. Phys. JETP* **34** 502
[10] Nasonov N 2001 *Phys. Lett.* A **292** 146
[11] Nasonov N, Noskov A 2003 *Nucl. Instr. Meth. In Phys. Res.* B **201** 67
[12] Blazhevich S V, Noskov A V 2006 *Nucl. Instr. and Meth.* B **252** 69
[13] Blazhevich S, Noskov A 2008 *Nucl. Instr. Meth. In Phys. Res.* B **266** 3770
[14] Blazhevich S, Noskov A 2009 *J. Exp. Theor. Phys.* **109** 901
[15] Andreyashkin M Yu , Zabaev V N, Ioshida K , Kaplin V V, Rozum E I, Uglov S R, Endo I 1995 *JETP Letters* **62** 791
[16] Andreyashkin M Yu , Zabaev V N et al 1997 *JETP Letters* **65** 594
[17] Nakayama K, Sekimura M, Yanase I, Endo I, Takashima Y, Kaplin V, Potylitsin A 1998 *Nucl. Instr. and Meth.* B **145** 236
[18] Takashima Y, Aramitsu K, Endo I, Fukumi A et al 1998 *Nucl. Instr. and Meth.* B. **145** 25
[19] Kaplin V V, Kuznetsov S I, Timchenko N A, Uglov S R, Zabaev V N 2001 *Nucl. Instr. and Meth.* B **173** 238
[20] Blazhevich S V, Zagorodnyuk R A, Noskov A V 2015 *Nucl. Instr. and Meth.* B **355** 114
[21] Blazhevich S V, Zagorodnyuk R A, Noskov A V 2014 *J. Exp. Theor. Phys.* **119** 641
[22] Blazhevich S V, Noskov A V 2014 *J. Exp. Theor. Phys.* **118** 550
[23] Bazylev V, Zhevago N 1987 *Emission From Fast Particles Moving in a Medium and External Fields* (Nauka, USSR, Moscow (in Russian)).

Cr/Sc multilayer radiator for parametric EUV radiation in "water-window" spectral range

S R Uglov[1], V V Kaplin[1], A S Kubankin[2], J-M André[3], K Le Guen[3], Ph Jonnard[4], S de Rossi[4], E Meltchakov[4] and F Delmotte[4]

[1] Tomsk Polytechnic University, pr. Lenina 30, Tomsk, 634050 Russia
[2] Belgorod State University, Belgorod, Russia
[3] CNRS UMR 7614, Laboratoire de Chimie Physique - Matière et Rayonnement Sorbonne Universites, UPMC Paris 06, 11 rue Pierre et Marie Curie, F-75231 Paris Cedex 05, France
[4] Laboratoire Charles Fabry, Institut d'Optique, Graduate School, 2 avenue Augustin Fresnel F-91127 Palaiseau, France

E-mail: uglov@tpu.ru

Abstract. The results of experimental investigation of parametric radiation generated by 5.7 MeV electrons in a multilayer structure consisting of 100 Cr/Sc bi-layers deposited on a Si_3N_4 membrane are presented. The multilayer structure was specially created for generation of parametric radiation with photon energy in "water-window" spectral range. First test measurements of angular distributions of radiation have been done and discussed.

1. Introduction

In a number of papers [1 – 6] it was shown that artificial periodic structures, as well as crystals, are suitable to generate tunable quasi-monochromatic X-rays by relativistic electrons. The energies E_{ph} of the photons emitted from the periodic structure lie in a narrow spectral range which is determined by the Bragg diffraction law. The mechanism of radiation generation in multilayer X-ray mirrors (MXM) is similar to the mechanisms of Parametric X-ray Radiation (PXR) and Diffracted Transition Radiation (DTR) generations in periodic crystal structures. Experiments [4 – 7] have been carried out for generation in MXM of X-rays with photon energies $E_{ph} = 6 - 15$ keV.

Recently in [10], it was experimentally shown that a multilayer mirror may be used for generation of quasimonochromatic radiation in the range of extreme ultraviolet (E_{ph} around 70 eV). So, by using structure with shorter period, one can get the quasimochromatic tunable source of photons in "water-window" spectral range. The "water-window" spectral range lies between the carbon and oxygen K absorption edges (284 and 543 eV, respectively). Radiation in this domain is required for many scientific and industrial applications, in particular, for the soft X-ray microscopy of biological objects.

This paper presents the construction of a multilayer radiator for generation of tunable radiation with photon energy in "water-window" spectral range and first results concerning the radiator test using the 5.7 MeV electron beam of the microtron M-5 of the Tomsk Polytechnic University. The multilayer radiator was specially prepared for generation of the parametric radiation (PR) in the "water window" spectral range with very thin substrate in order to decrease the contribution of background radiation.

Content from this work may be used under the terms of the Creative Commons Attribution 3.0 licence. Any further distribution of this work must maintain attribution to the author(s) and the title of the work, journal citation and DOI.
Published under licence by IOP Publishing Ltd

2. Cr/Sc radiator

In the experiment, the radiator consists of a pile of 100 Cr/Sc bi-layers with a period d = 2.34 nm caped by a 2.5 nm thick B_4C layer; the whole stack is deposited onto a 0.5 μm thick Si_3N_4 membrane. The thickness of Cr and Sc layers, a and b, respectively, are the same. Scheme of the Cr/Sc multilayer radiator is shown in figure 1.

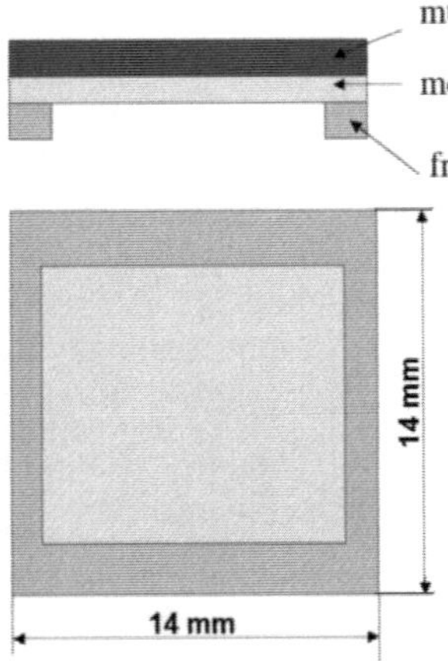

Figure 1. Scheme of Cr/Sc multilayer radiator. The thickness of B_4C cap layer is 2.5nm; Si frame is 500 μm thick.

The multilayer structure was prepared using the facility of the Laboratoire Charles Fabry by magnetron sputtering technique under constant argon pressure using plasma discharges by radio frequency power [11]; it was characterized by grazing incidence X-ray reflectivity GIXR and X-ray emission spectroscopy XES. Interfacial imperfection of Cr/Sc systems are mainly due to roughness with a typical rms roughness height of about 0.5 nm at each interface as indicated by GIXR; the analysis of Cr 3d valence states by XES does not evidence the presence of interfacial compounds resulting from the interdiffusion of the Cr and Sc layers [12].

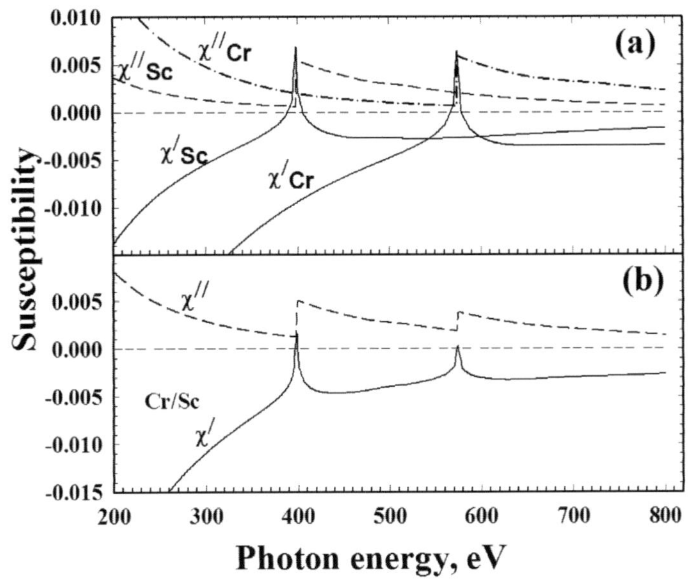

Figure 2. Dielectric susceptibilities: (a) Cr and Sc layers; (b) Cr/Sc bi-layer.

The global multilayer radiator (multilayer + substrate) was also tested using the X-ray diffractometer "Shimadzu XRD 6000". The width of rocking-curve measured using the Cu K X-rays was about 0.2° that is in about 10 times more than that of theoretical one. All these characterizations indicate that the optical quality of this radiator is yet very far from perfection which has likely affected its PR emissive properties as shown hereafter. The mechanical properties of the Si_3N_4 membrane acting as substrate have not yet been studied in details; nevertheless it is likely that a large part of the relatively poor optical quality of the radiator can be attributed to the substrate.

Real χ' and imaginary χ'' parts of the dielectric susceptibilities $\varepsilon_{a,b}(\omega) = 1 + \chi'(\omega) + i\cdot\chi''(\omega)$ of Cr, Sc layers and average one $\varepsilon_{Cr/Sc}(\omega)$ of Cr/Sc bi-layer are shown in figure 2. The value of $\varepsilon_{Cr/Sc}(\omega)$ was calculated as $(a\cdot\varepsilon_{Cr} + b\cdot\varepsilon_{Sc}) / (a + b)$. The reflectivity of our Cr/SC multilayer structure supposed to be ideal (no interfacial roughness and interdiffusion) was calculated using the Parratt-method [13] for a number of Bragg angles and photon energies and shown in figure 3. The calculation evidences the reflectivity enhancements just below the L edge of Sc (398 eV) at the Bragg angle $\theta_0 = 42°$ and around L edge of Cr (574 eV) at $\theta_0 = 27°$ as a result of the anomalous scattering [14]

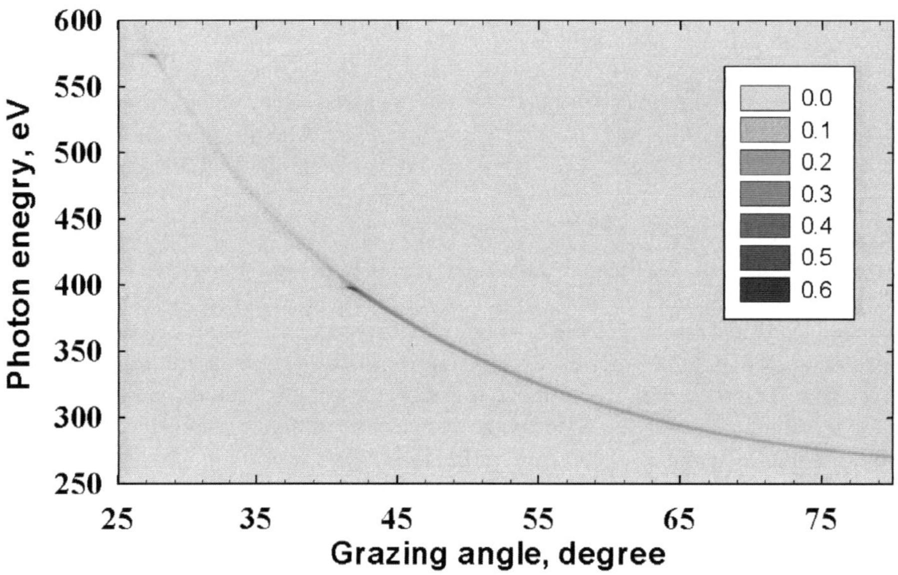

Figure 3. Reflectivity of ideal Cr/Sc mirror as function of Bragg angles and photon energies.

3. Theoretical background

The angular and spectral-angular densities of radiation were calculated using different theories [3, 5]. The calculation shows that the maximum PR intensity is emitted near the direction $\theta_{Dy} = 1/\gamma$ (γ is the Lorentz factor) with respect to the plane of diffraction. Therefore, the search for the PR effect should be carried out near the expected direction of maximum intensity of the radiation by scanning the detector through the angular distribution in a plane that is at an angle $\theta_{Dy} = 1/\gamma$ with respect to the diffraction plane **VN** (**N** is the normal to the target surface). As it is known, the back transition radiation (BTR) formed at the interaction of electrons with the input surface of the target is emitted also in the same direction. To suppress the BTR contribution, a filter (0.2 μm LiF + 0.1 μm Palyren) was used.

The angular distributions of 30 – 600 eV BTR and PR calculated for a number of Bragg angles θ_0 by taking into account the photon absorption in the filter are shown in figures 4(a) and 5. Figure 4(b) illustrates the influence of the filter on BTR intensity generated at $\theta_0 = 40°$ on upper Cr layer of the multilayer structure. The Cr surface was taken for calculation of the BTR contribution because the intensity of BTR from Cr surface is higher than that from B_4C, Si_3N_4 or Sc surfaces. The angular distributions of BTR were calculated using the Pafomov's theory [15].

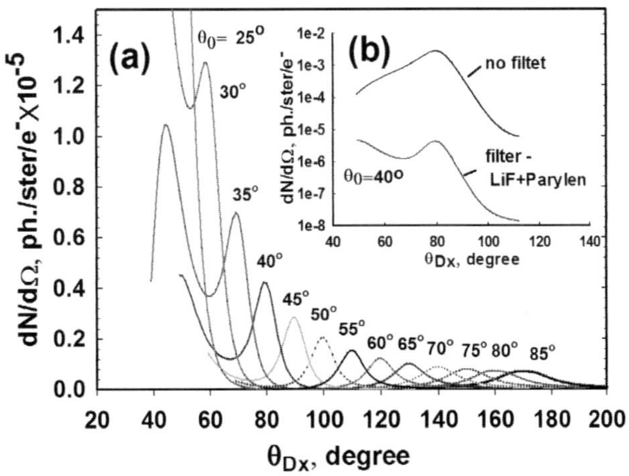

Figure 4. Angular distributions of 30 – 600 eV BTR: (a) θ_0 between 25° and 85°; (b) at $\theta_0 = 40°$, with and without a filter.

The angular distributions of PR were calculated using the theory [6]. The upper solid curve in figure 5 shows the dependence of photon energy versus $\theta_D = 2\theta_0$. As can be seen from a comparison of the data presented in figures 4(a) and 5, BTR contribution to overall radiation yield can be neglected at using the filter. Also is very important to note the difference of changing of the BTR and PR intensities via the angle θ_0. The BTR intensity increases with decrease of the angle θ_0, while PR intensity has a maximum around $\theta_0 \approx 40°$.

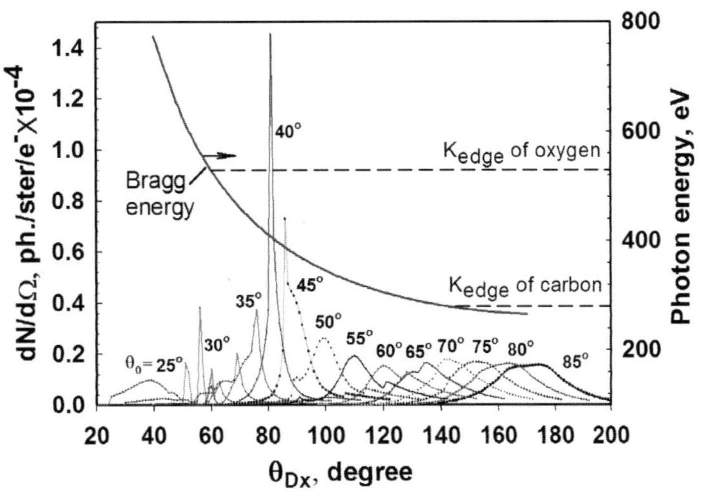

Figure 5. Angular distributions of PR for angles θ_0 between 25° and 85°.

4. Experimental procedure

Experiment on the angular distribution of the radiation generated by the 5.7 MeV electrons into the Cr/Sc radiator was carried out using the electron beam delivered by the microtron M-5 at Tomsk. The radiation was detected by a channel electron multiplier (CEM-model VEU-6). A detailed description of the installation can be found in [8 – 10]. The geometry of the interaction between the electron beam and the target and of the generation of radiation is shown in figure 6.

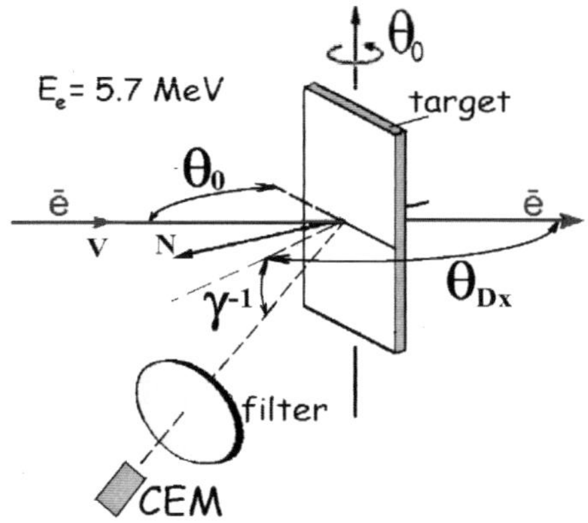

Figure 6. Scheme of geometry of the interaction of electron beam with the radiator and generation of radiation at experimental setup based on microtron M-5. Electron energy – 5.7 MeV; repetition rate – 25 Hz, pulse duration – 0.4 μs, electron current on target – 4 pC per pulse.

A series of angular distributions was measured for several angles between the electron beam and the radiator surface in the range of 35° – 72°. The distributions were measured in the diffraction plane **VN** and at the angle $\theta_{Dy} = 1/\gamma = 5.14°$ ($\gamma = 11.15$) with respect to this plane. To suppress the soft component of ordinary BTR, the measurements were carried out with the "LiF + Parylen" filter considered in section 3.

5. First experimental results and discussion

Angular distributions of radiation generated in the Cr/Sc structure are shown in figure 7 and 8. The black points in figure 7 show the angular distribution of density of radiation generated in the Cr/Sc structure at $\theta_0 = 41°$.

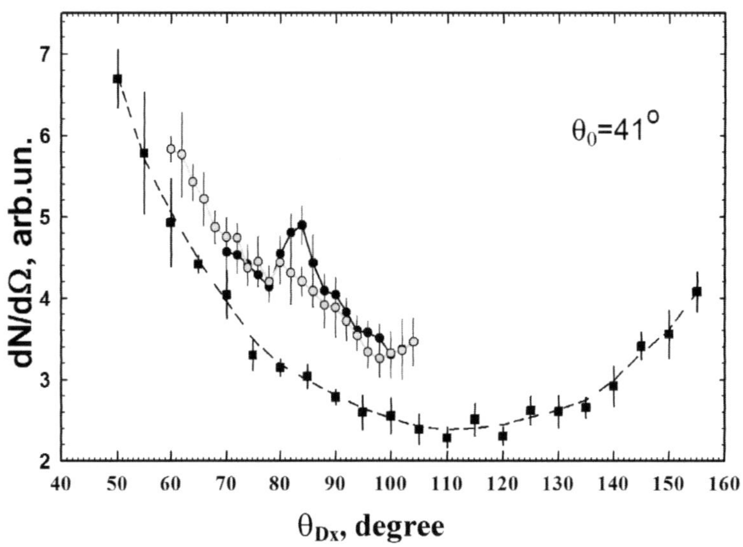

Figure 7. The black points are for radiation emitted from Cr/Sc structure at $\theta_0 = 41°$. The light points are for the Si_3N_4 substrate without Cr/Sc structure. Lower curve shows background level.

The light points are for radiation generated by the nude substrate (Si_3N_4 membrane) at the same angle $\theta_0 = 41°$. The black and light points in figure 8 correspond to the angular distribution of radiation generated in the Cr/Sc structure at $\theta_0 = 55°$ and $\theta_0 = 35°$, respectively. The lower dashed curves, shown in figures 7 and 8, present background level measured when the electron beam did not hit the target.

Comparison of the results of measurements, presented in figure 7 and 8, with the calculations presented in figures 4(a) and 5, shows that the experimentally observed trend of change of the intensity of the maxima of the angular distributions, namely the existence of the global maximum of radiation intensity near $\theta_0 = 41°$, is in qualitative agreement with the PR calculations.

The FWHM of the PR angular distribution is about 6° that is about in 2 times less than typical value of FWHM of BTR one. As can be seen from figure 7, the Cr/Sc structure generates radiation with intensity more 2 times larger than the Si_3N_4 membrane, but this value is almost 10 times less than that derived from model calculations for an ideal periodic structure.

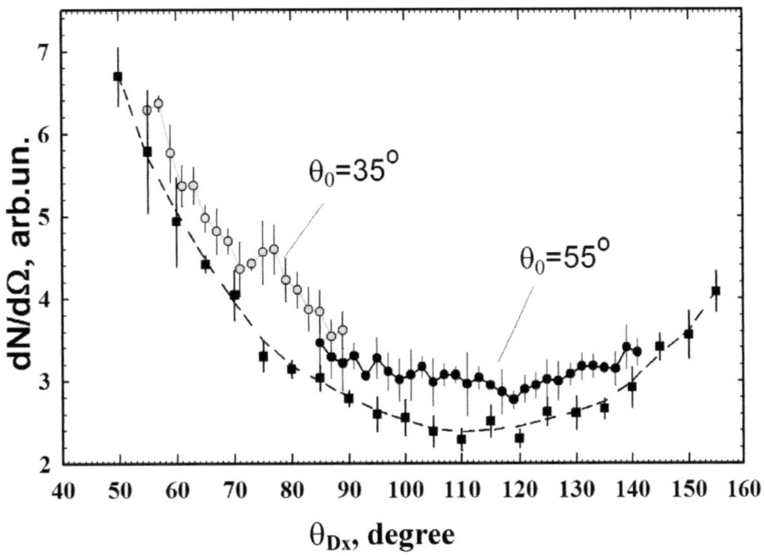

Figure 8. The black and light points illustrate the emission from Cr/Sc structure for the two angles $\theta_0 = 55°$ and $\theta_0 = 35°$, respectively. Lower curve shows background level.

In addition, due to the small thickness of the substrate, the electron beam is scattered weakly, even at high angles of inclination of the target, and completely falls into the aperture of the Faraday cup, that eliminates the need of correction of the value of electron current measured, that associated with a change in the thickness of the target when the target is rotated.

6. Conclusion

The presented results of the test experiment show the low yield of generated radiation. Probably, the reason of the essential difference between experimental and theoretical values of PXR yield results from interfacial roughness, mutual diffusion of layer materials and lack of flatness of the substrate because these factors can severely deteriorate efficiency of PXR generation, as it takes place concerning the decrease of reflectivity of real photons in X-ray optics due to imperfection of X-ray mirror. An improvement of the multilayer fabrication technique is necessary to minimize these radiator defects in order to create an effective source of radiation. Due to the thin (thickness – 0.5 μm) Si_3N_4 substrate of the multilayer structure, the contribution of background radiation generated in the target was less in about 3 times than that in experiments [9, 10] when a multilayer structures had the thick Si substrates. Additionally, the use of target with very thin substrates might be promising to increase the efficiency of this type of radiator when using cyclic accelerators because of possible multiple passes of electrons through them [16].

It should be noted that the structure composed of Cr and Sc layers is capable to generate the Cherenkov radiation because the values of $\chi'_{Cr/Sc}(\omega_{LCr})$ and/or $\chi'_{Cr/Sc}(\omega_{LSc})$ may exceed 0 near the L absorption edges of Cr and Sc (see figure 2b). By means of optimizing of the ratio of the thicknesses of Cr and Sc layers, the effect of diffracted Cherenkov radiation can be observed at the appropriate orientation of the structure with respect to an electron beam [17, 18].

Acknowledgements
The present research was partially supported by the Ministry of Education and Science (state contract No. NIR MON 3761), and by the Russian Foundation for Basic Research (project No. 14-02-01032). The French team was supported by the Agence Nationale pour la Recherche, project No. ANR-10-BLAN-0924.

References

[1] Bazylev V A, Zhevago N K 1987 *Radiation of fast particles in matter and in external fields* (Moscow: Nauka)
[2] Andre J-M, Barchewitz R, Bonnelle C and Pardo B 1993 *J. Opt. Paris* **24** 31
[3] Andre J-M, Pardo B and Bonnellel C 1999 *Phys. Rev.* E **99** 968
[4] Kaplin V V, Uglov S R, Zabaev V N, Piestrup M A, Gary C K, Nasonov N N and Fuller M K 2000 *Appl. Phys. Lett.* **76** 3647
[5] Nasonov N N, Kaplin V V, Uglov S R, Piestrup M and Gary C 2003 *Phys. Rev.* E **68** 036504
[6] Nasonov N N, Kaplin V V, Uglov S R, Zabaev V N, Piestrup M and Gary C 2005 *Nucl. Instrum. and Meth.* B **227** 41
[7] Kaplin V V, Uglov S R, Sohoreva V V, Bulaev O F, Voronin A A, Piestrup M, Gary C and Fuller M 2009 *Nucl. Instrum. and Meth.* B **267** 777
[8] Uglov S R, Zabaev V N, Kaplin V V and Kuznetsov S 2012 *J. of Phys.: Conf. Ser.* **357** 012012
[9] Uglov S R, Zabaev V N and Kaplin V V 2013 *Nucl. Instrum. and Meth.* B **309** 79
[10] Uglov S R, Kaplin V V, Sukhikh L G and Vukolov A V 2014 *JETP Letters* **100/8** 503
[11] Hardouin A 2007 *Ph. D thesis Université Paris Sud Orsay*
[12] Le Guen K, Maury H, Andre J-M, Jonnard Ph, Hardouin A et al. 2007 *Appl. Phys. Lett.* **91** 234104
[13] Parratt L G 1954 *Phys. Rev.* **95** (2) 359
[14] Marmoret R, André J-M 1983 *Appl. Opt.* 22 (1) 17
[15] Pafomov V E 1969 *Trudy FIAN* **44** (28)
[16] Kaplin V, Uglov S, Bulaev O, Goncharov V, Piestrup M and Gary C 2001 *Nucl. Instrum. and Meth.* B **175** 3
[17] Kubankin A, Nasonov N, Kaplin V, Uglov S, Piestrup M and Gary C 2006 *Radiation Physics and Chemistry* **75** 913
[18] Kubankin A, Likhachev V, Nasonov N, Rakitjansky A and Zhukova P 2006 *Nucl. Instrum. and Meth. B* **252** 124

Journal of Physics: Conference Series **732** (2016) 012018 doi:10.1088/1742-6596/732/1/012018

Characteristics of Smith-Purcell radiation in millimeter wavelength region

G A Naumenko[1], A P Potylitsyn[1], P Karataev[3], V Bleko[2], L G Sukhikh[1], M V Shevelev[1], Yu Popov[1]

[1]Tomsk Polytechnic University, Lenina Ave. 2a,Tomsk, Russia
[2]RASA Center in Tomsk, Tomsk Polytechnic University, Lenina Ave. 2a, Tomsk, Russia
[3]John Adams Institute at Royal Holloway, University of London, Egham, Surrey, TW20 0EX, United Kingdom

E-mail: naumenko@tpu.ru

Abstract. Investigations of the Smith-Purcell radiation (SPR) were began with non-relativistic electron beams with some unexpected experimental results. Further the experimental investigations were performed with relativistic electron beams for application to beam diagnostics. Large discrepancy between different theoretical models significantly increases the role of experimental studies of this phenomenon. In this report we present some problems and features of experimental investigations of SPR in millimeter wavelength region. The problems of prewave zone and coherent effects are considered. The shadowing effect, focusing of radiation using a parabolic SPR target and effect of inclination of target strips were investigated with moderately relativistic electron beam.

1. Introduction

The incoherent SPR arising when a charged particle passes over a periodic structure was observed for the first time in 1953 [1]. In view of a target periodicity the SPR will be monochromatic, i.e., there exists a stiff correlation between its wavelengths and the direction of radiation (the so-called Smith-Purcell dispersion relationship from [1]):

$$\lambda = \frac{d}{k}\left(\frac{1}{\beta} - \cos\theta\right),$$

where λ is the radiation wavelength, k is the diffraction order, d is the target period, β is the electron speed to that of light ratio, and θ is the observation angle (figure 1). Figure 2 shows the dependence of relation λ / d on observation angle for different k and $\beta \approx 1$.

Very exotic results have been obtained in experiment [2]. In this work using the beam of electron microscope, the fourth order was observed in dependence of infrared radiation intensity on the beam current. Unfortunately this effect was not confirmed yet.

In recent experiments [3, 4], the possibility of creating a monochromatic radiation source of the THz range on the basis of the Smith-Purcell radiation (SPR) has been demonstrated. The SPR from low relativistic electrons is also used in orotrons. The dispersion relationship is very useful for using of a coherent SPR (CSPR) in non-invasive beam diagnostics, namely for bunch length measurements

Content from this work may be used under the terms of the Creative Commons Attribution 3.0 licence. Any further distribution of this work must maintain attribution to the author(s) and the title of the work, journal citation and DOI.
Published under licence by IOP Publishing Ltd

[5]. For the non-relativistic electron energies ($E_e \leq 100\,\text{keV}$), the approach developed by van den Berg [6, 7] ensures a reasonable agreement with experiment [8, 9].

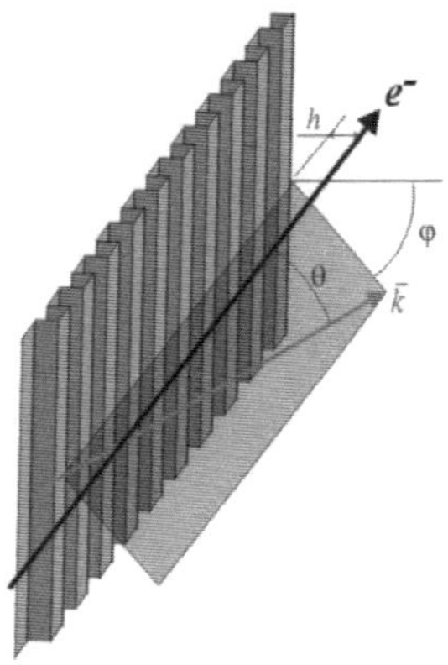

Figure 1. Designation of variables parameters.

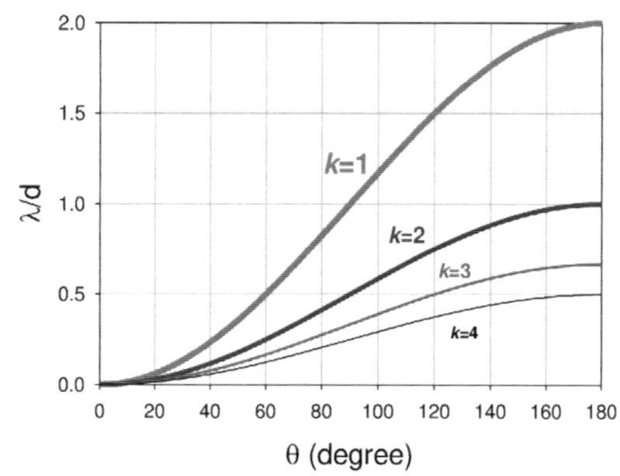

Figure 2. Dependence of relation λ / d on observation angle for different *k*.

In experiment [10], the possibility of applying the SPR of high relativistic electrons as a spontaneous radiation mechanism for a free-electron laser has been investigated. However until now no conventional model to calculate the characteristics of the SPR from high relativistic electrons is available.

In [11] the different models for SPR characteristics calculation were theoretically compared for high relativistic electrons:

a) resonant diffraction radiation (RDR) model. This model is used for gratings consisting of infinitely thin perfect conducting strips separated by vacuum gaps, including inclined strips [12];

b) surface current (SC) model (the SPR is considered as a radiation generated by the current induced by the field of a particle moving in vacuum on a perfect conducting periodic surface);

c) Van den Berg's (VdB) model (applicable for volume gratings).

There was shown that the predictions of most models differ by approx. 2 orders of magnitude for the electron energy 20 MeV [13] and by several orders for the electron energies E_e = 855 MeV [14]. The available experimental results do not provide an ultimate conclusion on the validity of one of these models.

We would like in this paper to take attention on the some features which take place during the experimental investigations of SPR in millimeter and sub-millimeter wavelength region.

2. Prewave zone effect

This effect takes place at a small distance *R* between a target and detector. In this case a target size contribution distorts the angular distribution of radiation, which is usually considered theoretically in a far field zone. This is well-known effect for a transition radiation (TR) [15], that appears when $R \leq \gamma^2 \lambda$, where γ is the Lorenz-factor of electrons and λ is the investigated wavelength of TR. However, it is not obvious to the SPR. The detailed analysis of the prewave zone effect for SPR was done in [11]. We may show that this effect for transversal size of SPR target $H \geq \gamma\lambda$ appears when

$R \gg \dfrac{L^2(1/\beta+\cos\theta)}{2d}+2\gamma^2 d\cdot(1/\beta-\cos\theta)$, where L is a longitudinal size of target, d is period of target and $\beta=\sqrt{1-1/\gamma^2}$.

For example:

a) for L = 200 mm; $\gamma=12$; $\theta=\pi/2$; d = 12 mm we obtain far field zone condition R > 5 m,

b) for L = 20 mm; $\gamma=2000$; $\theta=\pi/2$; d = 800 nm (KEK ATF conditions, optical region) R > 4 m.

It is clear, that it is problem (sometimes impossible) to provide measurements in far field zone. To exclude the prewave zone effect a parabolic telescope (see figure 3) with detector placed in focus of parabolic mirror may be used for the spectral and angular distributions measurement. This method was proposed and tested in [16] and gives the same angular distribution as in the far field zone.

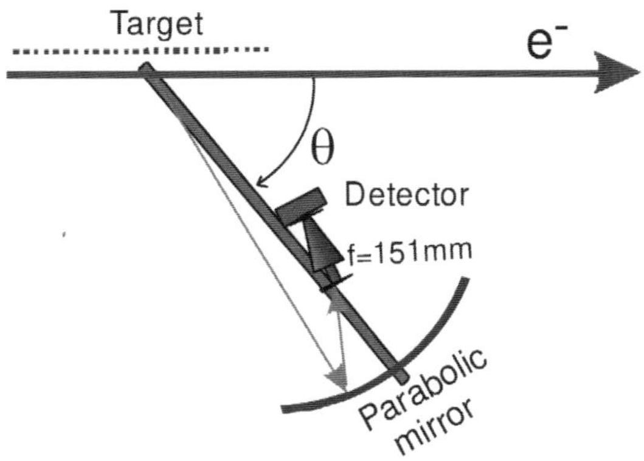

Figure 3. Scheme for suppression of the prewave zone effect.

3. Coherency factor

Another problem is the coherent effect of radiation of a bunched electron beam. The coherent radiation intensity of electron bunch may be presented as $W_{\text{coh}} \approx N_e \cdot f^2 \cdot W_{\text{incoh}}$. Here $N_e \gg 1$ is a bunch population, W_{incoh} is the incoherent radiation of electrons in bunch and f is the geometrical form-factor of a bunch. For SPR the longitudinal form-factor $f_{\parallel}$ gives contribution only, if $\sigma_{\perp} << \dfrac{\gamma\lambda}{2\pi} e^{\frac{2\pi h}{\gamma\lambda}}$ (see [31]), where $\sigma_{\perp}$ defines a transversal size of bunch and h is impact-parameter. For Gaussian approximation of longitudinal distribution of electron in bunch $f = f_{\parallel} = e^{-2\left(\frac{\pi\cdot\sigma_{\parallel}}{\beta\cdot\lambda}\right)^2}$, where $\sigma_{\parallel}$ defines a longitudinal size of bunch.

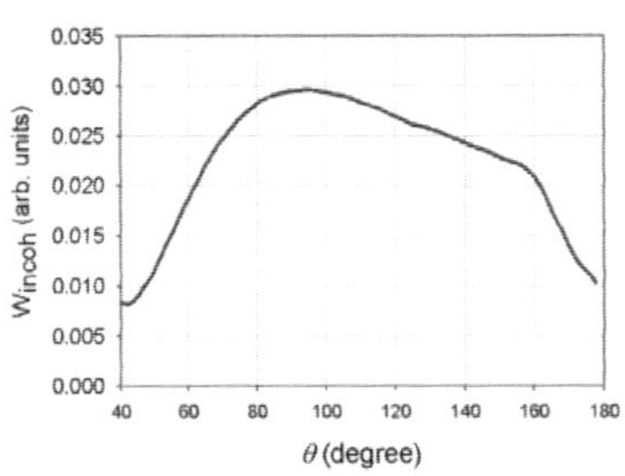

Figure 4. SPR polar distributions, calculated using surface current model for incoherent radiation.

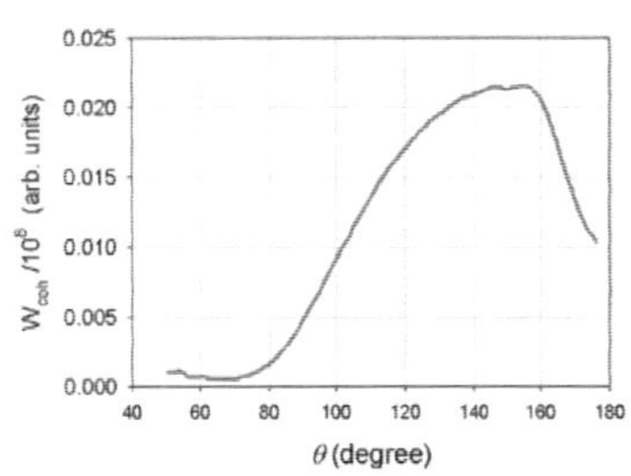

Figure 5. SPR polar distributions, calculated using surface current model for coherent radiation.

In figures 4 and 5 are presented the sample of SPR polar distributions, calculated using surface current model for incoherent and coherent radiation correspondently for next conditions: L = 160 mm; d = 8 mm; strip width is equal 4 mm; impact-parameter h = 10 mm; $\sigma_{\|} = 2$ mm; $N_e = 10^8$.

So, form-factors are crucial in the formation of the spectral and angular characteristics of SPR.

4. Main SPR properties, observed using the electron beam of Tomsk Polytechnic University microtron

The experiment was performed on the extracted electron beam of microtron PTI TPU with the following parameters: electron energy – 6.1 MeV; macro-pulse duration $-4\,\mu s$; the macro-pulse frequency $1 \sim 8$ Hz; the characteristic bunch length in the Gaussian approximation $\sigma_{\|} \approx 2$ mm; the bunch population N_e = 10^8, the number of bunches in macro-pulse – 10^4, the cross-section of the extracted beam $4 \times 2\,\text{mm}^2$, the angular divergence of the extracted beam – 0.08 rad. In figure 6 is presented the dependence of squared form-factor on the wavelength for these conditions.

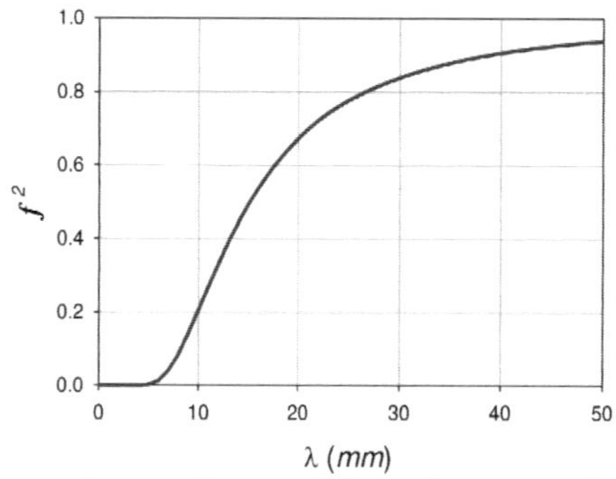

Figure 6. Dependence of squared form-factor on the wavelength.

Under these conditions for $\lambda > 9$ mm radiation of electron bunch is coherent. We may expect the radiation for $\lambda < 9$, but with a small intensity, which results large relative errors. Because of the coherency the radiation intensity is increased by N_e times (i.e., for the specified bunch population it is by 8 orders), making it available for the measurement using existing detectors operating at a room temperature.

For radiation registration, we used broadband detector DP-21M, manufactured in Semi-conductive Devices Research Institute (Tomsk, Russia) with parameters: efficiency in the wavelength region $\lambda = 3 \sim 16\,\text{mm}$ is estimated to be constant to a $\pm 15\%$ accuracy. The detector sensitivity is 0.3 V/mWatt. To suppress the accelerator RF system background the beyond cutoff waveguide with diameter 10 mm was mounted on the detector, which cuts a radiation having a wavelength $\lambda > 17$ mm. Thus, the spectral range of the measurement is $\lambda = 9 \sim 17$ mm.

The angular characteristics of the radiation were measured using the parabolic telescope (figure 3), which provides the measurements in terms of far field zone. The angular resolution of a telescope is 0.035 radians.

4.1. Comparison of different theoretical models of SPR

There are few theoretical models for calculate the SPR characteristics for the different grating profiles. We consider here three models: van den Berg's model (vDB) [6] (applicable for volume gratings), surface current model (SC) [17, 18] (the SPR is considered as a radiation generated by the current induced by the field of a particle moving in vacuum on a perfect conducting periodic surface), and resonant diffraction radiation model (RDR) [19] (for gratings consisting of infinitely thin perfect conducting strips separated by vacuum gaps). For comparison we use calculations and experimental results from [20] (figure 7). We can see in figure 7 a large discrepancy between radiation yields obtained using these three models. Comparing the absolute experimental data with the theoretical predictions [20] ascertain that the RDR and SC models.

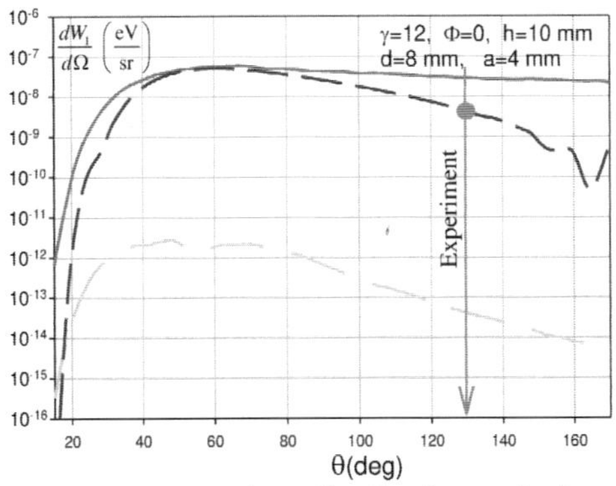

Figure 7. Angular distribution of the SPR intensity for a flat grating according to the RDR model (solid line), surface current model (a dashed curve) and for van den Berg's model (dash-dotted line). The point is the absolute experimental measurement

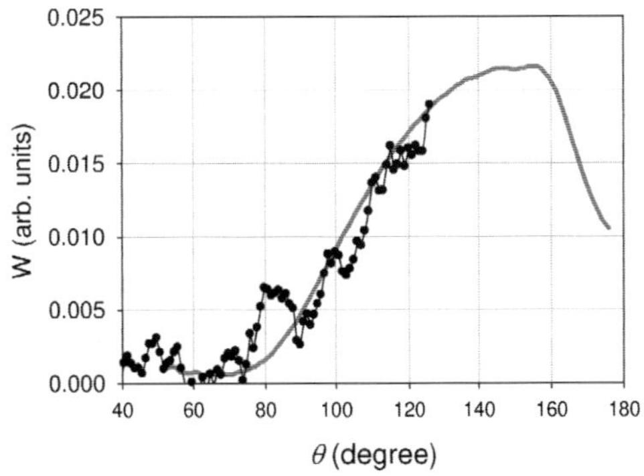

Figure 8. Polar angular distribution of measured coherent SPR. Dotted line is the experimental results. Solid line is calculation for far field zone using surface current model, normalized on the experimental results.

4.2. Angular dependence of coherent SPR

The polar angular distribution of measured coherent SPR (CSPR) from the flat target with parameters: $L = 200$ mm, $d = 12$ mm, strip width – 6 mm, impact-parameter $h = 10$ mm, is shown in figure 8.
Solid line depicts the calculation of CSPR intensity normalized on the experimental results, for far field zone using surface current model for bunch length $\sigma_{\|} \approx 2$ mm and for the same target parameters.

4.3. SPR focusing

Here we describe the possibility of CSPR focusing using a parabolic grating (figure 9). In this case we can obtain the focusing effect in a pre-wave zone without additional optic devices.

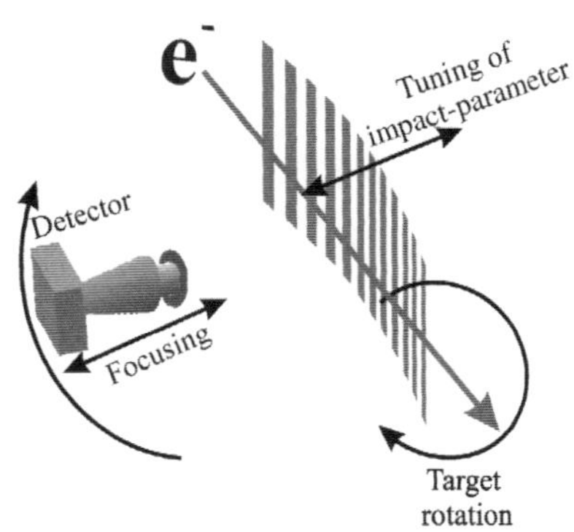

Figure 9. Scheme of CSPR focusing using a parabolic grating

In figure 10 is shown the dependence of radiation intensity on the distance between target and detector where we can see the maximum of radiation intensity in the focus of parabola. The comparison of measured azimuthally orientation distributions of CSPR yield from polar-focusing (red curve) and flat (blue curve) gratings in corresponding conditions is shown in figure 11.

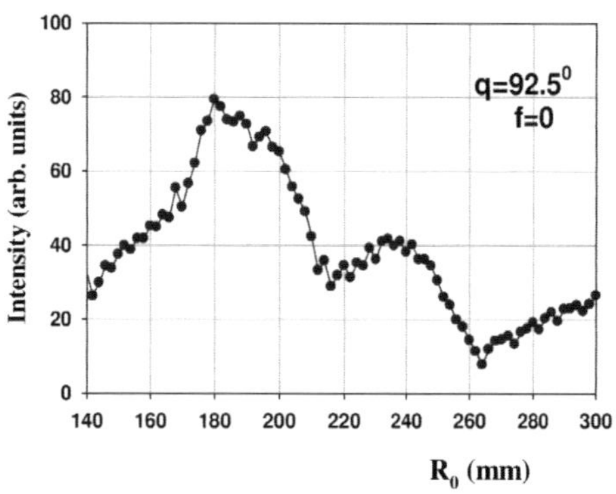

Figure 10. Experimentally measured distribution of CSPR yield vs. distance from the polar-focusing grating to detector.

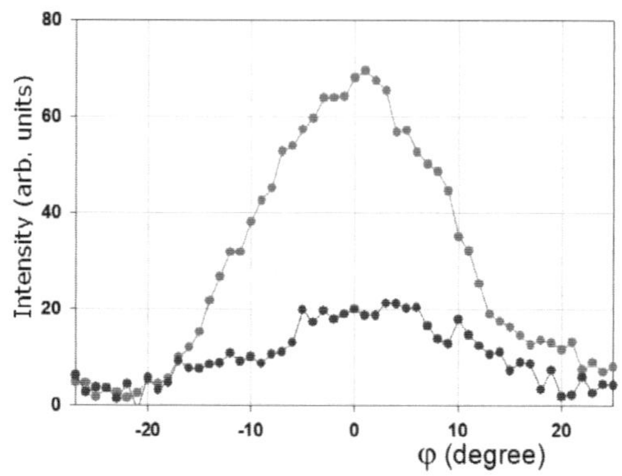

Figure 11. The comparison of experimentally measured azimuthally (at angle φ) orientation distributions of CSPR yield from polar-focusing (red curve) and flat (blue curve) gratings in corresponding conditions.

Here we can see that the focusing effect is limited due to increase of the impact-parameter in the center of parabolic target.

4.4. Shadowing effect in SPR

The shadowing of electron electromagnetic field in case of diffraction and transition radiation was considered theoretically and experimentally in number of papers [21] – [25]. In Pseudo-photon approach for ultra-relativistic electron the properties of electron field are very close to the properties of real photons. There is a region downstream to a conductive or absorbing screen where the Coulomb field is partly missing. In terms of paper [22] this effect is named “shadow effect”, and the term "semi-bare electron" has been introduced in [26, 27] to describe the similar effect in the framework of quantum electrodynamics for an electron scattered at a large angle.

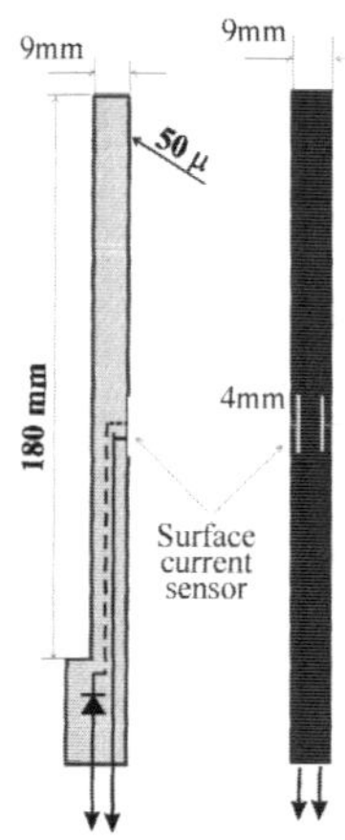

Figure 12. Active strip with surface current sensor.

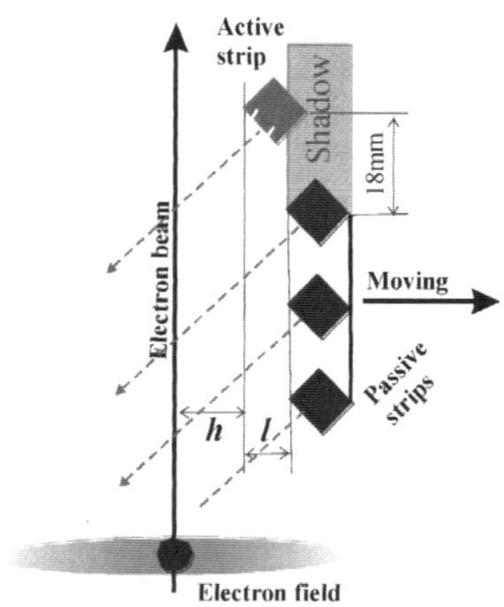

Figure 13. Scheme of shadowing measurements.

In [22] the shadowing effect for a Smith-Purcell radiation (SPR) was predicted by Prof. X. Artru. We apply in [28] the well-known technique, which is used in strip-line beam position monitors [29]. This sensor registered a surface current component perpendicular to the slit (see figure 12). The sensor was inserted in the element typical for a SPR target element (active element).

To be sure that we measure the concerned amount, we had measured the dependence of the surface current on impact-parameter (dots in figure 14). The solid line in figure 14 is the fit to experimental data.

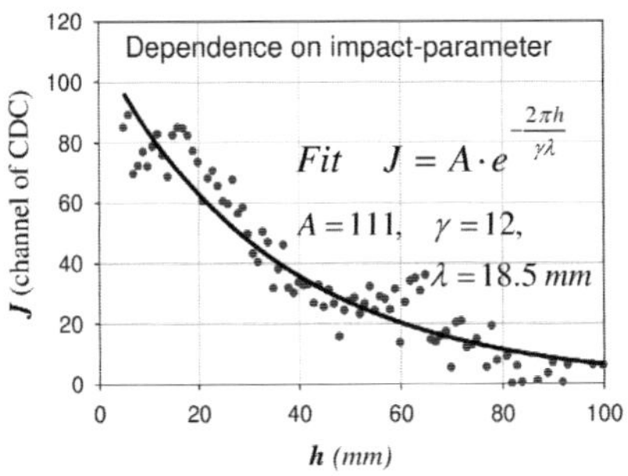

Figure 14. Dependence on impact-parameter. Dots are the experimental points. Solid line is fit shown approximation.

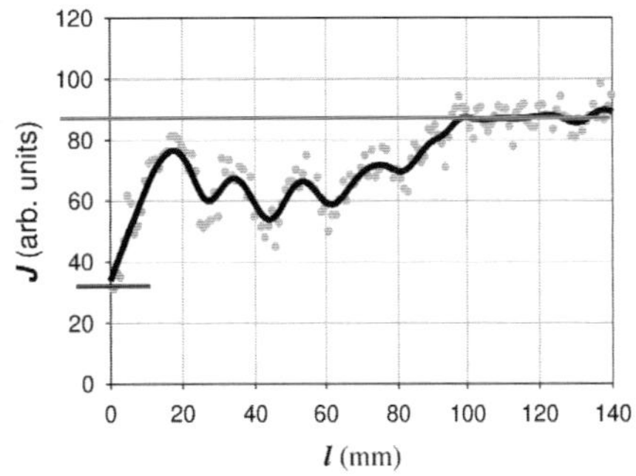

Figure 15. Dependence of surface current on the surface of active element on impact-parameter of the passive elements.

For measuring the shadowing effect we had inserted three grating elements of the same geometry but without a sensor (passive elements) upstream to the active element (see figure 13). In this geometry the active grating element corresponds to the fourth element in SPR target with width of element 9 mm and period 18 mm. The dependence of surface current on the surface of active element on impact-parameter of the passive elements (figure 15) shows the shadowing effect is equal 64%. This effect is of considerable importance to be taken into account in a calculation of SPR.

4.5. Effect of SPR target inclination (conical effect)

This item is devoted to the so named conical effect, predicted theoretically in [30]. According this prediction the maximum of azimuthal (φ) and polar (θ) distribution of SPR depends on an inclination angle ψ of a SPR target in respect to an electron beam (figure 16).

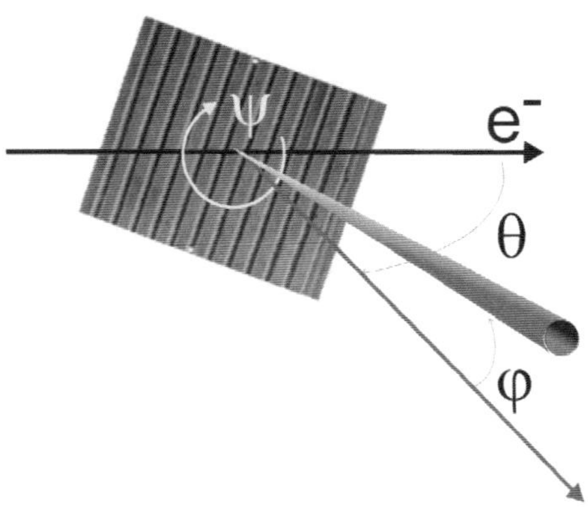

Figure 16. Designation of variable parameters in the experiment.

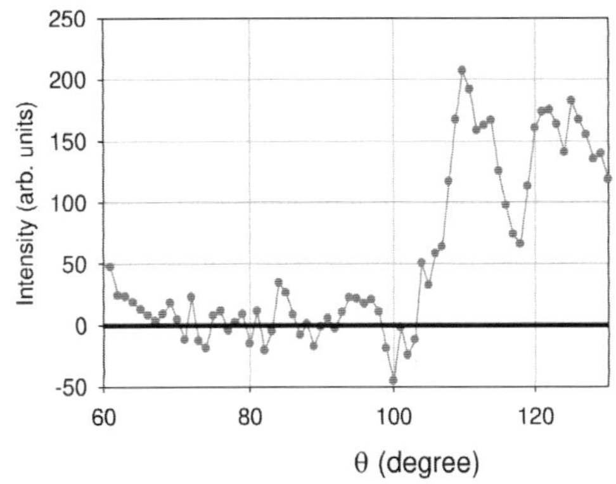

Figure 17. The dependence of radiation intensity on the polar angle θ for $\varphi = 0$ and $\psi = 0$.

The dependence of radiation intensity on the polar angle θ for $\varphi=0$ and $\psi=0$ in designation of variable parameters in the experiment shown in figure 16 is shown in figure 17. In figure 18 are shown the azimuthal dependences of SPR intensity in the point $\theta=110^{\circ}$ for $\psi=0$, $\psi=3^{\circ}$ and $\psi=5^{\circ}$. The dependences are normalized on the maximum of radiation intensity.

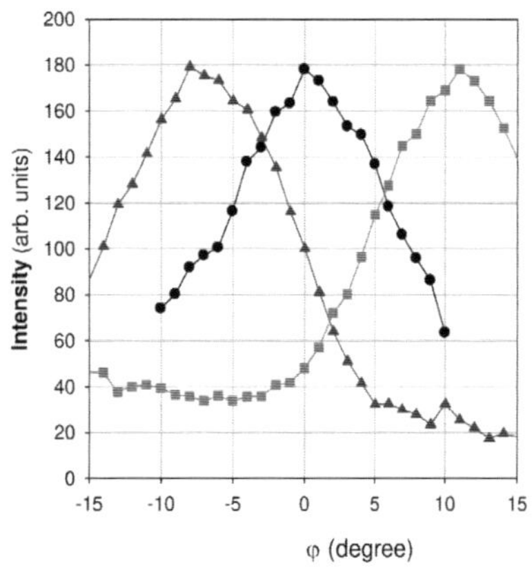

Figure 18. Azimuthal dependences of SPR intensity for $\psi=0$ (circles), $\psi=3^{\circ}$ (triangles), $\psi=5^{\circ}$ (squares).

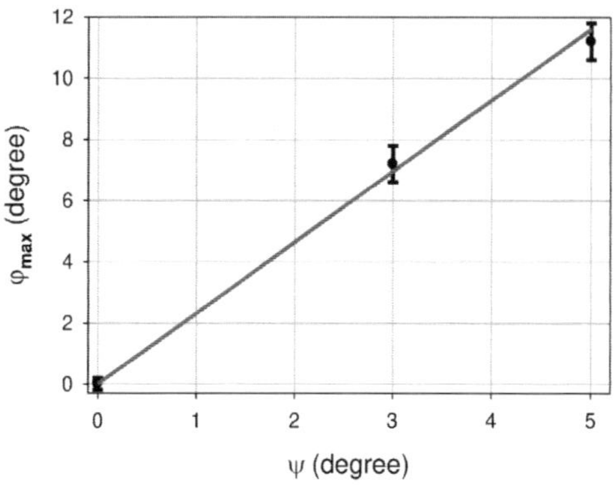

Figure 19. Dependence of maximum position φ_{max} on the orientation angle ψ of the target in respect to the beam direction. Line is the linear fit.

Using these results we can build the dependence of maximum position φ_{max} on the orientation angle ψ of the target in respect to the beam direction (figure 19). In figure 19 we can see the strong dependence of angular peak position φ_{max} on the orientation angle ψ of the target.

The effect of azimuthal inclination of radiation due to rotation of target may be used for direct noninvasive measurement of a beam direction at a given point of the beam trajectory.

5. Conclusion

The existing theoretical calculations of SPR properties show the large discrepancy between them and can be used for estimations only (except kinematical calculations), and should be tested experimentally. Prewave zone effect should be estimated, for each geometry in angular and spectral measurements. The coherent properties of radiation of a bunched electron beam may be used for bunch length measurement. A parabolic SPR target may be used for the focusing of SPR. The shadowing effect should be taken into account in theoretical estimations. The effect of azimuthal inclination of radiation due to rotation of target may be used for direct noninvasive measurement of a beam direction.

Acknowledgment

This work was supported by grant RFBR 15-52-50028

References

[1] Lampel M C 1999 *Advanced Accelerator Concepts: Eighth Workshop* (AIP, New York)
[2] Bakhtyari A, Walsh J E, and Brownell J H 2002 *Phys. Rev. E* **65** 066503
[3] Shibata Y, Hasebe S, Kimihiro Ishi K *et al* 1998 *Phys. Rev. E* **57** 1061
[4] Korbly S E, Kesar A S, Sirigiri J R and Temkin R J 2005 *Phys. Rev. Letters* **94** 054803
[5] Doucas G, Blackmore V, Ottewell B and Perry C 2006 Phys. Rev. ST AB **9** 092801

[6] Van den Berg P M 1973 *J. Opt. Soc. Am.* **63** 1588
[7] Van den Berg P M 1973 *J. Opt. Soc. Am.* **63** 689
[8] Gover A, Dvorkis P and Elisha U 1984 *J. Opt. Soc. Am. B* **1** 723
[9] Adischev Y N, Vukolov A V, Karlovets D V, Potylitsyn A P and Kube G 2005 *JETP Letters* **82** 174
[10] Backe H, Lauth W, Mannweiler H *et al* 2006 *NATO Workshop 'Advanced Radiation Sources and Applications'* (Springer, New York) 267
[11] Karlovets D V and Potylitsyn A P 2006 *Phys. Rev. ST AB* **9** 080701
[12] Potylitsyn A, Karataev P, Naumenko G 2000 Phys. Rev. *E* **61** (6) 7039
[13] Kesar A S 2005 *Phys. Rev. ST Accel. Beams* **8** 072801
[14] Kube G, Backe H, Euteneuer H *et al* 2002 *Phys. Rev. E* **65** 056501
[15] Verzilov V A 2000 *Phys. Lett. A* **273** 135
[16] Kalinin B N, Naumenko G A, Potylitsyn A P *et al* 2006 *JETP Letters* **84** 3 110
[17] Brownell J H, Walsh J and Doucas G 1998 *Phys. Rev.* E **57** 1075
[18] Brownell J H and Doucas G 2005 *Phys. Rev. ST Accel. Beams* **8** 091301
[19] Potylitsyn A P 1998 *Nucl. Instrum. Methods Phys. Res. Sect. B* **145** 60
[20] Naumenko G, Kalinin B, Saruev G *et al* 2007 *Proceedings of symposium FEL* Novosibirsk Russia WEPPH055
[21] Naumenko G, Artru X, Potylitsyn A *et al* 2010 *Journal of Physics: Conference Series* **236** 012004
[22] Artru X, Ray C 2007 *International Conference on Charged and Neutral Particles Channeling Phenomena II,* Proc. SPIE **6634**
[23] Naumenko G A, Potylitsin A P, Sukhikh L G *et al* 2009 JETP Letters **90** (2) 96
[24] Naumenko G, Artru X, Potylitsyn A, Popov Yu and Sukhikh L 2010 *Proceedings of symposium "Charged and neutral particles channeling phenomena - Channeling 2008"* 511
[25] Naumenko G, Potylitsyn A, Popov *et al* 2009 *ArXiv.org physics* arXiv:0901.2630
[26] Feinberg E L 1979 *Sov. Phys. Uspekhi* **22** 479
[27] Shul'ga N F and Syshchenko VV 2000 *Journal Physics of Atomic Nuclei* **63** 2018
[28] Naumenko G A, Potylitsyn A P, Popov Yu A, Shevelev M V 2011 *Nuovo Cimento C* **35** 4 305
[29] Sargsyan V 2004 *TESLA Report* **03**
[30] Sergeeva D Yu, Tishchenko A A and Strikhanov M N 2015 *Phys. Rev. ST* **18** 052801
[31] Naumenko G 2015 *Advanced Materials Research* **1084** 138

Simulation of Coherent Diffraction Radiation Generation by Pico-Second Electron Bunches in an Open Resonator

L G Sukhikh, A P Potylitsyn and D A Verigin

Tomsk Polytechnic University, Lenin avenue, 30, Tomsk, 634050, Russia

E-mail: `sukhikh@tpu.ru`

Abstract. In this report we present new approach for calculation of processes of diffraction radiation generation, storage and decay in an open resonator based on generalized surface current method. The radiation characteristics calculated using the developed approach were compared with those calculated using Gaussian-Laguerre modes method. The comparison shows reasonable coincidence of the results that allows to use developed method for investigation of more complicated resonators.

1. Introduction

Stimulated radiation is responsible for optical and infrared laser operation principle based on atomic properties of various materials. Later the term "laser" was used to specify light generated in undulators in electron accelerators known as Free Electron Lasers (FEL) [1, 2]. In this case electrons moving on a periodical path generate synchrotron radiation. The radiation traveling along the beam in the undulator starts interacting with the electrons stimulating the production of the radiation with the same wavelength and phase, which are the main properties of the laser light.

In the experiments [3, 4] the authors observed Stimulated Transition Radiation (TR) and Stimulated Synchrotron Radiation respectively, which, in contrast to undulator radiation, had a continuous spectrum. Authors of Ref. [3] used a special 4-mirror resonator in order to focus coherent TR generated by the electron bunch on the both surfaces of the target alternatively in the moment of passage of the subsequent electron bunch. In papers [4, 5] the authors proposed a concept of the so-called "broadband FEL" based on stimulated TR in the closed cylindrical resonator where resonator mirrors were used at the same time as TR targets. Afterwards the radiation is recirculated in the resonator. As a result every subsequent bunch generates radiation in the presence of radiation field from all preceding bunches stimulating the radiation production.

In Ref. [6] we proposed to generate intense THz radiation in an semi-confocal resonator to stimulate coherent diffraction radiation (DR) generated both by the flat mirror in forward direction (FDR) and by the concave mirror in backward direction (BDR). Due to DR non-invasive nature [7] there was no direct interaction of the electron beam with the mirrors resulted in fine background conditions and the beam parameters remaining almost the same as the initial ones. It was demonstrated that the radiation intensity stored in such a resonator growed up exponentially as a function of the number of bunches, which has an obvious analogy with

Content from this work may be used under the terms of the Creative Commons Attribution 3.0 licence. Any further distribution of this work must maintain attribution to the author(s) and the title of the work, journal citation and DOI.
Published under licence by IOP Publishing Ltd

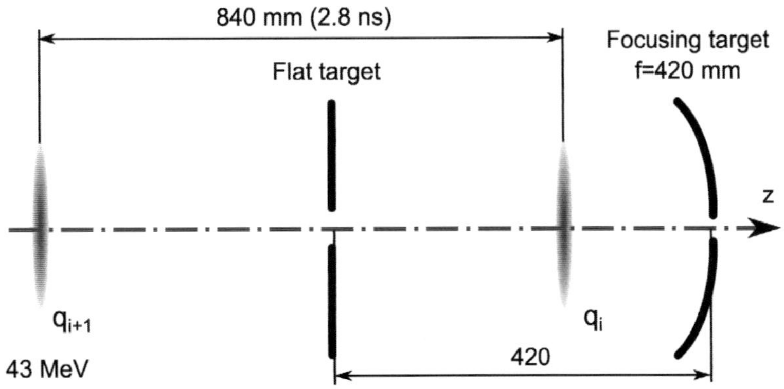

Figure 1. Simulation scheme.

undulator FELs, where the intensity grows up exponentially as a function of the number of undulator periods [6].

The aim of this paper is to show the analytical way for calculation of diffraction radiation characteristics in the open semi-confocal resonator taking into account radiation generation in the presence of external field, its storage and resonator free runs.

2. Theoretical model of DR generation in open resonator

Passing through openings in the first and second resonator mirrors point-like electron bunches with a charge q generate FDR and BDR, consequently. The first bunch passes through the opening in the first mirror generating FDR that travels along z-axis together with the electron. The slippage effect could be neglected due to a broadband DR spectrum. The bunch reaches the second mirror and generates BDR in presence of the FDR from the first mirror. Both BDR and reflected FDR travels back and reaches the first mirror at the moment when the second bunch arrives. Thus, both FDR and BDR are generated in the presence of the external electric field that was generated previously.

Let us formulate the problem as follows. Train of the point-like electron bunches with 43 MeV energy travels along z-axis passing through an open resonator that consists of flat and focusing mirrors (see figure 1). The distance between bunches in the train is equal to 840 mm (2.8 ns) while the distance between resonator mirrors is equal to 420 mm. The second mirror is paraboloidal one with radius of curvature equal to $R = 840$ mm (focal distance $f = 420$ mm). In the transverse plane both mirrors are assumed to be circles with a diameter equal to $a = 100$ mm. The openings in the mirrors are assumed to be round ones. The opening diameter in the focusing mirror is equal to $a_1 = 5$ mm, in the flat mirror – $a_0 = 5$ mm. The resonator parameters coincide with ones from the previous experiment [6].

For our calculation we assumed that both mirrors were perfect conductors. According to generalized surface current method [8] the electric field of FDR generated by the first bunch $\mathbf{E}_\mathbf{f}^{(1)}$ could be written as integration of the double cross-product:

$$\mathbf{E}_\mathbf{f}^{(1)}(\mathbf{r_0}, \omega) = \frac{1}{2\pi} \int_{S_1} dS_1 \, [\mathbf{n_1}, \mathbf{E_e}(\mathbf{r}, \omega)] \times \nabla_0 G(\mathbf{r_0}, \mathbf{r}, \omega), \tag{1}$$

where $\mathbf{r_0}$ is the observation point; $\mathbf{r}$ is the radiation point; ω is the radiation angular frequency ($\omega = 2\pi c/\lambda$, c is the speed of light); $\mathbf{n_1} = \{0, 0, 1\}$ is the surface normal to the first mirror; $\mathbf{E_e}(\mathbf{r}, \omega)$ is the electric field of the bunch; $\nabla_0 G(\mathbf{r_0}, \mathbf{r}, \omega)$ is the gradient of free space Green function. The integration is carried out over the surface of the flat mirror S_1, where $dS_1 = \rho d\rho d\phi$.

The electron bunch field $\mathbf{E_e}(\mathbf{r},\omega)$ could be written as follows [8]:

$$\mathbf{E_e} = \frac{q}{\pi c^2}\frac{\omega}{\beta^2\gamma}e^{i\frac{\omega}{\beta c}z}\left\{\frac{\vec{\rho}}{\rho}K_1\left[\frac{\omega\rho}{\beta c\gamma}\right] - \frac{i}{\gamma}\frac{\mathbf{V}}{\beta c}K_0\left[\frac{\omega\rho}{\beta c\gamma}\right]\right\}. \tag{2}$$

Here q is the bunch charge, $\beta = v/c$ is the electron reduced velocity, $i = \sqrt{-1}$, $\vec{\rho} = \{x = \rho\cos\phi, y = \rho\sin\phi\}$, $\mathbf{V} = \{0, 0, \beta c\}$, K_n is the modified Bessel function of the second kind of the n-th order.

The gradient of the free space Green function $\nabla_0 G(\mathbf{r_0},\mathbf{r},\omega)$ could be written as follows:

$$\nabla_0 G(\mathbf{r_0},\mathbf{r},\omega) = -(\mathbf{r}-\mathbf{r_0})\frac{\exp\left[i\frac{\omega}{c}|\mathbf{r}-\mathbf{r_0}|\right]}{|\mathbf{r}-\mathbf{r_0}|^3}\left(i\frac{\omega}{c}|\mathbf{r}-\mathbf{r_0}| - 1\right) \tag{3}$$

After the FDR is generated by the first bunch it travels in the resonator along z-axis. Both electron bunch and FDR reach the second mirror almost at the same time. In this case BDR is generated in the presence of external field. Thus, electric field of BDR $\mathbf{E_b^{(1)}}(\mathbf{r_0},\omega)$ could be written as follows:

$$\mathbf{E_b^{(1)}}(\mathbf{r_0},\omega) = -\frac{1}{2\pi}\int_{S_2} dS_2\left[\mathbf{n_2}, \left(\mathbf{E_e}(\mathbf{r},\omega) + \mathbf{E_f^{(1)}}(\mathbf{r},\omega)\right)\right] \times \nabla_0 G(\mathbf{r_0},\mathbf{r},\omega), \tag{4}$$

where $\mathbf{n_2} = -\dfrac{2f\left\{\frac{\rho\cos\phi}{2f}, \frac{\rho\sin\phi}{2f}, 1\right\}}{\sqrt{4f^2+x^2+y^2}}$ is the normal vector to the parabolic mirror surface [9]. The integration is performed over the surface of the parabolic mirror S_2, $dS_2 = \dfrac{\sqrt{4f^2+\rho^2}}{4f}\rho d\rho d\phi$.

After generation of BDR the first bunch leaves the resonator. The BDR travels back to the first mirror and reaches it almost in the same time with second bunch resulting in the generation of the FDR in the presence of the external field. In this case the formula to find field of FDR looks similar to equations (1),(4):

$$\mathbf{E_f^{(2)}}(\mathbf{r_0},\omega) = -\frac{1}{2\pi}\int_{S_1} dS_1\left[\mathbf{n_1}, \left(\mathbf{E_e}(\mathbf{r},\omega) + \mathbf{E_b^{(1)}}(\mathbf{r},\omega)\right)\right] \times \nabla_0 G(\mathbf{r_0},\mathbf{r},\omega), \tag{5}$$

Thus, the problem consists of the calculation of radiation spatial distributions on the mirror surfaces generated by the electron bunch train. In the case of i-th bunch the FDR and BDR could be calculated in the following way:

$$\begin{aligned}\mathbf{E_f^{(i)}}(\mathbf{r_0},\omega) &= \frac{1}{2\pi}\int_{S_1} dS_1\left[\mathbf{n_1}, \left(\mathbf{E_e}(\mathbf{r},\omega) + \mathbf{E_b^{(i-1)}}(\mathbf{r},\omega)\right)\right] \times \nabla_0 G(\mathbf{r_0},\mathbf{r},\omega),\\ \mathbf{E_b^{(i)}}(\mathbf{r_0},\omega) &= -\frac{1}{2\pi}\int_{S_2} dS_2\left[\mathbf{n_2}, \left(\mathbf{E_e}(\mathbf{r},\omega) + \mathbf{E_f^{(i)}}(\mathbf{r},\omega)\right)\right] \times \nabla_0 G(\mathbf{r_0},\mathbf{r},\omega)\end{aligned} \tag{6}$$

The equations (6) allow to calculate radiation characteristics in the resonator for all electron bunches one by one. After the bunch train leaves the resonator the radiation enhancement stops and decay process starts (resonator free runs). The decay of stored radiation could be calculated using the same formalism as described before but assuming that the bunch field is equal to zero.

There exists a significant difficulty that makes it almost impossible to carry out direct calculation of radiation field in the resonator using the proposed approach. Each radiation generation event adds double integration in the final result. For example, in order to calculate FDR from the second bunch (equation (5)) one needs to calculate six-dimensional integral. In order to overcome this problem the following assumption was made. Calculating BDR from the first electron bunch (equation (4)) one needs to know distribution of the FDR from the first bunch on the surface of the focusing mirror. Such a distribution could be calculated easily using

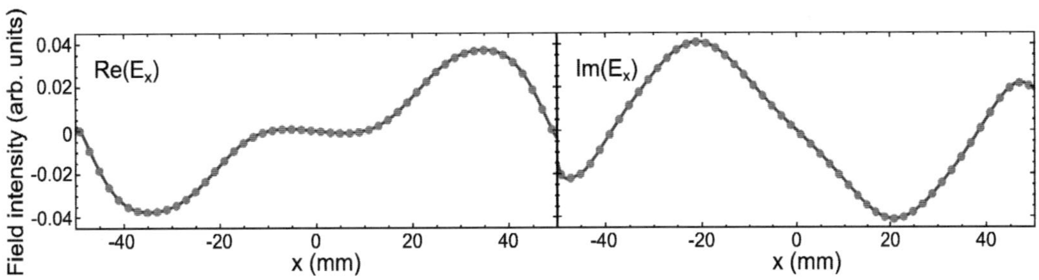

Figure 2. Comparison of calculated field distribution and its interpolation. Left plot – real part of E_x-component, right plot – imaginary part of E_x-component. Red circles – cross-section of calculated E_x-component, blue line – cross-section from interpolation 2D surface, $y = 0$, $\lambda = 3.5\,\text{mm}$.

equation (1) that allows numerical calculation of the spatial distribution of all three components of the electrical field on the mirror surface. Practically these field components are calculated in a finite number of points. Amount of such points should be enough for correct representation of the field. Thus, one can easily calculate the FDR field distribution in the specified number of points situated on the mirror surface and to interpolate obtained distribution assuming its smooth character. In this case the numerical calculation of BDR will include just double integration over the focusing mirror surface that could be easily carried out numerically.

Our calculation was carried out using "Wolfram Mathematica 7.0" software [10]. The FDR field distribution on the surface of focusing mirror was calculated in the range $-50 \le x, y \le 50\,\text{mm}$ with the step equal to $2\,\text{mm}$. The calculated field consists of 6 components, namely:

$$\mathbf{E}_\mathbf{f}^{(1)}(\mathbf{r_0}, \omega) = \{\Re(E_x) + i\Im(E_x), \Re(E_y) + i\Im(E_y), \Re(E_z) + i\Im(E_z)\} \tag{7}$$

Each field component was interpolated by 2D surface using spline interpolation procedure in "Wolfram Mathematica 7.0". Figure 2 demonstrates comparison of calculated field distribution (x-component of FDR field) and results of interpolation by 2D surfaces in the form of cross-sections along x-axis ($y = 0$) both for real and imaginary parts of the field. The calculation was carried out at $\lambda = 3.5\,\text{mm}$.

One can see in figure 2 that coincidence between calculated field distribution and its interpolation is good and obtained distribution of the field is a smooth function that allows to use proposed method for calculation of further reflections of the field in the resonator avoiding multifold integration.

The spatial distributions of DR on the mirror surfaces has azimuthal symmetry as it was expected from the problem geometry [7]. Both distributions have minimum in the center and one maximum, i.e. one could observe double-lobe distribution expected from the characteristics of DR. The position of maximum differs for flat mirror and concave mirror due to the fact that concave mirror naturally focuses the radiation on the surface of the flat mirror.

3. Gaussian-Laguerre modes in semi-confocal resonator

In a resonator electromagnetic wave is confined by its mirrors. In this case it could be described by superposition of plane waves or transverse electromagnetic modes which occurs in the resonator. Different solutions for modes are used for different types of symmetry. Gaussian-Laguerre modes (GLM) are used for calculation of transverse distributions of radiation stored in confocal resonators with axial symmetry. In order to apply these modes to our problem we "expanded" our semi-confocal resonator and assumed that we have confocal resonator that consists of two focusing mirrors with curvature radii equal to $R = 840\,\text{mm}$. The intermirror

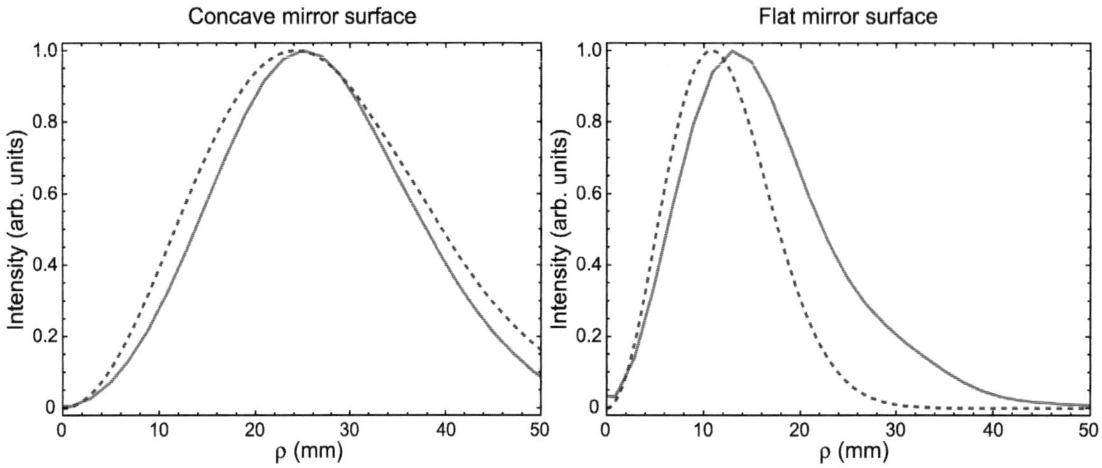

Figure 3. Comparison of radial distribution of GLM (blue dashed curve) and DR in the resonator after 15 electrons (red solid curve). The calculation was carried out at $\lambda = 3.5\,\mathrm{mm}$

distance is also equal to $L = 840\,\mathrm{mm}$. Using Gaussian-Laguerre modes one could estimate the radiation spatial distribution on the surface of focusing mirror and in the center of resonator where the radiation waist is situated.

According to [11] the amplitude of GLM m, n in the radiation waist could be described as following:

$$\Psi^0_{mn}(\rho,\varphi) = \left(\frac{\rho\sqrt{2}}{w_0}\right)^m \exp\left[-\frac{\rho^2}{w_0^2}\right] L_n^m\left(-\frac{2\rho^2}{w_0^2}\right) e^{im\varphi}. \tag{8}$$

Here $w_0 = \sqrt{\frac{L\lambda}{4\pi}}$, L_n^m is the generalized Laguerre polynomial.

Fresnel transformation of GLM allows to obtain amplitude of radiation at any distance z from the waist position inside resonator:

$$\begin{aligned}\Psi_{mn}(\rho,\varphi,z) = \frac{w_0}{w(z)}\left(\frac{\rho\sqrt{2}}{w(z)}\right)^m \times \\ \times \exp\left[-\frac{\rho^2}{w^2(z)} + \frac{ik\rho^2}{2R(z)} + im\varphi - i(2n+m+1)\arctan\left(\frac{4z}{L}\right)\right] L_n^m\left(\frac{2\rho^2}{w^2(z)}\right).\end{aligned} \tag{9}$$

Here $w(z) = w_0\sqrt{1+\left(\frac{4z}{L}\right)^2}$, $R(z) = z\left[1+\left(\frac{L}{4z}\right)^2\right]$, $k = 2\pi\lambda^{-1}$.

Due to the fact that the DR distribution has double-lobe structure [7] we took into account only GLM Ψ_{m0}. Figure 3 shows comparison of radial distribution of GLM and DR calculated using equations (6) for 15 point-like electron bunches on the surfaces of both mirrors. GLM on the surface of concave mirror was calculated in the form $|\Psi_{10}|^2$ assuming $z = 420\,\mathrm{mm}$. GLM on the surface of flat mirror was calculated in the form $|\Psi^0_{10}|^2$. During the calculation we keep $\lambda = 3.5\,\mathrm{mm}$.

In figure 3 one can see that radial distribution of DR calculated on the surface of concave mirror coincides rather well with Gaussian-Laguerre mode $|\Psi_{10}|^2$. Coincidence of radial distributions of DR and GLM $|\Psi^0_{10}|^2$ on the surface of flat mirror is not so good. The GLM distribution is narrower than DR distribution, however peak positions are close to each other. Thus, one could say that use of GLM gives reasonable agreement with DR approach and could be used for resonator investigation.

4. Quality factor of semi-confocal resonator

During experimental investigation of coherent DR generation in semi-confocal resonator its quality factor (Q-factor) was defined to be equal to $Q = 72.9$ [6]. The Q-factor was defined using the following procedure. After all electron bunches leaved the resonator the radiation stored in it decayed following exponential law. The part of radiation that came out the resonator from the opening in the flat mirror was measured as function of time and was observed as a peak structure with time distance between peaks equal to distance between electron bunches ($\Delta t = 2.8\,\mathrm{ns}$). In order to estimate Q-factor exponential fit of this signal tail was done by the function $A(t) = A_0 e^{-t/t_0}$, where A_0 is free scaling fit parameter and t_0 defines Q-factor in the following form [6]:

$$Q = 2\pi \frac{t_0}{\Delta t}. \tag{10}$$

In our theoretical investigation we tried to estimate the value of Q-factor for the semi-confocal resonator using both DR approach and Gaussian-Laguerre mode approach. During the experiment [6] the flat mirror consisted of glass substrate with reflecting coating and had the opening diameter $a_0 = 5\,\mathrm{mm}$. However, the reflecting layer was deposited not on the whole substrate surface. The area inside the circle with diameter 15 mm was not deposited and thus consisted of pure glass. Using both theoretical approaches we could not take into account this part with no reflecting coating. In order to estimate resonator Q-factor we assumed two cases. In the first case the opening diameter was assumed to be equal to 5 mm that should result in the upper limit of Q-factor. In the second case the opening diameter was assumed to be equal to 15 mm that should result in the lower limit of Q-factor.

In order to estimate Q-factor using DR approach we calculated a train consisted of 15 point-like electron bunches. The DR was calculated using equations (6) at radiation wavelength equal to $\lambda = 3.5\,\mathrm{mm}$ for both opening diameters. After each third bunch starting from third one 25 relaxation free runs were calculated using again equations (6) but assuming that electron field is absent ($\mathbf{E_e}(\mathbf{r}, \omega) = 0$). In order to calculate the part of radiation that comes out of the resonator the DR intensity distribution on the flat mirror surface was integrated over the opening area. This value was assumed to be a "signal" that was measured over time. Figure 4 show an example of radiation yield vs. time that was calculated for $a_0 = 15\,\mathrm{mm}$. Cavity free runs were calculated

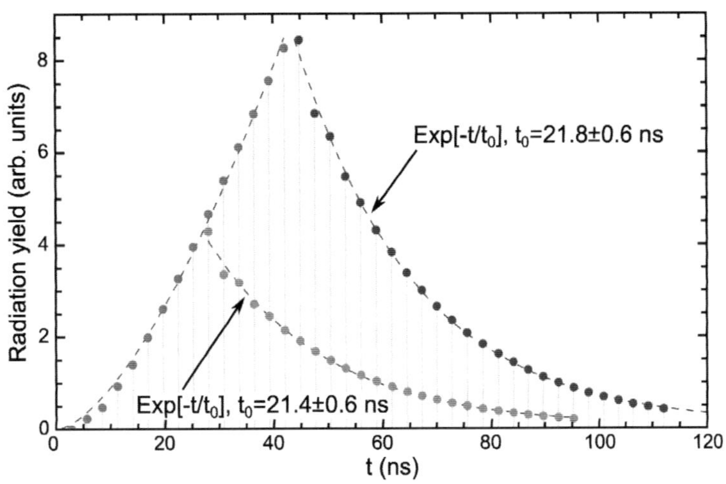

Figure 4. Example of time dependence of radiation yield. Red dots – radiation yield electrons, blue dots – resonator free runs after 15 bunches, green dots – resonator free runs after 9 bunches. The calculation was carried out at $\lambda = 3.5\,\mathrm{mm}$ and $a_0 = 15\,\mathrm{mm}$

for 9 electron bunches in the train and 15 bunches in the train.

In the case of DR calculation approach the resonator Q-factor was estimated exactly like it was done during the experiment [6]. The values of resonator Q-factor calculated at $a_0 = 5\,\mathrm{mm}$ are shown in figure 5 by blue triangles and at $a_0 = 15\,\mathrm{mm}$ – by red dots. The errors shown in figure 5 are errors caused by fit errors of parameter t_0. It is obvious that Q-factor should not depend on number of electron bunches in the train. In order to demonstrate this the value of Q-factor in figure 5 was plotted versus number of bunches, i.e. resonator free runs were calculated after each third bunch. The resonator Q-factor value $Q = 72.9$ measured experimentally is shown by black solid line.

Cavity Q-factor could be also calculated using GLM as following:

$$Q = 2\pi \frac{1}{c_\Sigma}, \tag{11}$$

where c_Σ is the total losses of energy in the resonator. In our case we took into account geometrical losses (c_G) and diffraction losses (c_D). According to [12]:

$$c_\Sigma = 1 - (1 - c_D^f)(1 - c_D^c), \tag{12}$$

where indexes f, c define flat and concave target, respectively.

According to [13], diffraction losses could be estimated as:

$$c_D = 1 - (1 - c_G)^2 \tag{13}$$

Geometrical losses could be calculated due to the fact that we know spatial distributions of

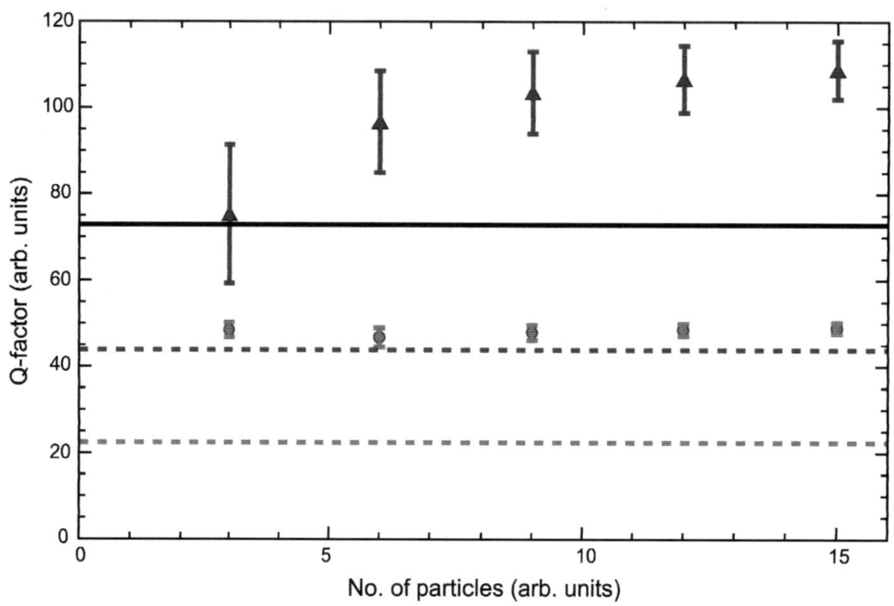

Figure 5. Dependence of the resonator Q-factor on number of electrons. Red dots – DR approach, $a_0 = 15\,\mathrm{mm}$; blue triangles – DR approach, $a_0 = 5\,\mathrm{mm}$; red dashed line – GLM approach, $a_0 = 15\,\mathrm{mm}$, blue dashed line – GLM approach, $a_0 = 5\,\mathrm{mm}$; black solid line – $Q = 72.9$, experimental result [6]

radiation intensity on the mirror surfaces [14]:

$$
\begin{aligned}
c_G^f &= \frac{\int_0^{a_0/2} \rho d\rho \int_0^{2\pi} |\Psi_{10}^0|^2 d\varphi + \int_{a/2}^{\infty} \rho d\rho \int_0^{2\pi} |\Psi_{10}^0|^2 d\varphi}{\int_0^{\infty} \rho d\rho \int_0^{2\pi} |\Psi_{10}^0|^2 d\varphi}, \\
c_G^c &= \frac{\int_0^{a_1/2} \rho d\rho \int_0^{2\pi} |\Psi_{10}|^2 d\varphi + \int_{a/2}^{\infty} \rho d\rho \int_0^{2\pi} |\Psi_{10}|^2 d\varphi}{\int_0^{\infty} \rho d\rho \int_0^{2\pi} |\Psi_{10}|^2 d\varphi},
\end{aligned}
\tag{14}
$$

where $a_0 = 5\,\mathrm{mm}$ or $a_0 = 15\,\mathrm{mm}$ (see further), $a_1 = 5\,\mathrm{mm}$, $a = 100\,\mathrm{mm}$.

Combining equations (11)–(13) one could obtain resonator Q-factor value. In figure 5 the calculated values of Q-factor are shown by red dashed line at $a_0 = 15\,\mathrm{mm}$ ($Q = 22.5$) and by blue dashed line at $a_0 = 5\,\mathrm{mm}$ ($Q = 43.9$).

5. Conclusion

In conclusion we would like to note that the well-known theory of Gaussian beams in open resonators and developed computer code allowing to simulate DR characteristics in a semi-confocal resonator give the coinciding results for transversal field components. This fact confirms the validity of the code based on the generalized surface current model [8]. The unified formulation for DR generation in resonator mirrors and the radiation storage can be applied for calculation of DR characteristics at any resonator configuration including the confocal one. The preliminary estimations of Q-factor for such a resonator show a possibility to achieve much higher values of Q-factor than in the semi-confocal resonator considered above.

Acknowledgments

The work was partially supported by the Russian Ministry of Education and Science within the program "Nauka", Grant No. 3.709.2014/K an by RFBR grant No. 15-52-50028.

References

[1] Dattoli D, Renieri A and Torre A 1993 *Lectures on Free Electron Laser Theory and Related Topics* (Singapore: World Scientific)
[2] Gover A, Dyumin E 2006 *Proceedings of Free Electron Laser conference FEL 2006 (Berlin, Germany)* p 1
[3] chi-Lihn H, Kung P, Settakorn C, Wiedemann H and Bocek D 1996 *Phys.Rev.Lett.* **76** 4163
[4] Shibata Y et al 1997 *Phys.Rev.Lett.* **78** 2740
[5] Sasaki S, Shibata Y, Ishi K, Ohsaka T, Kondo Y, Hinode F, Matsuyama T and Oyamada M 2002 *Nucl. Instr. Meth. A* **483** 209
[6] Aryshev A et al 2014 *Nucl. Instr. Meth. A* **763** 424
[7] Potylitsyn A, Ryazanov M, Strikhanov M and Tishchenko A 2010 *Diffraction Radiation from Relativistic Particles* (Berlin: Springer)
[8] Karlovets D V and Potylitsyn A P 2009 *Phys.Lett.A* **373** 1988
[9] Naumenko G A, Cha V A, Kalinin B N, Popov Yu A, Potylitsyn A P, Saruev G A and Sukhikh L G 2008 *Nucl. Instr. Meth. B* **266** 3733
[10] http://www.wolfram.com/
[11] Allen L, Beijersbergen W, Spreeuw R J C and Woerdman J P 1992 *Phys. Rev. A* **45** 8185
[12] Kubarev V V 2000 *Kvantovaya Elektronika* **30** 826
[13] Kubarev V V, Persov B Z, Vinokurov N A, Davidov A V 2004 *Nucl. Instr. Meth. A* **528** 199
[14] Gurov M G and Dmitriev A K 2009 *Russian Physics Journal* **52** 464

Coherent forward and backward diffraction radiation of relativistic electrons in a dielectric targets

G A Naumenko, M V Shevelev and Yu A Popov
Tomsk Polytechnic University, Tomsk, Russia

E-mail: naumenko@tpu.ru

Abstract. During the interaction of the relativistic electrons field with a dielectric target various types of electromagnetic radiation, such as Cerenkov radiation, diffraction radiation, transition radiation can be generated. In this report we present the results of experimental studies of the diffraction radiation generated by relativistic electrons in a dielectric target at the interface vacuum-insulator and insulator-conductor in the millimeter wavelength range. The experimental results show that the component of the diffraction radiation of relativistic electrons at the interface insulator-conductor, for any significant refractive index of insulator, is suppressed. The analysis of the results from different points of view was done.

1. Introduction

The interaction of relativistic particles with the material of the dielectric target can generate various types of electromagnetic radiation, such as Cerenkov radiation (ChR), diffraction radiation (DR), the transition radiation (TR). The nature of the radiation depends essentially on the characteristics of the materials that define the properties of the boundary between the media. In the case of perfect dielectrics, the main characteristic is a real part of the permittivity or refractive index. DR, by definition, is a radiation generated when the particles move near the target. It should be noted, that ChR might be generated, when particles move close to a target, not crossing it. For example, in [1] the forward DR and ChR in such conditions were simultaneously registered.

It is shown in [2] that TR and DR are of the same nature – radiation generated due to polarization of the target material by the electromagnetic field of moving electron. In the case of DR, only part of the electron field crossing the target is involved in the process of target material polarization. The theoretical analysis and experimental research on planar dielectric TR targets are well known [3, 4]. In this case, mainly TR on the upstream and downstream surface of the target is considered. The resulting radiation has characteristic properties, depending on the thickness of the target, its dielectric constant, and the electron energy [5]. The theoretical model of the generation of TR and DR for a charged particle moving near the edge of the screen of finite size and a finite conductivity with frequency dispersion has been developed in [6, 7].

In the case of relativistic electrons, usage of pseudo-photon approximation [8] is useful for an intuitive understanding of the processes during an interaction of the electron field with a substance. In [9] this viewpoint is called "the virtual photon method". In this case, the longitudinal component of the electron field is negligible (i.e., the field is nearly transversal), and the speed is nearly equal to the light speed. So, the field of the electron in the interaction with the matter has locally nearly the same properties as the radiation field. During an interaction, a target "can not distinguish" the electron field

Content from this work may be used under the terms of the Creative Commons Attribution 3.0 licence. Any further distribution of this work must maintain attribution to the author(s) and the title of the work, journal citation and DOI.
Published under licence by IOP Publishing Ltd

from radiation field and gives the same response. From this point of view, DR and TR represent refraction and reflection of an electron field at the interface of target substance.

Next, we consider radiation produced during the passage of an electron through a dielectric plate of finite thickness as well as radiation produced during the moving of an electron near edge of a target. For definiteness hereinafter the surface at the side of the electron source will be called an upstream surface, and the opposite surface will be called a downstream surface. Moreover, forward radiation will be called a radiation in semisphere in the direction of the electron motion, and the backward radiation will be called a radiation in the semisphere towards the electron source. In case of DR electrons are not scattered by the target material. This simplifies the procedure of the experiment and does not require taking into account the calculations of scattering. Therefore, we focus our attention at the DR in a dielectric targets.

Intuitively, if we screen the downstream surface of the flat dielectric target (figure 1) using a conductive plate, which is a mirror in the investigated wavelength range, then a backward DR (pseudo-photon field of the electron, reflected from the mirror) on the border between two media (insulator-metal) can be generated.

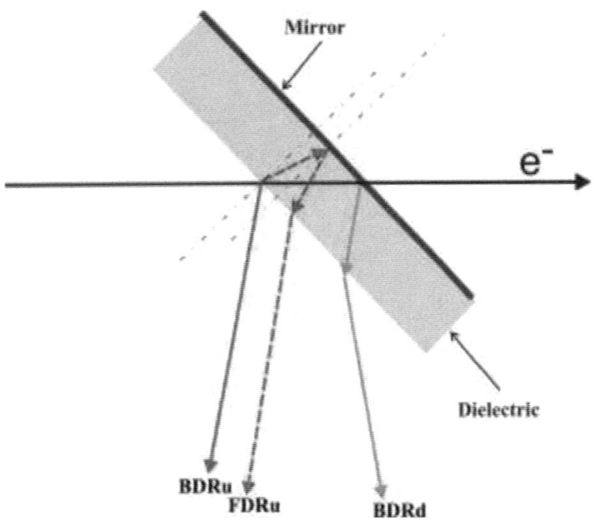

Figure 1. Scheme of DR components separation on the planar dielectric target.

During the passage of relativistic electrons above the target, in geometry shown in figure 1, the following radiation components can be generated:

a) backward DR from the upstream surface of the target (BDRu);

b) forward DR from the upstream surface (FDRu), which is reflected from a mirror, refracted on the upstream surface, and propagates out of the target;

c) backward DR at the insulator-metal border (BDRd), which is also refracted at the upstream surface and propagates out of a target;

d) ChR, which is reflected from the mirror and by the refraction on the upstream surface of the target will be released.

This geometry is interesting because all of these components can be measured in the same conditions. However, it should be noted that BDRu and FDRu have the same direction and cannot be separated in the angular distribution.

But, if we use not a flat target, but a dielectric wedge (figure 2), we can separate these components.

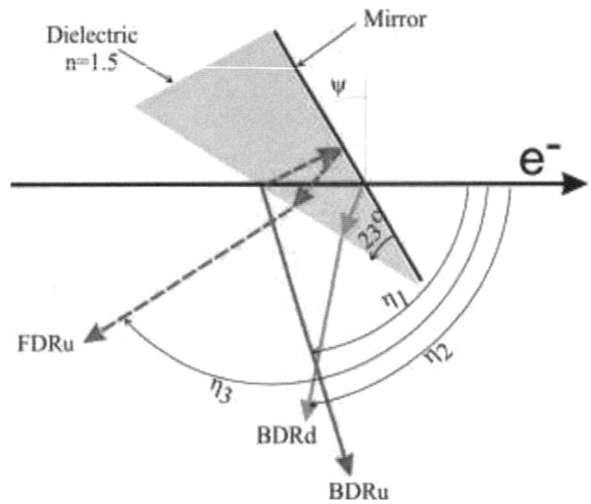

Figure 2. Scheme of separation of the DR components from the dielectric wedge.

The exact theoretical solution of this problem for the described geometry of DR target, which can be used in the experiment, is very difficult due to a complicate border conditions. The goal of the experiment, which will be described below, is to measure and compare the intensity of DR components a, b and c.

2. Experiment

The experiment was performed on the extracted electron beam of microtron PTI TPU with the following parameters: electron energy – 6.1 MeV, macro-pulse duration – $4\,\mu s$, the macro-pulse frequency $1-8$ Hz, the characteristic length of the electron bunch in the Gaussian approximation $\sigma_z = 2.4$ mm, the bunch population $Ne = 10^8$, the number of bunches in macro-pulse – 10^4, the cross-section of the extracted beam – 4×2 mm, the angular divergence of the extracted beam – 0.08 rad.

Under these conditions for $\lambda > 9$ mm radiation of electron bunch is coherent [10]. Because of the coherency the radiation intensity increases by *Ne* times (i.e. for the specified bunch population it is by 8 orders) in comparison with incoherent radiation, making it available for the measurement using existing detectors operating at room temperature. For radiation registration, we used broadband detector DP-21M, manufactured in Semi-conductive Devices Research Institute (Tomsk, Russia) on the basis of a broadband antenna with a threshold diode. The detector ensures the detection of radiation at room temperature with a sensitivity of approximately 0.3mV/W in wavelength region $\lambda = 0.5-25$ mm with accuracy $\pm 15\%$.

To suppress the accelerator RF system background the beyond cut off waveguide with diameter 10 mm was mounted on the detector, which attenuates a radiation having a wavelength $\lambda > 17$ mm. Thus, the spectral range of the measurement is $\lambda = 9-17$ mm.

The angular characteristics of the radiation were measured using the parabolic telescope (figure 3) which is a detector, located at the focus of a parabolic mirror, placed on the rod. The rod may be rotated around an axis passing through the center of the radiating face of the wedge. In this case, the measured angular radiation characteristics coincide with the characteristic measured in the far field zone [11] for comparison with theoretical calculations.

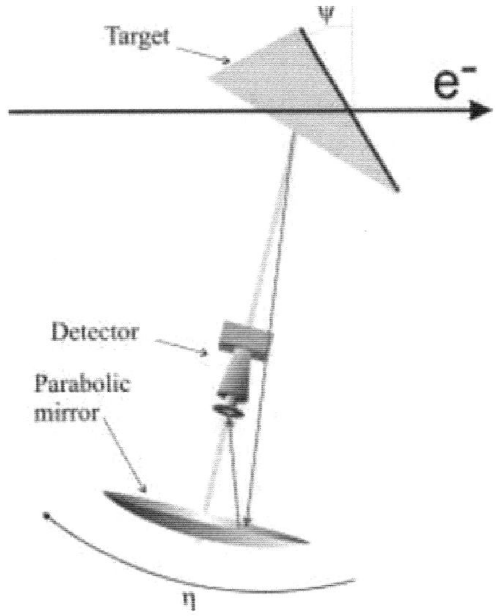

Figure 3. Scheme of the experiment.

The angular resolution of a telescope is 0.035 radians. If the measurement of the angular distribution with this resolution is performed in the horizontal plane (in the plane of figure 3), we will register only the vertical component of the radiation polarization. So we can eliminate the contribution of ChR in the angular dependence, because this component of ChR propagates in a vertical plane at the Cherenkov angle in respect the horizontal plane. Thus, we can measure the radiation characteristics of only components DR, types a, b and c.

For investigation we chose the wedge from paraffin with the opening angle of 23 degrees and with the refractive index of n = 1.5. The size of the diagonal of the wedge was 200 mm. The refractive index of this target was measured using a beam from millimeter wavelength emitter. The target has been tested on absorption of radiation in investigated wavelength region. It was found that absorption is less than 5%. Impact-parameter of the target was 15 mm. Orientation of the wedge is chosen so that the peaks of all three components of the angular distribution of the radiation separated. For this purpose, using the pseudo-photon point of view, the emission angle η from the target for each radiation component as a function of the target orientation angle ψ was calculated (figure 4).

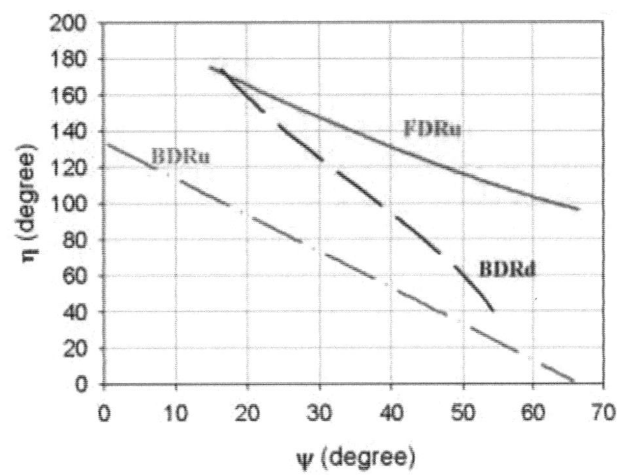

Figure 4. Calculated dependences of a propagation angle η of DR components on orientation angles ψ of the target.

Technically the range of observation angles η is limited between $95° - 145°$.

TR and DR are special cases of polarization radiation, which is a radiation of target material, polarizing by a Coulomb field of a relativistic electron. In case of DR in contrast to TR, only part of electron field takes part in polarization of target material, but the process of radiation is the same. The same nature of TR and DR for conductive targets has been shown convincingly in [2]. So we can suppose the same similarity for radiation in dielectric targets. Unfortunately, no analytical solutions using appropriate theoretical models of DR for combined dielectric–metal targets with geometry used in our experiment were in theory considered. Therefore we use for estimations the exact solution from [4] for forward and backward TR in dielectric targets. These estimates showed that for $n = 1.5$ the intensity of backward radiation is negligible in comparison to the forward radiation. So we will not be interested in the radiation component BDRu (see figure 2) and we chose for the measurement the target orientation angles $\psi = 35°$ and $\psi = 40°$. In this case both FDRu and BDRd may be observed in the same η - scan in our limits of $\eta = 95° - 145°$.

Figure 5 shows the measured angular dependences of the radiation intensity for these angles of orientation.

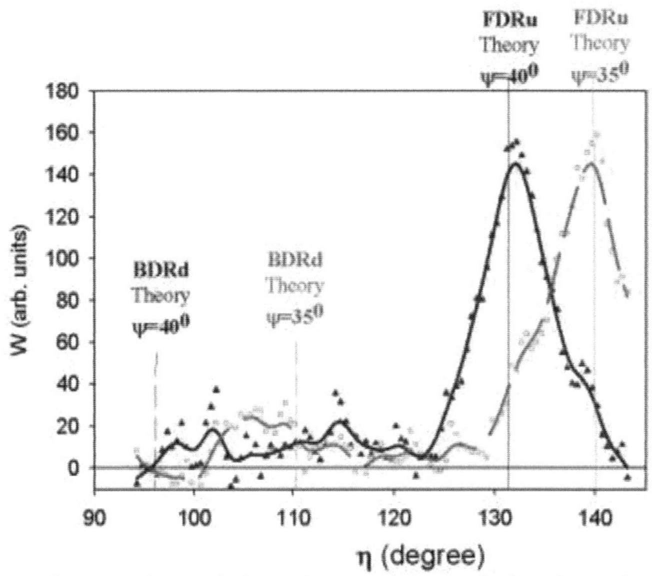

Figure 5. The experimental dependence of the radiation intensity W on the observation angle η for different angles ψ of the target orientation. Dots are the experimental values and lines represent the smoothed experimental data. Vertical lines represent calculated positions of the peak maximum.

These results show that the component BDRd is practically absent outside the statistical error. That is, the pseudo-photon electron field in the investigated spectral range is not reflected from the mirror. This result is totally unexpected and requires careful analysis.

3. Discussion

As was mentioned above, no solutions using appropriate theoretical models of DR for combined dielectric–metal targets with geometry used in our experiment were in theory considered. Using the approximation of similarity of TR and DR mentioned above, we attempt to use for the qualitative analysis of these results the expression (2.35) from [4] for the backward transition radiation at the interface between two semi-infinite media:

$$W(\theta_1)=\frac{q^2v^2}{\pi^2c^3}\sin^2\theta_1\cos^2\theta_1\sqrt{\varepsilon_1}\left|\varepsilon_2-\varepsilon_1\right|^2$$
$$\times\left|\frac{\left[1-\left(v^2/c^2\right)\varepsilon_1+\left(v/c\right)\sqrt{\varepsilon_2-\varepsilon_1\sin^2\theta_1}\right]}{\left[1-\left(v^2/c^2\right)\varepsilon_1\cdot\cos^2\theta_1\right]\left[1+\left(v/c\right)\sqrt{\varepsilon_2-\varepsilon_1\sin^2\theta_1}\right]\left(\varepsilon_2\cos\theta_1+\sqrt{\varepsilon_1\left(\varepsilon_2-\varepsilon_1\sin^2\theta_1\right)}\right)}\right|^2.$$

Here all the notations are the same as they were in the source. We are interested in the case where the second medium is an ideal metallic mirror, i.e. $\varepsilon_2\to\infty$. Then, this expression takes the form:

$$W(\theta_1)=\frac{q^2}{\pi^2c}\frac{\beta^2\sqrt{\varepsilon_1}\sin^2\theta_1}{\left|1-\beta^2\varepsilon_1\cos^2\theta_1\right|^2},\tag{1}$$

where $\beta=v/c$.

In the vacuum $(\varepsilon_1=1)$ from (1) we obtain the expression for the TR:

$$W(\theta_1)=\frac{q^2}{\pi^2c}\frac{\beta^2\sin^2\theta_1}{\left|1-\beta^2\cos^2\theta_1\right|^2}.$$

On the other hand for $n_1=(\beta\cos\theta_1)^{-1}$, the expression (1), where $n_1=\sqrt{\varepsilon_1}$ is a refractive index, has a pole of Cherenkov radiation, generated in the target and reflected from the mirror. Analyzing the angular distribution of radiation from (1) we can see that at the insulator-metal transition the TR will be generated only if $1\le n_1<(\beta\cos\theta_1)^{-1}$ (we do not consider here a negative refractive index). For our experiment this condition reduces to $1\le n_1<1.003$. That is, our target with $n_1=1.5$ in these conditions may generate at the metal-insulator transition in the studied spectral range Cherenkov radiation only, which is suppressed in the experiment due to selected polarization of detected radiation. As the principle of reflection of electron field from conductive screen for right angle of incidence and for an inclined angle of incidence of electrons is the same, we may expect that for inclined angle of incidence of electrons the backward TR will be also negligible. Hence, given the same nature of generation of TR and DR, it can be argued that the experimental results are consistent with the classical theory.

The problem of Maxwell's equations is that we may obtain in some cases the accurate final decisions, but these solutions do not provide information about the physics of the processes occurring in the interaction of relativistic particles with matter (in a sense – "black box"). For example, in [14] is shown the solution of Maxwell's equations for evolution of electromagnetic field generated by a relativistic electron downstream to the conductive screen. However, no information about a nature of this radiation may be extracted from this solution.

As mentioned, the pseudo-photon method may be useful for the intuitive understanding of the processes taking place in the interaction of an electron field with a substance. According to this approach the field of the electron in the interaction with the matter has locally nearly the same properties as the radiation field. A forward DR and TR on the upstream surface of the target is a refraction of the electron field on this surface. So the direction of propagation of an electron field like a direction of propagation of usual plane waves in the medium with a refractive index $n_1>1$, is different from the direction of the electron velocity. Hence it follows logically, that the field of electron and the electron itself are separated, and the electron moves without some part of its Coulomb field in the investigated spectral range. The separate electron field propagates further in the form of real photons. This field is reflected from the mirror, refracted on the upstream surface of the target, and will be registered as FDRu. The similar case of separation of an electron field and an electron itself, when electrons are scattered for large angles, is considered in the work of E.L. Feinberg [12], which introduces the concept of "half-naked electron" (electron deprived of part of its Coulomb field). This view is also used in [15, 16]. In this case, at the input of the insulator-metal transition (downstream face of the target) in the geometry shown in figure 2, the electron become "half-naked" in the investigated spectral range, and cannot generate backward DR on this transition. That is, the radiation BDRd should be absent, as is observed in the experiment.

In this paper we considered the problem of interaction of electron Coulomb field with combined dielectric-metal target from two points of view: classical viewpoint and pseudo-photon viewpoint. There is no contradiction between these points of view and experimental results. However the pseudo-photon point of view allows us to understand more in detail the processes occurring in the interaction of relativistic electrons with matter. In [12] the interpretation of the processes based on "half-naked electron" is considered as a possible, but not the only interpretation of the mechanism of interaction of a charge field with the substance. However, in [13, 14] was shown that opposite interpretation of these phenomena is baseless. Thus, the interpretation based on the concept of "half-naked electron" is the only one.

The importance of this concept in consideration of the interaction of the Coulomb field of relativistic charged particles with matter is that this viewpoint raises a number of questions about the nature of the charge and its Coulomb field and it can be considered as an example for the application of this approach to other types of interactions (such a strong interactions in [12]).

Acknowledgment
This work was supported by the Ministry of education and science of the Russian Federation Grant number 3761.

References

[1] Naumenko G A, Potylitsyn A P, Shevelev M V and Popov Yu A 2011 *JETP Lett.* **94** 258
[2] Potylitsyn A P 1998 *Nucl. Instrum. and Methods* B **145** 169
[3] Denisov S P 1997 *Soros. Obrazov. Zh.* **3** 124
[4] Ginzburg V L and Tsytovich V N 1990 *Transition Radiation and Transition Scattering* (Bristol: Adam Hilger)
[5] Ter-Mikaelian M L 2001 *Phys.-Usp.* **44** 571
[6] Karlovets D V and Potylitsyn A P 2009 *JETP Lett.* **90** 326
[7] Karlovets D V 2011 *J. Exp. Theor. Phys.* **113** 27
[8] Ter-Mikaelian M L 1972 *High Energy Electromagnetic Processes in Condensed Media* (New York: Wiley-Interscience)
[9] Jackson J D 1998 *Classical Electrodynamics*, 3rd ed (New York: J. Willey & Sons)
[10] Aleinik A N, Aryshev A S, Kalinin B N *et al* 2002 *JETP Lett.* **76** 337
[11] Kalinin B N, Naumenko G A, Potylitsyn A P *et al* 2006 *JETP Lett.* **84** 110
[12] Feĭnberg E L 1979 *Sov. Phys. Usp.* **22** 479
[13] Naumenko G A, Potylitsyn A P, Shevelev M V, Soboleva V V and Bleko V V 2013 *JETP Lett.* **98** 593
[14] Naumenko G A, Popov Yu A and Shevelev M V 2012 *J. Phys: Conf. Ser.* **357** 012005
[15] Shul'ga N F and Syshchenko V V 2000 *J. Phys. of Atomic Nuclei* **11** 2018
[16] Artru X and Ray C 2007 *Proc. SPIE: Int. Conf. on Charged and Neutral Particles Channeling Phenomena II* vol 6634 (Rome, Italy) p 66340Y-1

Computer simulation of electron-positron pair production by channeling radiation in amorphous converter

S V Abdrashitov[1,2], O V Bogdanov[1,2], S B Dabagov[3,4,5], Yu L Pivovarov[1] and T A Tukhfatullin[1]

[1] National Research Tomsk Polytechnic University, Lenin Ave. 30, 634050, Tomsk, Russia
[2] National Research Tomsk State University, Lenin Ave. 36, 634050, Tomsk, Russia
[3] INFN – Laboratori Nazionali di Frascati, Via E. Fermi 40, I-00044 Frascati (RM), Italy
[4] RAS PN Lebedev Physical Institute, Lenin Ave. 53, 119991, Moscow, Russia
[5] NRNU MEPhI, Kashirskoe highway 31, 115409, Moscow, Russia

E-mail: abdsv@tpu.ru

Abstract. We consider the radiator-converter approach at 200 MeV channeled electrons (the SPARC_LAB LNF facility energies) for the case of using W crystalline radiator and W amorphous converter. A comparison of the positron production by the axial channeling radiation and the bremsstrahlung is performed. The positron stopping in the convertor is studied by means of computer simulations. It is shown that for the maximum yield of positrons the thickness of the W amorphous converter should be taken 0.35 cm in the case of using the axial channeling radiation resulting to total yield of positrons 5 10^{-3} e^+/e^- and 0.71 cm in the case of using the bremsstrahlung resulting to total yield of positrons 3.3 10^{-3} e^+/e^-.

1. Introduction

The problem of the positron beam generation remains of interest during last decade, in connection with the physics of slow positrons [1], relativistic positronium atom beams [2–4], modern research of electron-positron plasma [5] and with the search for the effective positron source for the electron–positron colliders [6–9]. The positrons are usually produced via conversion of photons with an energy greater than 1 MeV into e^--e^+ pair in the field of a nucleus. Several schemes are suggested for the intense positron beam generation. Basically, the initial high energy electron beam is used for the photon beam generation in several different ways: bremsstrahlung (BS) [10], channeling radiation (CR) [11–12] or coherent bremsstrahlung (CB) [10, 13–14], the Compton scattering [15], and even undulator radiation [16]. A new technique involving the high-energy laser to irradiate solid gold and platinum targets [17] seems to be very promising for the neutral electron-positron plasma production but poorly applicable as a positron source for colliders. A comparison of these schemes is given in details in [18]. Most of all of these schemes are used as the source of the photons a multi-GeV electron beam: 1.2 GeV at the Institute for Nuclear Study, Tokyo, Japan [11, 13]; 3 and 8 GeV at KEK, Tsukuba, Japan [14] and electron beams on the SPS CERN transfer lines, Geneva, Switzerland [10, 15]; 46.6 GeV at the SLAC, USA [16].

Content from this work may be used under the terms of the Creative Commons Attribution 3.0 licence. Any further distribution of this work must maintain attribution to the author(s) and the title of the work, journal citation and DOI.
Published under licence by IOP Publishing Ltd

However, the features of the positron production by sub-GeV electron beam remain not described in detail. One of the important advantage of sub-GeV electron beam for the positron production is the avoiding the thermal degradation of the crystal wherein the CR photons are generated. The comparison of the $1 \div 3$ MeV positron yield from a thin amorphous W converter of 0.1 mm thickness produced by BS, by <100> axial and (110) planar CR in a 10 μm W crystal for the positron energy range of was carried out recently in [18]. Obviously, the positron energy spectrum after the W converter depends not only on incident photon spectrum, but also on positron energy loss in a radiator on the way to leave it. Therefore, in continuation of [18], here we perform a computer simulation of the positrons stopping in an amorphous W converter for the cases when one uses the <100> axial CR and BS from 200 MeV electrons. The principal scheme of hybrid positron source is shown in figure 1.

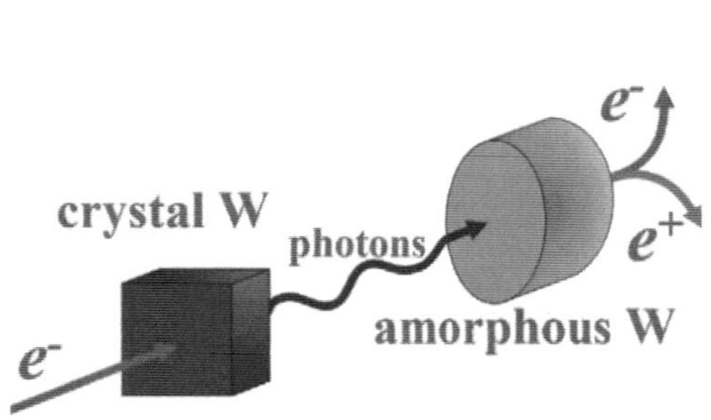

Figure 1. The scheme of hybrid positron source using CR from primary electron beam.

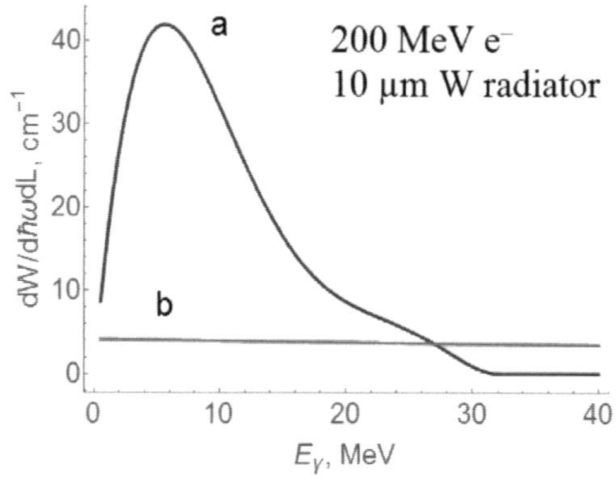

Figure 2. (a) CR intensity spectrum from 200 MeV electrons <100> axial channeling in W; (b) BS intensity spectrum from 200 MeV electrons in amorphous W.

2. Bremsstrahlung and channeling radiation spectra

In our simulations, the thickness of a W crystal (radiator) was10 μm, that is less than the dechanneling length for 200 MeV electrons [19]. The general properties of CR are well described in [20–21]. The realistic simulations of CR spectra can be performed using different models. For example, it can be done in the frame of binary collision model [22]. The faster way to simulate the CR intensity spectra (energy radiated during penetration through a crystal per unit crystal length) in the case of axial CR (figure 2a) from 200 MeV electrons in W is to apply the BCM-1.0 code [23]. This code was recently applied to simulate the orientation dependence of the CR total yield [18, 24]. The BS spectrum in an amorphous W (figure 2b) was calculated according to the Schiff formulae [25].

For the SPARC_LAB primary 200 MeV electron beam the brilliant peak of axial CR is located near 6 MeV, thus allowing creation of positrons with maximal kinetic energy up to 5 MeV. The total yield of BS is twice greater than that of axial CR [24], but the axial CR spectrum is almost 20 times narrower than the BS one and the maximum value of it is almost 10 times greater than the maximum for BS spectrum (figure 2).

3. Electron-positron pair production in a thick converter

For calculation of the e^--e^+ pair production by a photon in an atomic field we shall use the Bethe-Heitler formula [26–28].

The spectrum of positrons generated by CR from electrons in thin converter can be determined in the following way:

$$\frac{dN_P}{dE_P} = n \cdot L_C \int \frac{1}{E_\gamma} \frac{dW}{dE_\gamma dL} \cdot L \cdot \frac{d\sigma(Z, E_P, E_\gamma)}{dE_P} dE_\gamma, \quad (1)$$

where n is the number of atoms per volume unit of W convertor, L_C is the convertor thickness, Z is the atomic number of the converter material, E_P is the total energy of positron, E_γ is the energy of the photon, $dW / dE_\gamma dL$ is the photon intensity spectra, L = 10 μm – the thickness of the radiator and $d\sigma(Z, E_P, E_\gamma) / dE_P$ – the Bethe-Heitler e^--e^+ pair production cross-section by a photon [29].

The yield of positrons from conversion of CR or BS into e^--e^+ pair in thin convertor of thickness L_C is determined by the expression:

$$N_P = n \cdot L_C \iint \frac{1}{E_\gamma} \frac{dW}{dE_\gamma dL} \cdot L \cdot \frac{d\sigma(Z, E_P, E_\gamma)}{dE_P} dE_\gamma dE_P = \int \frac{dN_P}{dE_P} dE_P. \quad (2)$$

Computer simulations shows that the total yield of positrons (2) produced in a thin 0.1 mm W convertor by axial CR from 200 MeV electrons in a 10 μm W crystal is about 3.5 10^{-4} e^+/e^- and 1.6 10^{-4} e^+/e^- in the case of BS. The more detail on results of simulation of photoproduction of positrons in a thin W amorphous converter by CR and BS are presented in in [18].

Let divide a thick W converter into N thin layers of thickness 0.1 mm. The e^--e^+ pair production by the radiation in the thin layer of the converter numbered i is calculated according to (1) neglecting to the radiation attenuation and positron energy losses in the one layer under consideration. The attenuation of the radiation passed through converter from the first layer to the layer i-1 is described in terms of linear attenuation coefficient. The main contributions to radiation attenuation are coherent and incoherent scattering, pair production and photoelectric absorption and defined using XCOM: Photon Cross Sections Database [30]. The positrons energy losses in the converter from the i+1 layer to the Nth layer are described in terms of continuous slowing down approximation (CSDA). For the positron energies under consideration, CSDA ranges of the positrons in a matter practically coincide with one for the electrons [31]. CSDA ranges for the electrons are calculated using the CASINO code [32].

Our computer simulation carried out for the N=100 W layers shows (figure 3) that the total yield of positrons reaches a maximum value about 5.0 10^{-3} e^+/e^- at the thickness of converter equal to 0.35 cm in the case of photoproduction by axial CR from 200 MeV electrons in W crystal, and 3.3 10^{-3} e^+/e^- at 0.71 cm in the case of using BS.

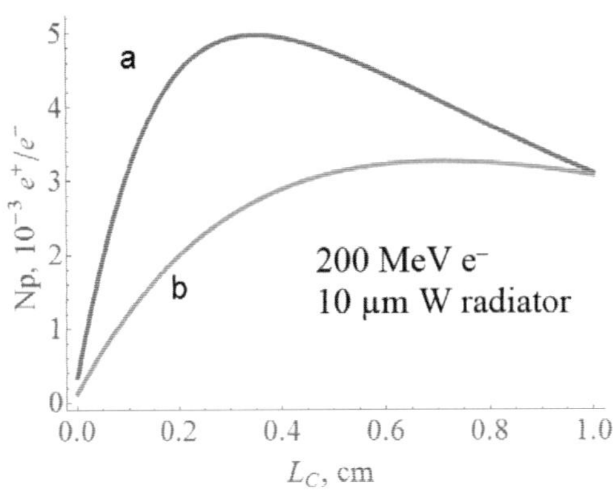

Figure 3. Total yield of positrons produced by the radiation from 200 MeV electrons in W as the function of the converter thickness: (a) axial CR; (b) BS

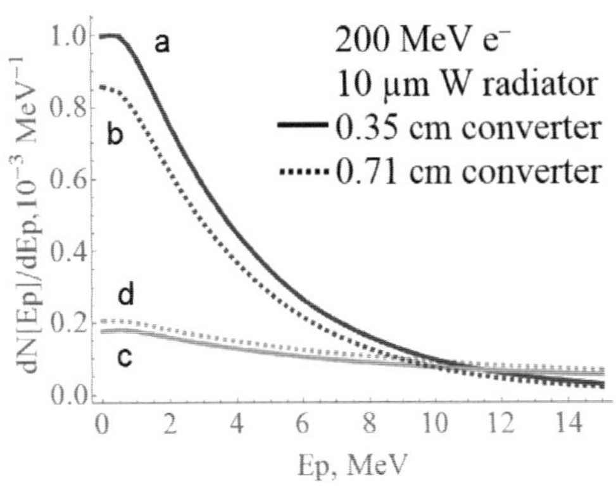

Figure 4. The energy spectra of positrons, produced by the radiation from 200 MeV electrons in W: (a, b) axial CR; (c, d) BS. Solid lines correspond to converter thickness 0.35 cm; dashed – 0.71 cm.

The energy spectrum of positrons produced by the axial CR from 200 MeV electrons in W is narrower than that produced by the BS (figure 4) for all thicknesses of converter. Moreover, in the case of axial CR it is possible to produce up to 5 time greater amount of low energy positrons than using the BS (figure 4a and 4d). These characteristics determine the benefits of the axial CR compared to BS.

4. Conclusions

Our previous studies [18] on hybrid positron source using channeling radiation and a thin W amorphous convertor are extended here to the case of more thick radiators. Two stages of simulations have been used: the first one is to calculate CR spectrum from initial 200 MeV electron beam in a 10 μm W radiator. The second stage is to take into account the created positrons energy loss in a downstream amorphous W convertor, in order to determine the final energy spectrum of positrons leaving a convertor.

The main new result is, that suggested hybrid scheme in the case of 0.35 cm amorphous W converter lead to the maximum total yield of positrons produced by axial CR from 200 MeV electrons in W crystal about 5.0 10^{-3} e^+/e^-, and 2.8 10^{-3} e^+/e^- in the case of using BS. The converter of greater thickness equals 0.71 cm allows to maximum total yield of positrons 3.3 10^{-3} e^+/e^- in the case BS and 4.1 10^{-3} e^+/e^- in the case of CR.

That means, for the SPARC_LAB electron beam parameters [33] and suggested hybrid scheme, in the case of 200 MeV electrons one can expect the maximum positron yield of about 2.3 $10^6 \div 10^7$ e^+/s in the case of BS (0.71 cm W converter) and about 3.5 $10^6 \div 10^7$ e^+/s in the case of <100> CR (0.35 cm W converter).

These numbers characterizing the total yield of positrons at SPARC_LAB electron beam can be increased using CR in a thicker radiator. In this case the influence of dechanneling on CR should be taken into account [34] as well as the contribution of coherent and incoherent bremsstrahlung [35] at channeling conditions.

Acknowledgements

The work was supported by the National Research Tomsk Polytechnic University (Excellence Program NR TPU), by the grant from the President of the Russian Federation Number MK-5202.2015.2, by the Ministry of Education and Science of the Russian Federation (Contract number 1.676.2014/K, program "Nauka") by The Tomsk State University Academic D.I. Mendeleev Fund Program in 2015-2016. The work was supported by Russian Fund for Basic Research according, project No. 16-32-00464 m_a.

The authors are grateful to Prof. X. Artru for fruitful discussion.

References

[1] Saitoh H, Stanja J, Stenson E V, Hergenhahn U, Niemann H, Sunn Pedersen T, Stoneking M R, Piochacz C and Hugenschmidt C 2015 *New J. Phys* **17** 103038
[2] Meshkov I N 1997 *Phys. Part. Nucl.* **28** 2 198
[3] Kunashenko Yu P 2005 *Nucl. Instr. and Meth.* B **229** 2 219
[4] Kunashenko Yu P and Pivovarov Yu L 2014 *J. of Phys.: Conf. Ser.* **517** 012007
[5] Zank G P and Greaves R G 1995 *Phys. Rev.* E **51** 6079
[6] Heuer R-D 2002 *Int. J. Mod. Phys.* A **17** 3469
[7] Biagini M E, Alesini D, Boni R *et al* 2010 The SuperB project: accelerator status and R&D *Proc. 2nd International Particle Accelerator Conference: IPAC'11 (San Sebastian, Spain, 4–9 September 2011)* pp. 3684-3686
[8] Rinolfi L, Chehab R, Dadoun O, Kamitani T, Strakhovenko V and Variola A 2013 *Nucl. Instr. and Meth.* B **309** 50
[9] Artru X, Chaikovska I, Chehab R *et al* 2015 *Nucl. Instr. Meth.* B **355** 60

[10] Chehab R, Cizeron R, Sylvia C *et al* 2002 *Phys. Lett.* B **525** 41
[11] Yoshida K, Goto K, Isshiki T *et al* 1998 *Phys. Rev. Lett.* **80** 7 1437
[12] Dolgikh V A, Kunashenko Yu P and Pivovarov Yu L 2003 *Nucl. Instr. Meth.* B **201** 253
[13] Kalinin B N, Naumenko G A, Potylitsin A P *et al* 1998 *Nucl. Instr. and Meth.* B **145** 209
[14] Inoue M, Takenaka S, Yoshida K *et al* 2001 *Nucl. Instr. and Meth.* B **173** 104
[15] Artru X, Chehab R, Chevallier M, Strakhovenko V M, Variola A and Vivoli A 2008 *Nucl. Instr. Meth.* B **266** 3868
[16] Alexander G, Barley J, Batygin Y *et al* 2008 *Phys. Rev. Lett.* **100** 210801
[17] Liang E, Clarke T, Henderson A *et al* 2015 Scientific Reports **5** 13968
[18] Abdrashitov S V, Bogdanov O V, Dabagov S B, Pivovarov Yu L and Tukhfatullin T A 2015 *Nucl. Instr. Meth.* B **355** 65
[19] Backe H, Kunz P, Laulh W *et al* 2008 *Nucl. Instr. Meth.* B **266** 3835
[20] Baier V N, Katkov V M and Strakhovenko V M 1998 *Electromagnetic Processes at High Energy in Oriented Single Crystals* (Singapore: World Scientific) p 554
[21] Dabagov S B and Zhevago N K 2008 *Riv. Nuovo Cimento* **31** 491
[22] Vyatkin E G, Pivovarov Yu L and Vorobiev S A 1987 *Nucl. Phys.* B **284** 509
[23] Bogdanov O V, Fiks E I, Korotchenko K B, Pivovarov Yu L and Tukhfatullin T A 2010 *J. of Phys.: Conf. Ser.* **236** 012029
[24] Abdrashitov S V, Bogdanov O V, Pivovarov Yu L and Tukhfatullin T A 2014 *Journal of Surface Investigation. X-ray, Synchrotron and Neutron Techniques* **8** 3 494
[25] Koch H W and Motz J W 1959 *Rev. Mod. Phys.* **31** 920
[26] Akhiezer A I and Berestetsky V B 1965 *Quantum Electrodynamics* (New York: Interscience Publishers)
[27] Kalinovskii A N, Mokhov N V and Nikitin Yu P 1989 *Passage of high-energy particles through matter* (New York: American Institute of Physics)
[28] Heitler W 1957 *The Quantum Theory of Radiation* (London: Oxford University Press)
[29] Dabagov S B, Abdrashitov S V, Bogdanov O V, Pivovarov Yu L and Tukhfatullin T A 2016 *Journal of Surface Investigation. X-ray, Synchrotron and Neutron Techniques* **10** 1 254
[30] Berger M J, Hubbell J H, Seltzer S M *et al* 2010 *XCOM: Photon Cross Section Database (version 1.5)* National Institute of Standards and Technology, Gaithersburg, MD.
[31] Salvat F, Fernandez-Varea J M, Sempau J 2011 Penelope-2011: A Code System for Monte Carlo Simulation of Electron and Photon Transport *Nuclear Energy Agency workshop proc. (Barcelona, Spain, 4–7 July 2011)* (Paris, OECD NEA)
[32] Drouin D, Couture A R, Joly D, Tastet X, Aimez V and Gauvin R 2007 *Scanning* **29** 3 92
[33] Ferrario M, Alesini D, Anania M *et al* 2013 *Nucl. Instr. Meth.* B **309** 183
[34] Bogdanov O V and Dabagov S B 2012 *J. of Phys.: Conf. Ser.* **357** 012029
[35] Bazylev V A and Zhevago N K 1987 *Radiation by Fast Particles in Matter and in External Fields* (in Russian) (Moscow: Nauka)

Multi-photon absorption in the channeling of electrons in an external field

VYaralov
Alikhanyan National Laboratory (Yerevan Physics Institute), Brothers Alikhanyan 2, Yerevan 0036, Armenia

E-mail: yaralov@mail.yerphi.am

Abstract. Following the methods developed for atom ionization by alternating electric field the probability of multi-photon absorption of photons of the strong external laser field by channeled electron (extraction of electron from the channel) have been calculated for different strengths of the monochromatic external field. The emission spectra of 54 MeV electron channeled in diamond crystal planes (110) are shown for different values of the resonant laser field of a frequency close to the transition frequency in the channel taking into account multi-photon absorption. It is shown that the multi-photon phenomena give some contribution to the total level width.

1. Introduction

The investigation of the interaction of a channeled electron and laser radiation is a natural development works associated with channeling [1-7]. In [8-11] the results of the theoretical predictions of splitting the spectrum of channeling radiation in a strong external field were presented. In quantum optics the effect of resonance external field on the spectrum of two-level atoms was sufficiently well studied both theoretically and experimentally. The spectrum is split into 3 peaks. The phenomenon is called dynamic Stark effect or "triplet Mollow" [12-14]. Nowadays this effect is being investigated by using ultra-short laser pulses with a field strength that exceeds the atomic [15-16]. Basically all the results were obtained in the semi-classical theory - a two-level quantum electron in a classical external field. The electron is described by the quantum-statistical method using the density matrix. For planar channeling the theoretical calculation is reduced to one-dimensional problem in the transverse direction with respect to the electron beam. In the theory of channeling were taken into account the effects leading to the departure of an electron from its level (de-channeling). That is: the scattering of electrons by atomic deviations from the ideal crystal structure; the scattering by the atom electrons; interaction with external field or vacuum field. The strong external field is another reason that reduces the lifetime of the considered two-level system. The number of photons is proportional to the square of the field strength. If the field is sufficiently strong, electron simultaneously interacts with many photons. The electron at the same time absorbs few resonant photons and escape from a potential well. The total energy of the photons must to exceed the extraction potential.

2. Multi-photon phenomena in the channeling

Photo-ionization of atoms was an object of investigation for a long time using different methods [17-20]. The theory developed in [17] combines multi-photon absorption with the tunnel effect. There is a certain probability of transition of the electron through the barrier. The given theory shows that the tunnel effect and multi-photon absorption are phenomena of the same nature. The input parameter is $\gamma = \omega\sqrt{2mI}/eF$ where ω, F – the frequency and strength of the external field, m, e – electron mass and charge, I – ionization potential (energy of the level). In the case of $\gamma << 1$ the general formula describes tunneling and for $\gamma >> 1$ it is multi-photon absorption.

Content from this work may be used under the terms of the Creative Commons Attribution 3.0 licence. Any further distribution of this work must maintain attribution to the author(s) and the title of the work, journal citation and DOI.
Published under licence by IOP Publishing Ltd

In the final state the interaction of the electron with the positively charged atomic ion was not taken into account. However this approach can be applied to calculate the probability of removal of an electron from a negative ion as well as to our case of channeling. In the final state only the external field affects the electron. There is no interaction between the ejected electron and averaged crystal potential. In [18] zero-radius-potential method is applied to calculate multi-photon absorption. For the short-range potential (smaller than the wavelengths of particles) the specific profile of potential is not essential and it is replaced by a one-dimensional delta function. Again, in the final state there is no interaction between the ejected electron and the positive residue. This one-dimensional model would be perfect for a plane-channeled electron. However, in the given model, the energy level of the electron in the channel (extraction potential) is significantly different from the real one. For a satisfactory agreement it is necessary to approximate the potential with the help of several delta functions. Therefore the calculations were made using formulas of work [17]. The probability of the process per unit time is as follows

$$W = \omega\left(\frac{\tilde{I}}{\hbar\omega}\right)^{1.5}\left(\frac{\gamma}{(1+\gamma^2)^{0.5}}\right)^{2.5} S\left(\gamma,\frac{\tilde{I}}{\hbar\omega}\right)\exp\left\{-\frac{2\tilde{I}}{\hbar\omega}\left[\operatorname{arsh}\gamma-\gamma\frac{(1+\gamma^2)^{0.5}}{1+2\gamma^2}\right]\right\}$$

$$S\left(\gamma,\frac{\tilde{I}}{\hbar\omega}\right)=\sum_{n=0}^{\infty}\exp\left\{-2A_n\left(\operatorname{arsh}\gamma-\frac{\gamma}{\sqrt{1+\gamma^2}}\right)\right\}\Phi\left\{\left[\frac{2\gamma}{\sqrt{1+\gamma^2}}A_n\right]^{0.5}\right\}$$

$$\tilde{I}=I\left(1+\frac{1}{2\gamma^2}\right)\qquad A_n=\operatorname{int}\left(\frac{\tilde{I}}{\hbar\omega}+1\right)-\frac{\tilde{I}}{\hbar\omega}+n\qquad \Phi(z)=\int_0^z \exp(y^2-z^2)\,dy$$

where I – extraction potential, i.e. the level energy taken with the opposite sign, $\tilde{I}$ – effective potential. Their difference is equal to the vibration energy of the electron in an external field. "int" means the integer part of the fraction.

3. Calculation results

All necessary parameters were calculated by the usual standard techniques developed in the theory of channeling. The calculated average potential of the plane-channeled electron is well approximated by an inverted parabola $V(x) = V_0\frac{x}{L}\left(2-\frac{x}{L}\right)$, where V_0 is well depth, L is half of the inter-planar spacing. The crystal field strength is $\frac{2V_0}{eL}\left(1-\frac{x}{L}\right)$, e is the electron charge. Let us consider 54 MeV electron channeled in the (110) plane of the diamond crystal. In this case in ħ = c = 1 system of units $V_0 = 23.6$ (eV), $L = 3.2{\times}10^{-4}(\mathrm{eV})^{-1}$, $e = 0.0854$. We are considering the radiation at the transition between the levels 4-3. The numbering of levels starts with the number 1. Transversal level energies $E_\perp$ are respectively: -7.325, -3.908 (eV). The maximum value of the crystal field strength is $F_{max}= 1.73{\times}10^6\ (\mathrm{eV})^2 = 7.5{\times}10^9$ (V/cm). The crystal field strength corresponding to level of 3.908 (eV) is equal to $6.85{\times}10^9$ (V/cm). These values greatly exceed the strength of the external field. Therefore we do not consider changes in the structure of the transverse electron levels. The transition frequency $\Delta E_\perp$ between selected levels is equal to 3.417 (eV). The frequency and wavelength of the external control field must be equal respectively $8.26{\times}10^{14}$(1/s), 363(nm). The required parameters of the laser field, however others. Let the electron moves at a speed of $v \approx c$ against the laser wave $E\cos[(n\omega/c)x+\omega t]$, where n is crystal refractive index. Suppose that at time t the electron is at the point $x = ct$. The wave at the same point and at the same time takes the form $E\cos[(n\omega/c)ct+\omega t] = E\cos[\omega(n+1)t]$. Thus the laser field acts on the electron with an effective frequency of $\omega(1+n)$. If the laser wave is directed at an angle of θ to the direction of the electron, the effective frequency is equal of $\omega(1+n/\cos\theta)$. Changing the angle θ one can change and select the desired effective frequency of the control field in a wide range. Bloch-momentum averaged parameters considered a two-level system are given in table 1 in ħ = c = 1 system of units.

Table 1. The parameters of two-level quantum system.

$\Delta E_{\perp}$(eV)	D(1/eV)	G_{scat}(eV)	G_{spont}(eV)
3.417	0.75×10^{-5}	0.255	3×10^{-8}

D is dipole momentum of the transition. The probability of spontaneous transition G_{spont} per unit time is much lesser than the probability of scattering G_{scat}. Therefore further spontaneous emission is not taken into account.

Let us now consider effective field for the selected pair of levels. Relevant parameters are given in table 2.

Table 2. The parameters of the interaction of channeled electron with effective laser field.

External field strength (V/cm)	Intensity (W/cm^2)	Rabi frequency (eV)	Rabi frequency (1/s)	Width due multi-photon absorption (eV)	Multi-photon absorption frequency (1/s)
0.5×10^8	3.3×10^{12}	0.17	0.258×10^{15}	0.023	0.35×10^{14}
10^8	1.3×10^{13}	0.35	0.532×10^{15}	0.063	0.96×10^{14}
2×10^8	5.3×10^{13}	0.69	1.048×10^{15}	0.114	1.73×10^{14}
3×10^8	1.3×10^{14}	1.04	1.58×10^{15}	0.116	1.76×10^{14}

Two-level channeled electron under the influence of the resonant external field oscillates between the levels with the Rabi frequency which is of the order of magnitude greater than the frequency of the multi-photon absorption. Keldysh parameter γ for various values of the field strength is respectively: 1.1; 0.55; 0.28; 0.18. Thus these external fields correspond to an intermediate region between a purely tunnel effect and a multi-photon absorption.

The spectra calculated with multi-photon absorption are shown in figure 1.

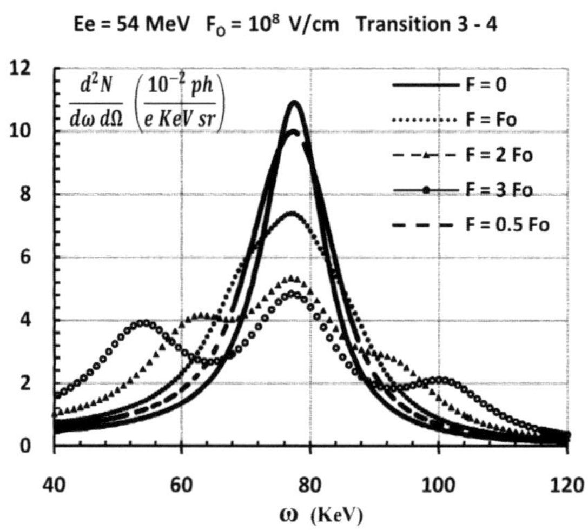

Figure1. The spectral distributions of the channeling radiation in external intensive field, taking into account multi-photon absorption for various strengths of the external resonance field F.

Discussion and conclusion

The field parameters listed in table 2 correspond to the selected two-level system (table 1). As noted above, the laser frequency should be lower than the transition frequency between the transverse channeled electron levels to provide these conditions. Due to the crystal polarization, strength of the laser field, on the contrary, should be higher than shown in table 2. Furthermore, only the force component perpendicular to the direction of electron counts. At the direction of the laser field at an angle it also leads to the need to increase the field strength. One can get the required field in the crystal with help such adaptations. Made calculations show that multi-photon absorption is making some contribution to the broadening of the level (less than the scattering) and is not a fundamental limitation on the possibility of observing the splitting of the emission spectra in the channeling of electrons in a crystal in a strong field. It would be interesting to test experimentally these predictions, but the main objective is to enhance radiation. Although the increase in the intensity of the control field leads to reducing of the central peak, its distribution is strictly centered on the frequency of the control field and is coherent with the control wave. If bunch of electrons is a small area compared to the wavelength of the control field, the electrons will radiate coherently.

Finally, it is appropriate to refer to the physical meaning of channeling radiation. Spontaneous radiation was omitted in the calculation. Allowance for spontaneous emission does not change the overall yield of radiation because of its smallness. The trajectory of the classical electron completely defines the radiation, regardless of the reasons for changing his movement. Similarly, the time-dependent density matrix of the channeling electron completely determines its radiation regardless of the reasons causing the change in its state. It turns out that the scattering plays a positive role in channeling radiation. This is the reason for the observed radiation. Multi-photon absorption destroying two-photon system plays only a negative role.

References

[1] Kumakhov M A 1986 *Izluchenie Kanalirovanikh Chastits v Kristallakh* (Moscow: Atomizdat)
[2] Bazilev V A and Zhevago N K 1987 *Izluchenie Bistrikh Chastits v Veshchestve i vo Vneshnikh Polyakh* (Moscow: Nauka)
[3] Gouanere M, Sillou D, Spighel M *et al* 1982 *Nucl. Inctr. and Method* **194** 225-8
[4] Genz H, Groening L, Hoffmann-Stascheck P *et al* 1996 *Phys. Rev.* B **53** 8922-36
[5] Dabagov S B, Beloshitsky V V and Kumakhov M A 1993 *Nucl. Instr. and Method* B **74** 368
[6] Dabagov S B and Zhevago N K 2008 *Rivista del Nuovo Cimento* **32** 491
[7] Beloshitsky V V and Kumakhov M A 1978 *J. Eksp.Teor. Fiz.* **74** 1244
[8] Yaralov V 2004 *Proc. Int. Conf. on Laser Physics–2004* Ashtarak Armenia 93-6
[9] Avakian R O, Ispirian K A and Yaralov V 2006 *Nucl. Instr. and Method* B **252** 20-25
[10] Yaralov V 2014 *J. of Phys: Conf. Ser.* **517** 012031
[11] Ispirian K A, Ispiryan M and Yaralov V 2015 *Nucl. Instr. and Method* B **355** 203
[12] Mollow B R 1969 Power spectrum of light scattered by two-level systems *Phys. Rev.* **188** 1969-75
[13] Mollow B R 1970 Resonant scattering of radiation from collision-damped two-level systems *Phys. Rev.* A **2** 76-80
[14] Mollow B R 1981 Theory of intensity dependent resonance light scattering and resonance fluorescence *Progr. in Opt.* **19** 1-43
[15] Cavaletto S M, Buth C, Zolt H *et al* 2012 Resonance fluorescence in ultrafast and intense x-ray free-electron-laser pulses *Phys. Rev.* A **86** 033402
[16] Ulhaq A, Weiler S, Roy C *et al* 2013 Detuning-dependent Mollow triplet of a coherently-driven single quantum dot *Opt. Exp.* **21** 4382-95
[17] Keldysh L V 1964 Ionization in the field of a strong electromagnetic wave *J. Eksp. Teor. Fiz.* **47** 1945 - - 57
[18] Perelomov A M, Popov V S and Terentiev M V 1966 Ionization of atoms in an alternating electric field *J. Eksp.Teor. Fiz.* **50** 1393
[19] Delone N B and Krainov V P 2000 *Multiphoton processes in atoms* (Berlin:Springer)
[20] Popov V S 2004 Tunnel and multiphoton ionization of atoms and ions in an intense laser field (Keldish theory) *Usp. Fiz. Nauk.* **174** 921

Quasicharacteristic radiation of relativistic electrons at orientation motion in lithium halides crystals along charged planes and axes

N V Maksyuta, V I Vysotskii, S V Efimenko
Taras Shevchenko National University of Kiev, Kiev 01601, Ukraine

E-mail: maksyuta@univ.kiev.ua

Abstract. The paper deals with the investigation of the orientation motion of relativistic electrons in charged (111) planes and charged [110] axes of lithium halides ionic crystals of LiF, LiCl, LiBr and LiI. On the basis of these investigations the spectra of quasicharacteristic radiation for the electron beams with various Lorentz-factors both in planar and axial cases have been calculated numerically.

1. Introduction

The given paper is a continuation of some papers dedicated to the investigation of the processes of the channeling and quasicharacteristic radiation (QCR) of relativistic electrons in charged planes and axes of LiM crystals having NaCl-type structure. Thus, in [1] the interaction potential of relativistic electrons with the charged (111) planes of LiH crystal has been calculated. More detailed research of similar potentials at M = H, D is given in [2] where a comparative analysis of analytically calculated QCR spectra with experimentally measured in [3] has been also done. The paper [4] gives analogous calculations for the case of M = F and the comparison of analytically calculated QCR spectra with experimental data given in [5] for three values of Lorentz-factor have been also made.

The given paper investigates the channeling of relativistic electrons in main charged (111) planes and main charged [110] axes of lithium halides of (M = F, Cl, Br, I). For such regime of motion some corresponding wave functions, energy levels of the channeling motion and QCR spectra were calculated by numerical solutions of Schrödinger equations.

2. The calculation of interaction potentials of electrons with the charged (111) planes of lithium halides crystals

Describing two 1s-electrons by a wave function of the form $\psi_1(r)=\sqrt{Z^*/\pi a_0}\exp\left(-Z^* r/a_0\right)$ and one 2s-electron (localized near lithium nucleus with the probability of $1-\alpha$) by a wave function of the form $\psi_2(r)=1/\sqrt{32\pi a_0}(2-r/a_0)\exp(-r/2a_0)$, we find the following expression for one-particle potential of Li^+ ion:

$$\varphi_+(r)=e\left\{2\left(\frac{Z^*}{a_0}+\frac{1}{r}\right)\exp\left(-\frac{2Z^* r}{a_0}\right)+\frac{(1-\alpha)f(r)}{a_0^3}\exp\left(-\frac{r}{a_0}\right)+\frac{\alpha}{r}\right\}. \quad (1)$$

Here a_0 is the Bohr radius; $Z^*=Z_{Li}-\beta=43/16$ is the screened charge of nucleus of Li^+ ion; $\beta=5/16$ is the screening parameter regarding Li^+ nucleus and calculated by a variation method taking into account the fact that every of 1s-electrons partially screens nucleus of Li^+ ion; α is the degree of ionicity of the link, $f(r)=r^2/8+ra_0/4+3a_0^2/4+a_0^3/r$.

Content from this work may be used under the terms of the Creative Commons Attribution 3.0 licence. Any further distribution of this work must maintain attribution to the author(s) and the title of the work, journal citation and DOI.
Published under licence by IOP Publishing Ltd

One-particle potentials for negatively charged ions of F^-, Cl^-, Br^- and I^- may be presented by Thomas – Fermi approximation [6]

$$\varphi_-(r)=\frac{e(Z+\alpha)}{r}\chi\left(\frac{r}{R}\right)-\frac{\alpha e}{r}, \tag{2}$$

where Z is the nucleus charge of M^- ion, $R\approx 0.8853a_0Z^{-1/3}$ is the screening radius of Thomas – Fermi, $\chi(x)=\sum_j\alpha_j\exp(-\beta_jx)$. For used in the given paper Moliere, Barrette and Firsov approximations, the constants α_j and β_j correspondingly equal to: $\alpha_j=\{0.1,0.55,0.35\}$, $\beta_j=\{6,1.2,0.3\}$ [6]; $\alpha_j=\{0.4,0.6\}$, $\beta_j=\{2.984,0.474\}$ [7]; $\alpha_j=4j\sinh^2(1.7\beta)\exp(-3.4\beta j)$, $\beta_j=2\beta j$, $j=1,2,\ldots$ [6]. Evidently the potentials (1), (2) satisfy the following boundary conditions:

$$\varphi_+(r)\underset{r\to 0}{\to}3e/r\,,\ \varphi_-(r)\underset{r\to 0}{\to}Ze/r\,,\ \varphi_\pm(r)\underset{r\to\infty}{\to}\frac{\pm\alpha e}{r}. \tag{3}$$

Note that the first two conditions in (3) follow from the requirement that the potentials should be pure Coulomb near nuclei when their charges are not screened by electrons [8].

The potentials of electro-neutral stems of $V_{0\pm}(x)$ of the charged planes of A^+- (111) and B^-- (111) (in the formula (1) the first and the second components and in formula (2) the first component are used) are calculated in a traditional way [6].

The calculation of the potential created by long-range Coulomb components based on the method given in [9]. The sense of this method is the following. At first the Coulomb potentials of ionic crystals (the last components in (1), (2) formulas) are substituted by the screened Coulomb potentials, namely, $\varphi_\pm(r)=\pm\alpha e/r\lim\limits_{b\to\infty}\exp(-r/b)$. Then after these potentials averaging along the planes, we come to the following expression for the summary potential of the full ensemble of the charged planes of (111) of the ionic A^+B^- crystal:

$$V_\pm(x)=2\pi b\alpha eg\lim_{b\to\infty}\sum_{n=-\infty}^{\infty}\left\{\exp\left(-\frac{|x-nd|}{b}\right)-\exp\left(-\frac{|x-nd-d/2|}{b}\right)\right\}, \tag{4}$$

where $d=a/\sqrt{3}$ is the distance between the same charged crystallographic of (111) planes, $g=4/a^2\sqrt{3}$ is the density of atoms in (111) planes, a is the lattice constant of the crystal. If first in the expression (4) pass to the limit there arises a conditionally convergent series. The methods summations of similar series for the ionic crystals of finite dimensions have been proposed in [10]. This paper deals with another approach which may be more correct from a mathematical point of view. First fulfill a decomposition of the potential (4) by one-dimensional vectors $\vec{g}_n=(2\pi n/d)\vec{e}_x$ of the reciprocal lattice (see, for example, [11]). After this we pass to the limit and come to the next expression for $V_\pm(x)$:

$$V_\pm(x)=\frac{4d\alpha eg}{\pi}\sum_{n=0}^{\infty}\frac{\cos[2\pi(2n+1)x/d]}{(2n+1)^2}. \tag{5}$$

At last, we confirm that the expression (5) is the expansion into a Fourier series of the following periodic function:

$$V_{\pm}(\xi) = -4V_0 \sum_{n=-\infty}^{\infty} \left[|\xi - n| - \frac{1}{4} \right] \theta\left[\frac{1}{2} - |\xi - n| \right], \tag{6}$$

where $\xi = x/d$ is the relative coordinate, $\theta(x)$ is the Heaviside function, $V_0 = V_{\pm}(n) = \pi d \alpha e g/2$. Further the potential (6) must be averaged over thermal oscillations (with using of thermal oscillations amplitudes of u_+ and u_- correspondingly in the planes of A^+ and B^-). For example, figure 1 (a) presents the graph of interaction potential of $U_{\pm}(\xi) = -e\langle V_{\pm}(\xi)\rangle_{u_+,u_-}$, calculated at T = 300 K for LiF (dash-dot curve). The same figure presents by the solid curve the graph of the interaction potential of $U_{0\pm}(\xi) = -e\langle V_{\pm}(\xi)\rangle_{u_+,u_-}$ with electro-neutral shells of (111) planes of LiF crystal, calculated at the same temperature and with usage of Firsov approximation. The graph of the total interaction potential of $U(\xi) = U_{0\pm}(\xi) + U_{\pm}(\xi)$, presented by the curve 1 in figure 1 (b), confirm that a long-range Coulomb interaction potential of $U_{\pm}(\xi)$ essentially transforms the interaction potential of $U_{0\pm}(\xi)$: in the planes of Li^+- (111) the depths of potential wells increase abruptly but in F^-- (111) planes they, in the contrary, essentially decrease. Evidently, analogous conclusions may be done for other lithium halides crystals. The graph of total interaction potentials of $U(\xi)$ at T = 300 K for LiCl, LiBr and LiI crystals (calculated also by Firsov approximation) are presented by the curves 2, 3 and 4 correspondingly.

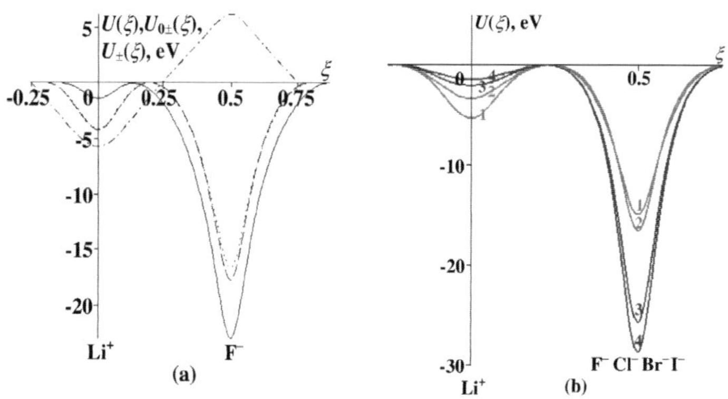

Figure 1. Interaction potentials graphs of electrons channeling along (111) planes: (a) of LiF crystal calculated using Moliere (dot curve) and Barrett (dash curve) approximations; (b) of LiF (curve 1), LiCl (curve 2), LiBr (curve 3) and LiI (curve 4) crystals calculated using Firsov approximation. Figure 1a presents by a solid curve the graph of the interaction potential of electrons with electro-neutral stems of the charged planes in LiF and also dash-dot curve shows a graph of the interaction potential with Coulomb components of the same planes. All the graphs are calculated at T = 300 K.

Note that in the calculation of the considered interaction potentials with B^- planes other approximations may be used. For example, figure 1 (a) presents by a dot curve the potential $U(\xi)$ in the case of LiF crystal calculated by means of Moliere approximation [1] and a dash curve by means of a Barrette approximation correspondingly [7]. However, as further analysis shows, QCR spectra for a LiF crystal coincide with the ones got experimentally (see [5]) only in the case of the channeling of relativistic electrons in potential wells calculated using Firsov approximation. This is just the basis for the choice of this approximation for all other lithium halides crystals.

3. The spectra of quasicharacteristic radiation for the channeling electrons in (111) planes of lithium halides crystals

The method of calculation of QCR was the following. The one-dimensional Schrödinger equation with a relativistic mass (see [6]) reduced to the Sturm – Liouville problem. Then this problem was solved by means of corresponding program from IMSL library of Fortran 90 in the result of which transverse energies of ε_k and their corresponding wave functions $\psi_k(x)$ for all lithium halides crystals at the same Lorentz-factor values which was chosen in [5] for the calculation of QCR spectra in LiF crystal was found. Numerical values of energies ε_k are shown in table 1.

Table 1. Numerical values of transverse energies of ε_k, arising at the channeling of electrons with Lorentz-factors of $\gamma_1 \approx 34$, $\gamma_2 \approx 60.7$ and $\gamma_3 \approx 107.1$ along (111) planes of lithium halides crystals.

ε_k		**0**	**1**	**2**	**3**	**4**	**5**	**6**	**7**	**8**	**9**
	γ_1	-9.90	-2.98	-2.40	-0.04						
LiF	γ_2	-10.96	-4.49	-3.45	-0.90	-0.53					
	γ_3	-11.83	-6.45	-3.85	-2.70	-1.37	-0.56	-0.02			
	γ_1	-11.74	-4.13	-2.10	-0.45	-0.07					
LiCl	γ_2	-12.79	-6.41	-2.38	-2.22	-0.66	-0.17				
	γ_3	-13.62	-8.44	-4.49	-2.61	-1.78	-1.16	-0.34	-0.15		
	γ_1	-19.88	-10.37	-3.92	-1.13	-0.55					
LiBr	γ_2	-21.15	-13.47	-7.52	-3.30	-1.31	-0.81	-0.20			
	γ_3	-22.13	-16.04	-10.94	-6.80	-3.65	-1.48	-1.46	-0.51	-0.27	-0.01
	γ_1	-22.57	-12.64	-5.57	-1.37	-0.74					
LiI	γ_2	-23.88	-15.94	-9.59	-4.85	-1.72	-0.88	-0.21	-0.02		
	γ_3	-24.89	-18.63	-13.27	-8.81	-5.27	-2.63	-1.00	-0.91	-0.27	-0.06

Figure 2 (a), (b) as illustration presents energy levels and their corresponding wave functions calculated in LiF crystal for the electrons with Lorentz-factor of $\gamma = 60.7$. The arrows in figure 2 (a) present the dipole radiation transitions contribute mainly in QCR spectrum (see figure 3).

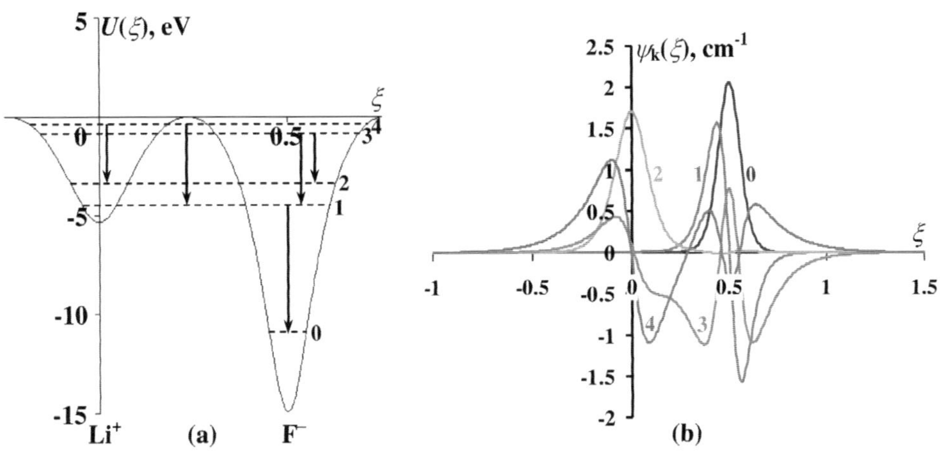

Figure 2. (a) Schematic presentation of energy level systems and radiation dipole transitions arising between them in the case of electrons channeling with a Lorentz-factor of $\gamma = 60.7$ along (111) planes in LiF crystal, (b) system of wave functions corresponding to these energy levels.

From the graphics of wave functions in figure 2 (b) is seen that spontaneous dipole transitions of $1 \rightarrow 0$, $3 \rightarrow 1$ and $3 \rightarrow 2$ are connected with the electrons localized in potential wells of F^-- planes. The transition of $4 \rightarrow 2$ is realized on the same electrons which localized mainly in Li^+- planes. At last, the transition of $4 \rightarrow 1$ relates to so-called non-direct transitions, since a channeling electron in the result of the loss of the transverse energy changes a region of localization on a spontaneous QCR.

Further, on the basis of the table 1 and using their corresponding wave functions spectral distributions of electron QCR energy were calculated numerically in a dipole approximation on a unit of the path length in correspondence with the following formula:

$$G(\omega) = \frac{1}{\hbar c}\frac{dI}{d\omega} = \frac{e^2 d^2 \omega}{\hbar c^4} \sum_{k=1}^{n-1} \sum_{l=0}^{k-1} \xi_{kl}^2 \Omega_{kl}^2 f\left(\frac{\omega}{2\gamma^2 \Omega_{kl}}\right) \langle P_k(\vartheta) \rangle, \qquad (7)$$

where $f(x) = (1 - 2x + 2x^2)\theta(1-x)$, $\xi_{kl} = \int_{-\infty}^{\infty} \psi_k(\xi)\xi\psi_l(\xi)d\xi$ is the matrix element of a dipole transition between energy levels of k and l, $\Omega_{kl} = (\varepsilon_k - \varepsilon_l)/\hbar$ is the frequencies of transitions between these levels, $\langle P_k(\vartheta) \rangle$ is the initial population (averaged by means of a normal distribution function $g(\vartheta) = (2\pi\vartheta_0^2)^{-1/2} \exp(-\vartheta^2/2\vartheta_0^2)$, where ϑ_0 is an angular dispersion) of k-th energy level by an electron beam incident on a crystal at an angle of ϑ to some channeling planes. Figure 3 shows the graphs of the dependencies (7) arising in lithium halides crystals at the channeling of electron beams with Lorentz-factors of γ= 34, 60.7, 107.1 ($\vartheta = 0$ mrad, $\vartheta_0 \approx 0.3$ mrad).

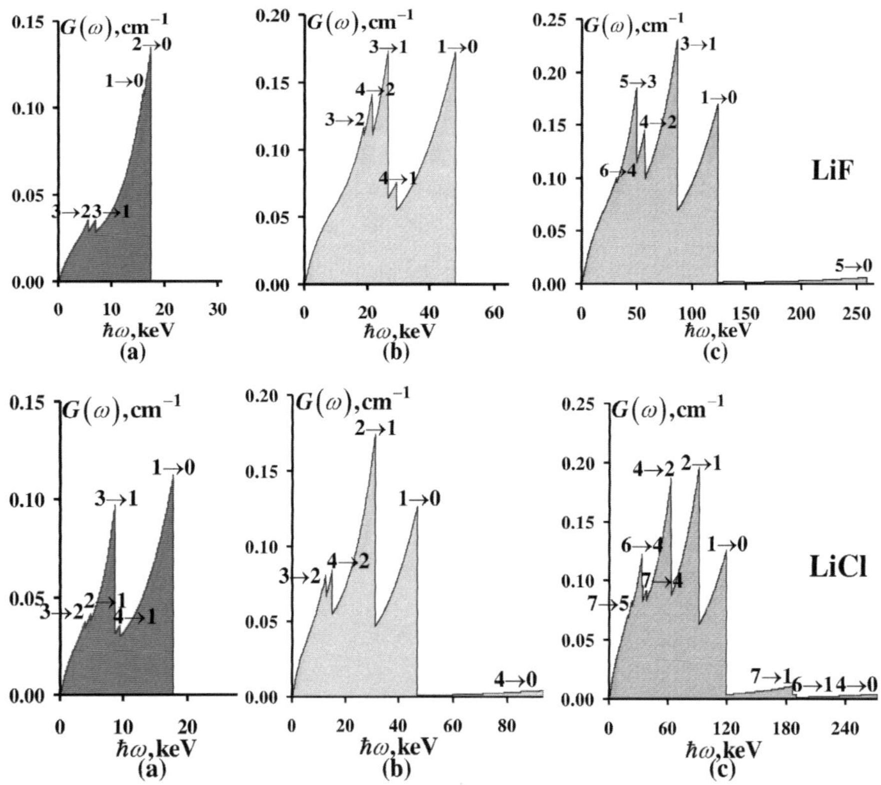

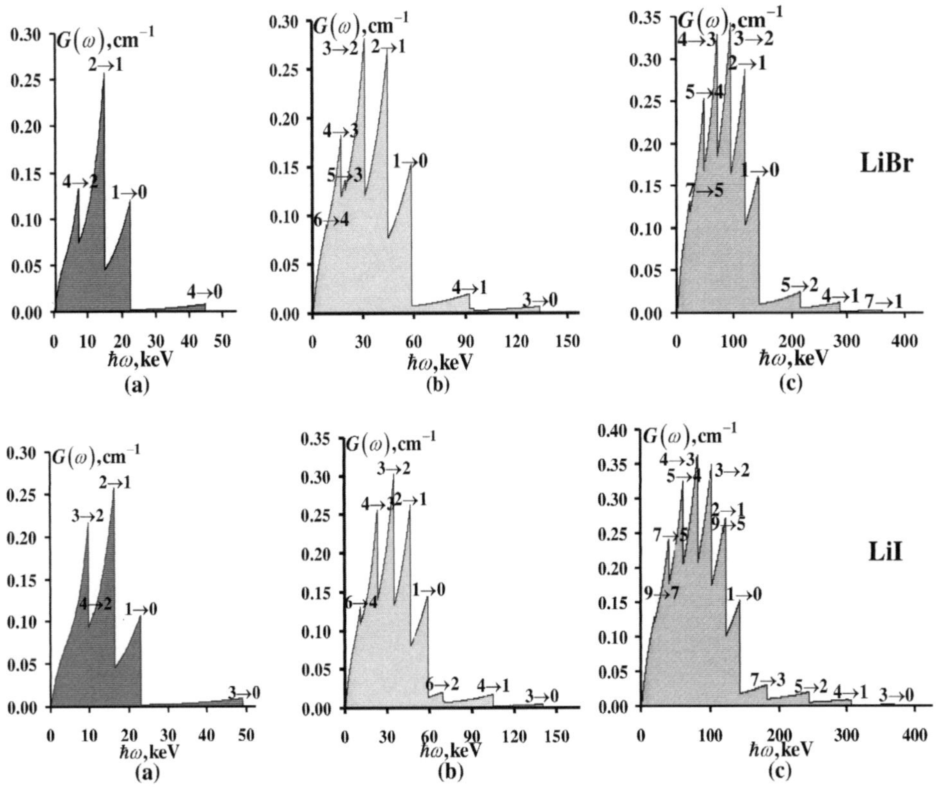

Figure 3. Spectral distributions of QCR energy by an electron per unit path of its motion at a zero angle to (111) planes of LiF, LiCl, LiBr and LiI crystals (the calculations are made for Firsov approximation) accordingly with Lorentz-factors of: (a) $\gamma = 34$, (b) $\gamma = 60.7$ and (c) $\gamma = 107.1$ (an angular dispersion $\vartheta_0 \approx 0.3\,\text{mrad}$). All the graphs present identification between the arising spectral peaks of QCR and displayed dipole radiation transitions.

As it is seen from figure 3 channeling electrons energy increases and also the transition from a crystal LiF to a crystal LiI causes a rise of spectral QCR peaks and their intensities (after a small falling in the case of LiCl crystal) for LiBr and LiI crystals increase approximately in one and a half, two times comparing with the intensities in LiF crystal.

4. The spectra of quasicharacteristic radiation at the channeling of electrons along the charged [110] axes of lithium halides crystals

The interaction potentials (quickly falling with a distance) of $U_{0\pm}(\xi,\eta)$ of relativistic electrons with electro-neutral stems of the charged [110] axes of ionic lithium halides crystals are calculated in a traditional way [6]. The interaction potential of $U_{\pm}(\xi,\eta)$ of the channeling electrons with a Coulomb components of all the ensemble of opposite charged axes is found analogically as in a planar case, namely, it is calculated as the result of the interaction potentials summation (with axes grouping by electro-neutral shells shown in figure 4 (a) (more detailed argumentation for the axial case is shown in [12]). As illustration in figure 4 (b) the 3D-graph is built on one two-dimensional period of $-1/4 \le \xi \le 3/4$, $|\eta| \le 1/4$ in relative coordinates of $\xi = x/a_x$, $\eta = y/a_y$ ($a_x = a/\sqrt{2}$, $a_y = a$) averaged by thermal oscillations of long-range interaction potential of $U_{\pm}(\xi,\eta)$ the electrons with the system of the charged [110] axes of LiF crystal. Figure 4 (c) presents 3D-graph of the total interaction potential of $U(\xi,\eta)=U_{0\pm}(\xi,\eta)+U_{\pm}(\xi,\eta)$ of the electrons with [110] axes in LiF crystal. It's seen, that in near-barrier regions a potential relief in $\eta=n/2$ planes, where $n=0,\pm1,...$, has minimax points.

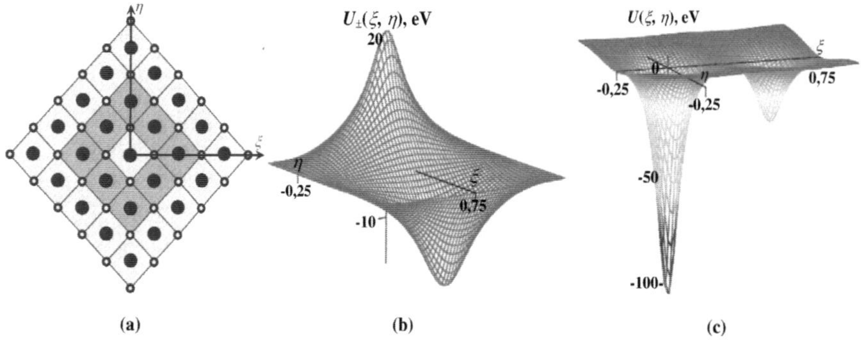

Figure 4. (a) Grouping of the charged axes of A^+B^- ionic crystals of NaCl-type by electro-neutral shells (light circles indicate A^+ axes and dark accordingly B^- axes), (b) 3D-graph of a long-range interaction potential of $U_{\pm}(\xi, \eta)$ of electrons with the system of charged [110] axes of LiF crystal, (c) 3D-graph of total interaction of $U(\xi, \eta)$ of the electrons with [110] axes in LiF crystal. The graphs are calculated at T = 300 K.

In figure 5 (a) in the section of $\eta = 0$ total interaction potentials of electrons with [110] axes for all lithium halides crystals are presented. Note that depths of potential wells in positive and negative charged [110] axes of these crystals correlate in the same way as in the planar case. As the result of numerical solution of the Schrödinger equation (more detailed description see in [12]) for all these potential wells there were found some transverse energy levels and corresponding wave functions for the channeling electrons with a Lorentz-factor of $\gamma = 8$. Analogically planar case on the basis of these data QCR spectra were calculated numerically for dispersion-less electron beams moving regarding [110] axes of lithium halides crystals at zero angles. The graphs of these spectral densities with the indication of dipole radiation transitions are shown in figure 5 (b).

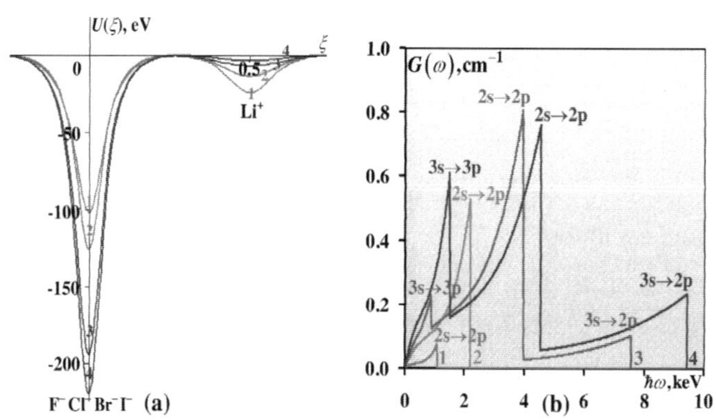

Figure 5. (a) Interaction potentials graphs of electrons channeling along [110] axes in LiF (curve 1), LiCl (curve 2), LiBr (curve 3) and LiI (curve 4) crystals calculated at T = 300 K using Firsov approximation, (b) spectral distributions of QCR energy by the electrons with Lorentz-factors of $\gamma = 8$ per unit path at its flying regarding [110] axes of LiF (curve 1), LiCl (curve 2), LiBr (curve 3), LiI (curve 4) crystals at a zero angle (there is an identification of spectral lines with corresponding radiation transitions).

As it is seen from figure 5 (b), even for a small value of Lorentz-factor the following tendency is evident: at the transition from LiF crystal to LiI one a quantity of radiation transitions increases abruptly and spectral QCR peaks shifting in high-frequency region rise by the intensity. Note that this results correlate with experimental data of orientation motion of electrons with longitudinal energies up to 5 MeV along different axial channels of crystal silicon (see [6, 13]).

5. Conclusions

By using the methods of the calculation of interaction potentials of the charged particles to the charged planes and axes in ionic crystals of NaCl-type there were got analytically interaction potentials of electrons with (111) planes and [110] axes in lithium halides of LiF, LiCl, LiBr and LiI. Wherein the interaction potentials calculation in crystallographic planes built of negative halides ions was made in the framework of Firsov approximation (as it follows from experimental data it is more optimal for LiF crystal).

On the basis of the numerical solutions of Schrödinger equations both in planar and axial cases the systems of energy levels and their corresponding wave functions arising at the channeling of relativistic electrons with various Lorentz-factors were calculated (they coincided with those numerical values which were chosen in [5] and [13]).

Proceeding from these data the QCR spectra in dipole approximation were got numerically both in planar and axial cases. At planar channeling with the transition from LiF crystal to LiI one, QCR energy increases but it is concentrated in the same spectral regions (it is observed for all Lorentz-factors). For an axial case this fact does not take place.

References

[1] Vysotskii V I, Kuz'min R N, Maksyuta N V 1987 *Sov. Phys. JETP* **66** 1150
[2] Maksyuta M V and Vysotskii V I 2012 *Bulletin of the University of Kiev. Series "Radio Physics and Electronics"* **17** 4
[3] Berman B L, Kephart J O, Datz S *et al* 1996 *Nucl. Instr. and Meth. in Phys. Res.* B **119** 71
[4] Maksyuta M V and Vysotskii V I 2012 *Bulletin of the University of Kiev. Series "Radio Physics and Electronics"* **17** 14
[5] Swent R L, Pantell R H, Park H *et al* 1984 *Phys. Rev.* B **29** 52
[6] Bazylev V A and Zhevago N K 1987 *Fast Particles Radiation in a Substance and in External Fields* (Moscow: Nauka)
[7] Barrrett J H 1979 *Phys. Rev.* B **20** 3535
[8] Gombas P 1951 *Statistical theory of an atom* (Moscow: Foreign Literature Publ. House)
[9] Maksyuta N V 2014 *Thesis of the Reports of XLIV Int. Conf. on Physics of Charged Particles Interaction with Crystals.* (Moscow) 19
[10] Korkhmazian N A, Korkhmazian N N, Babadjanyan N E 2003 *J. Tech. Phys.* **73** 1
[11] Ohtsuki Y-H 1983 *Charged Beam Interaction with Solids*, Taylor & Francis Ltd London and New York
[12] Maksyuta N V, Vysotskii V I, Efimenko S V 2015 *Nucl. Instrum. Methods.* B **355** 90
[13] Andersen J U, Bonderup E, Laegsgaard E *et al* 1982 *Nucl. Instrum. Methods.* **194** 209

Radiation of the magneto-crystalline undulator

V Epp[1] and V V Kaplin[2]

[1] Tomsk State Pedagogical University, ul. Kievskaya 60, 634061 Tomsk, Russia
[2] Tomsk Polytechnic University, pr. Lenina 30, 634050 Tomsk, Russia

E-mail: epp@tspu.edu.ru

Abstract. The radiation at grazing incidence of relativistic positively charged particles on the crystal surface in the presence of magnetic field is studied theoretically. The magnetic field is supposed to be parallel to the surface. Dependent on the initial conditions the particle can be captured in the channeling mode and perform periodic oscillations along the surface of the crystal. The spectrum, angular distribution and polarization of radiation are calculated. The emission spectrum of a single particle is discrete and it extends up to very large numbers of harmonics. If the magnetic field is much weaker than the electric field of atoms, the frequency range of radiation of the particle beam does not depend on magnetic field and is defined solely by the energy of the particles and by the surface averaged potential, though the frequency of the first harmonic is defined only by the magnetic field. In case of channeled positrons the characteristic energy of the emitted photons is of order $10\gamma^{3/2}$ (eV), where γ is the particle relativistic factor. The main part of radiation is bound to a narrow cone and is polarized largely orthogonal to the surface of the crystal.

1. Introduction

New sources of X-ray and gamma radiation, consisting of the particle accelerators and precisely oriented mono-crystals are created on the basis of the channeling phenomenon. There is interest in developing a radiation source using the periodically deformed crystals. See, for example, the recent papers [1, 2]. The channeled positrons are bound to the bent atomic crystal planes, and emit monochromatic radiation of a frequency which depends on the particles energy and the period of the crystal plate bending. Such a "crystalline undulator" was first proposed in [3, 4], where the deformation of the crystal was proposed to be performed by an ultrasonic wave. There are also a number of projects of multi-crystal undulators where the particle is deflected successively in the opposite directions at passing through a series of mutually oriented ultrathin crystals (see, for example, [5]). The radiation generated in the crystal undulators can be harder than one can get in a magnetic undulator at the same energy of the positrons, since the period of the deformed crystal may be much smaller than the period of undulator magnetic field.

A "magneto-crystalline undulator" which produces electromagnetic radiation by means of charged particles moving near the flat surface of a crystal in a magnetic field parallel to the surface was proposed in [6]. The positively charged particles at grazing incidence to the surface of the crystal are reflected from the surface and returning back by the magnetic field, if the field is orthogonal to the average velocity of the particles. The quantum energy levels of the particle transverse motion at the surface channeling in a magnetic field, and the possible frequency of radiation were also found in [7].

Content from this work may be used under the terms of the Creative Commons Attribution 3.0 licence. Any further distribution of this work must maintain attribution to the author(s) and the title of the work, journal citation and DOI.
Published under licence by IOP Publishing Ltd

Interaction of the particles with the crystal surface in the presence of a magnetic field can be observed when the crystal is mounted on the beam path in the chamber of a circular accelerator. The dynamics of the beam in an accelerator at its interaction with a radiator which is thin in profile and extended along the beam, in case of grazing incidence geometry is not investigated in accelerator physics. An effect of asymmetric generation of the X-ray radiation produced at grazing interaction of 33 MeV electrons with 50 mm Si plate in a magnetic field have recently been observed in experiments reported in [8].

Here we report our theoretical investigation on radiation at the surface channeling of fast positrons. It is assumed that the crystal is placed in a uniform magnetic field parallel to the crystal surface. The initial velocity of the particle is directed so that the Lorentz force exerted by the magnetic field is directed to the surface of the crystal. The result of the action of two competing forces – the force caused by the magnetic field and the repulsive force of the crystallographic plane, is that the particle oscillates in the vicinity of the surface.

2. The particle dynamics

Let the axis Z be directed along the vector of the magnetic field, and the axis Y be orthogonal to the surface of the crystal directed outward (figure 1). The axis X lies in the crystal surface so that the vector $\boldsymbol{v}$ of the particle initial velocity is lying in the plane XY. The crystallographic plane forming the surface of the crystal has the coordinates $y = 0$. In order to improve the

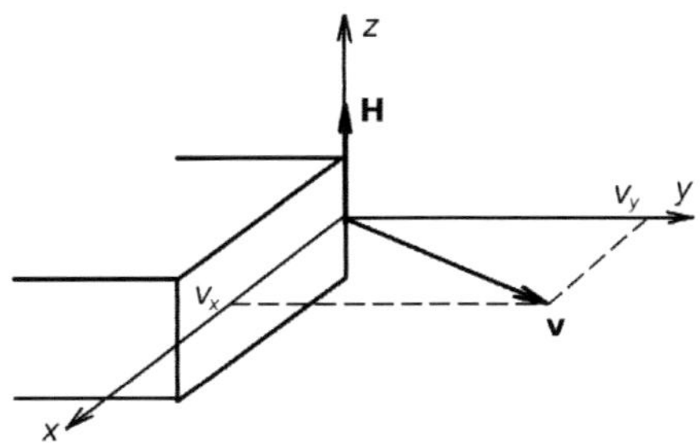

Figure 1. The coordinate system

possibility of analytical solution of the problem we simplify the potential of the electric field of the crystallographic plane as follows:

$$U(y) = U_0 e^{-y/a}.$$

where U_0 is the averaged potential of atomic plane, and a is the screening radius.

The electric $\boldsymbol{E}$ and magnetic $\boldsymbol{H}$ fields in the halfspace $y > 0$ can be written as

$$\boldsymbol{H} = (0, 0, H), \quad \boldsymbol{E} = (0, E(y), 0), \quad E(y) = -\frac{dU(y)}{dy} = \frac{U_0}{a} e^{-y/a},$$

where H is a constant. Relativistic equations of motion of a particle in this field have the form

$$\frac{\mathrm{d}p_x}{\mathrm{d}t} = \frac{e}{c}\dot{y}H, \quad \frac{\mathrm{d}p_y}{\mathrm{d}t} = eE(y) - \frac{e}{c}\dot{x}H. \qquad (1)$$

Here p_x, p_y are projections of the particle momentum on the coordinate axes:

$$p_i = \gamma m_0 \dot{x}_i, \quad x_i = x, y, \quad \gamma = \frac{1}{\sqrt{1-\beta^2}}, \quad \beta^2 = \frac{\dot{x}^2 + \dot{y}^2}{c^2},$$

m_0 and e are the mass and the charge of the particle, c is the speed of light. We assume that $\dot{y}$ is much less then $c\gamma^{-1}$. Integration of the equations (1) gives

$$\dot{x} = \omega y + V_1, \tag{2}$$

$$\frac{m\dot{y}^2}{2} = \mathcal{E}_0 - \frac{eU_0}{a}e^{-y/a} - m\omega\left(\frac{\omega y^2}{2} + V_1 y\right). \tag{3}$$

Here $m = m_0\gamma$, $\omega = eH/mc$, $\mathcal{E}_0$ and $V_1 = v_{x0} - \omega y_0$ are constants of integration. The last equation describes the one-dimensional motion of a particle in direction of the Y axis in the field with effective potential energy

$$U_{ef} = \frac{eU_0}{a}e^{-y/a} + m\omega\left(\frac{\omega y^2}{2} + V_1 y\right).$$

The integral of motion $\mathcal{E}_0$ plays the role of "transverse" energy of the particle. It is convenient to introduce the dimensionless quantities

$$V = \frac{U_{ef}}{eU_0}, \quad \chi = \frac{y}{a}, \quad \eta = \frac{aH}{U_0} = \frac{amc^2}{ReU_0},$$

where $R = c/\omega$ is a parameter that determines the strength of the magnetic field. It is approximately equal to the particle orbit radius in this field. The parameter η is the ratio of the force acting on the particle by the magnetic field to the electrostatic repulsion force of the crystallographic plane. The dimensionless potential V with this notations takes the form

$$V = e^{-\chi} + \chi\eta\beta_0 + \frac{a}{R}\beta_0\eta\chi(\chi - 2\chi_0), \tag{4}$$

where $\chi_0 = y(0)/a$. Since we are only interested in periodic motion in the Y-direction, we can take as a starting point, without loss of generality, the point where y has a maximum. Then $v_{0y} = 0$, $v_{0x} = v_0 = c\beta_0$, where v_0 is the initial velocity of the particle. We put $\beta_0 = 1$ where it is not critical. The value of a/R is much smaller than unity. The value of η in normal conditions is also small. For example, if $a = 1$ Å, $H = 10^4$ Oe, $eU_0 = 30$ eV, we have $\eta = 10^{-2}$. Consequently, the terms containing a/R in the formula for the potential energy can be neglected. As a result, the interaction potential takes the form

$$V = e^{-\chi} + \chi\eta. \tag{5}$$

A plot of this function is shown in figure 2. The minimum of potential lies in the plane with coordinate $\chi = \chi_m$ and has a value of V_m:

$$\chi_m = -\ln\eta, \quad V_m = \eta(1 - \ln\eta). \tag{6}$$

In the considered case $\eta \ll 1$ we have: $\chi_m \gg 1$ and $V_m \ll 1$. This means that the coordinate y_m of the potential minimum is located at a considerable distance from the atomic plane ($y_m \gg a$). This makes it possible to split the space above the surface of the crystal into two areas: $y < y_m$ and $y \geq y_m$. We assume that potential in region $y < y_m$ is represented by the potential of atomic plane, and in the area $y \geq y_m$ by potential of the magnetic field. Such combined potential has the form

$$V(\chi) = \begin{cases} e^{-\chi}, & \chi \leq \chi_m, \\ \eta\chi, & \chi \geq \chi_m. \end{cases} \tag{7}$$

Hence, we have to solve the equations of motion in two separate regions: $\chi \le \chi_m$ and $\chi \ge \chi_m$ and take the final coordinates in the first region as the initial coordinates for the second region. The solution of equations (2) and (3) in the region $\chi \ge \chi_m$ has the form

$$x = v_0 t, \quad y = y_0 - \frac{1}{2}R(\omega t)^2. \tag{8}$$

And in the region $\chi \le \chi_m$

$$x = v_0 t, \quad y = 2a \ln \frac{\cosh \alpha(t - t_0)}{\sqrt{\varepsilon}}, \tag{9}$$

with

$$\alpha = \omega\sqrt{\frac{y_0 R}{2a^2}}, \quad \varepsilon = \frac{\mathcal{E}_0}{eU_0} \tag{10}$$

and integration constant t_0, defined by condition of the trajectory continuity in the plane $y = y_m$:

$$t_0 = t_1 + \frac{1}{\alpha}\operatorname{Arth}\frac{\sqrt{\chi_0^2 - \chi_1^2}}{\chi_0}, \quad t_1 = \sqrt{\frac{2a(\chi_0 - \chi_1)}{c\omega}}, \tag{11}$$

where t_1 is the time moment when the particle crosses the plane $y = y_m$. Actually, if we neglect the short-range potential in region $y > y_m$, the particle moves in magnetic field along an arc of a circle. But in accepted approximation this arc is equivalent to osculating parabola (8). One period of the trajectory of a particle moving under the law (8) and (9) is shown in figure 3.

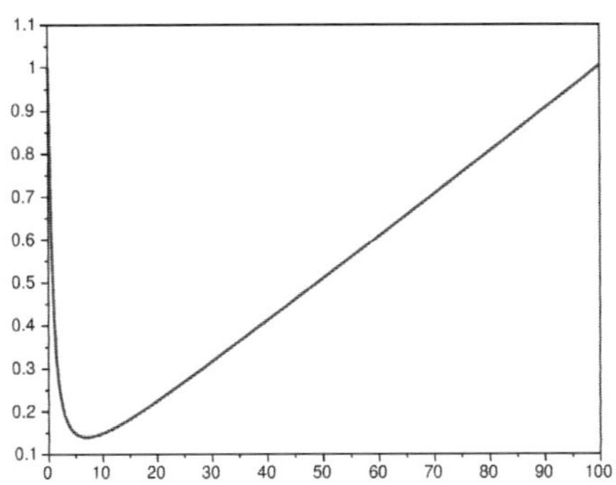

Figure 2. Potential $V(\chi)$, plotted according to equation (5) for $\eta = 0.01$.

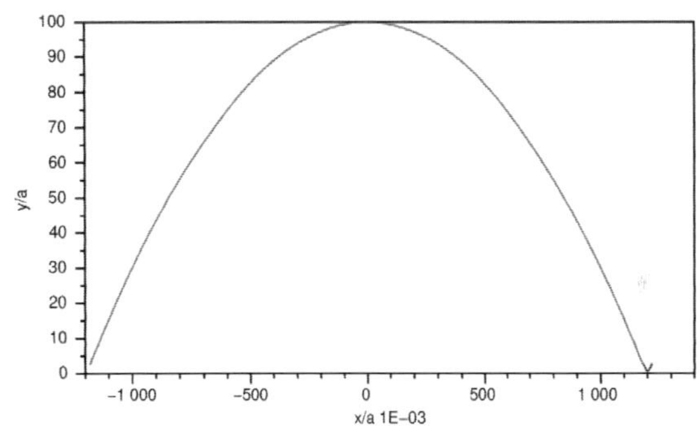

Figure 3. One period of trajectory of a charged particle. $H = 50$ kOe, $mc^2 = 40$ MeV, $eU_0 = 30$ eV, $a = 1$ Å, $\varepsilon = 0.5$.

The trajectory period $T = 2t_0$ in approximation $\eta \ll 1$ consists mainly of the time spent by the particle in the magnetic field: $T \approx 2t_1$. This follows from equations (10) and (11) — as $\alpha \gg \omega$, the particle spends much less time $2(t_0 - t_1)$ in the electric field of the surface atomic layer than in the magnetic field. It is also seen in figure 3.

3. Radiation

Let us denote by $\omega_0 = 2\pi/T$ the frequency of the particle oscillation in y direction. Since the particle motion is periodic, we use the formulae obtained in the theory of undulator radiation (see., e.g. [9, 10]). In particular, the spectral and angular distribution of the energy $d\mathcal{E}$, emitted in a solid angle $d\Omega$, in the frequency range $d\tilde{\omega}$ is defined by expression

$$\frac{d\mathcal{E}}{d\Omega d\tilde{\omega}} = \frac{e^2\gamma^6(1 - \boldsymbol{\beta}\boldsymbol{n})}{c\omega_0^2}\frac{\sin^2 \pi\nu N}{\sin^2 \pi\nu}(\rho_\sigma + \rho_\pi)|\dot{\boldsymbol{\beta}}(\nu)|^2, \tag{12}$$

where $\boldsymbol{n}$ is the unit vector in the direction of radiation, N is the number of complete periods of the particle oscillations, $\dot{\beta}(\nu)$ is the Fourier transform of acceleration, ρ_σ and ρ_π are the functions defining pattern of mutually orthogonal polarization components of radiation. Namely, ρ_σ defines the angular distribution of radiation polarized parallel to the plane of the particle trajectory. The frequency of the first harmonic is determined by period of the trajectory shifted by the Doppler effect, and ν indicates the number of harmonic. We define the direction of radiation $\boldsymbol{n}$ and the unit vectors of linear polarization $\boldsymbol{n}_\sigma$ and $\boldsymbol{n}_\pi$, in a spherical coordinate system as shown in figure 4. Functions ρ_σ and ρ_π in this case are of the form

$$\rho_\sigma = \frac{8(1-\psi^2\cos 2\phi)^2}{(1+\psi^2)^5}, \quad \rho_\pi = \frac{8\psi^4\sin^2 2\phi}{(1+\psi^2)^5}, \tag{13}$$

where $\psi = \vartheta\gamma$. A detailed analysis of the polarization components ρ_σ and ρ_π and its graphical representation can be found in [9, 10]. Calculating the Fourier components of the particle acceleration and substituting the result into expression (12), we obtain

$$\frac{d\mathcal{E}}{d\Omega d\tilde{\omega}} = \frac{2e^2\gamma^4\omega^2(1+\psi^2)}{c\omega_0^2}\left|\frac{\xi\nu}{\sinh\xi\nu} - \frac{\sin\omega_0 t_1\nu}{\omega_0 t_1\nu}e^{i\pi\nu}\right|^2 \frac{\sin^2\pi\nu N}{\sin^2\pi\nu}(\rho_\sigma+\rho_\pi), \tag{14}$$

where $\xi = \pi^2 a/y_0$. The first term containing the hyperbolic sine is responsible for radiation in the electric field of the crystallographic plane. Since $\xi \ll 1$, this term is significantly different from zero at high harmonic numbers $\nu \sim \xi^{-1} \gg 1$. The emission spectrum (14) consists of narrow lines at the integer values of ν. The envelope of the spectrum is shown in figure 5.

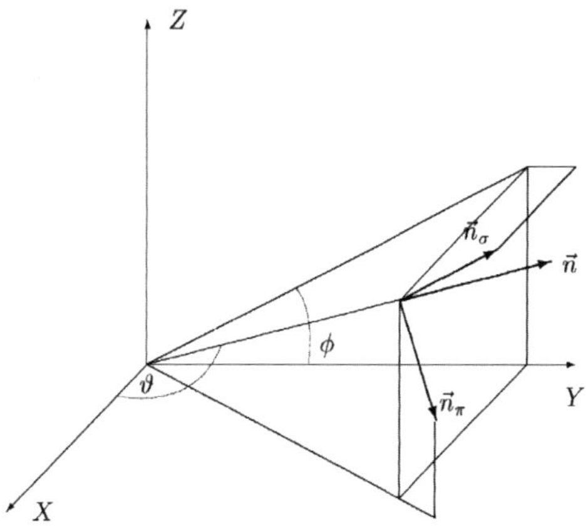

Figure 4. Unit vectors in the spherical coordinate system.

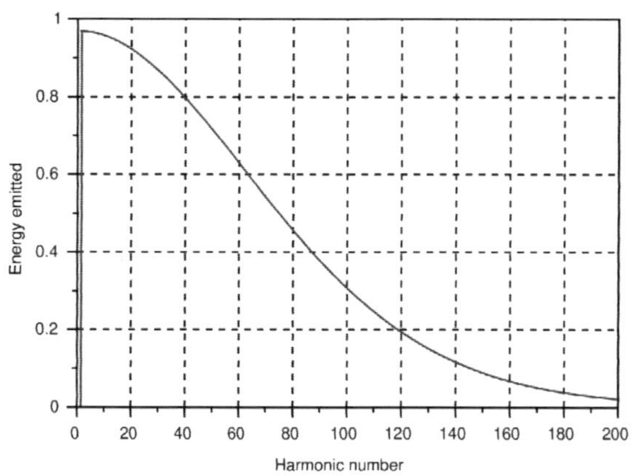

Figure 5. The radiation spectrum of a single particle. $H = 10$ kOe, $mc^2 = 40$ MeV, $eU_0 = 30$ eV, $a = 1$ Å, $y_0 = 500$ Å.

The angular spectral density of radiation of a beam of charged particles can be obtained by averaging of expression (14) over the transverse energy of the particles. Suppose that the beam of particles is of uniform distribution and insides on the crystal parallel to its surface. We assume that the particles perform very large number of oscillations, and proceed to the limit $N \to \infty$. The spectrum in this case becomes discrete and ν takes only integer values. The second term in (14) vanishes, as $\Omega_0 t_1 n \approx \omega_0 t_0 n = \pi n$, and the angular spectral density becomes

$$\frac{d\mathcal{E}}{d\Omega d\tilde{\omega}} = \frac{2e^2\gamma^4 L\sqrt{y_0}(1+\psi^2)}{\pi^2 c R^{3/2}} \frac{\xi^2\nu^2}{\sinh^2\xi\nu}(\rho_\sigma+\rho_\pi),$$

where $L = 2N\sqrt{2y_0 R}$ is the channeling length (the projection of the trajectory on the x axis). Next we average this equation over the initial coordinate y_0 of the particle. Using notation $x = \xi\nu = \nu\pi^2 a/y_0$ we obtain the spectral and angular distribution of the emitted energy of the beam per particle

$$\frac{\overline{\mathrm{d}\mathcal{E}}}{\mathrm{d}\Omega\mathrm{d}\tilde{\omega}} = \frac{2e^2\gamma^4 L\sqrt{2b}(1+\psi^2)}{\pi^2 c R^{3/2}}(\rho_\sigma + \rho_\pi)S(z), \tag{15}$$

where b is the maximum possible value of the initial position, in which the particle is captured into channeling. The value of b can be found from condition that the transverse energy of the particle is less than the potential barrier $V = 1$. It follows then from equation (7) that $b = a/\eta$. The emission spectrum is determined by the spectral function $S(z)$

$$S(z) = z^3 \int\limits_z^{kz} \frac{\mathrm{d}x}{x^2 \sinh^2 x}, \quad z = \frac{\pi\tilde{\omega}\sqrt{aR\eta}(1+\psi^2)}{2^{3/2}c\gamma^2}, \quad k = \sqrt{\frac{b}{y_m}}. \tag{16}$$

The graph of the spectral function $S(z)$ for $k = 10$ is shown in figure 6. The emission spectrum

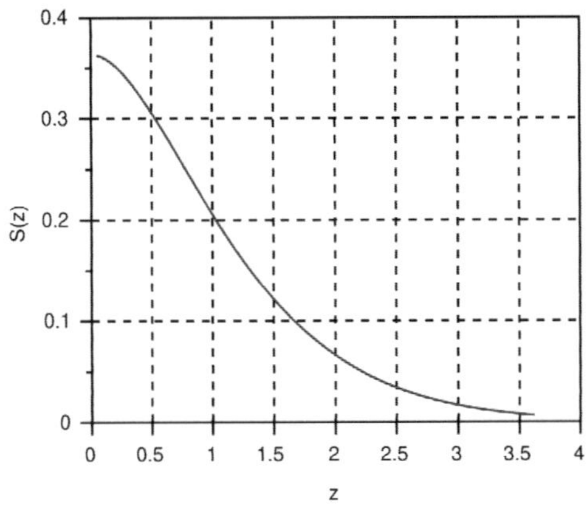

Figure 6. The spectral function $S(z)$ for $k = 10$.

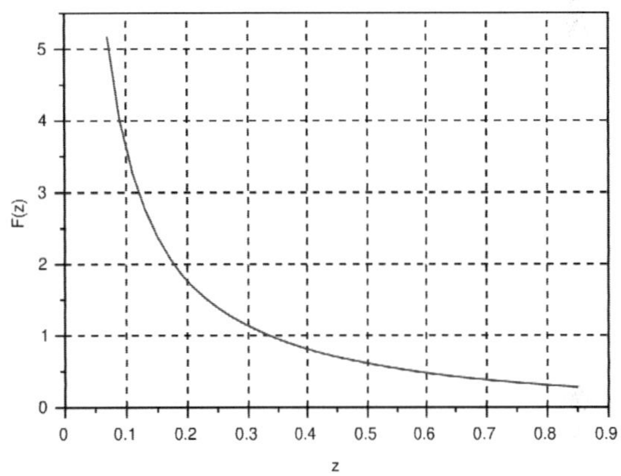

Figure 7. The normalized photonic spectrum $F(z)$ for $k = 10$.

extends from zero to a frequency $\tilde{\omega}_{cr}$ corresponding to the values of $z \sim 1$ or

$$\tilde{\omega}_{cr} \sim \frac{c\gamma^2}{\sqrt{aR\eta}} = \frac{c}{a}\sqrt{\frac{eU_0}{m_0 c^2}}\gamma^{3/2}. \tag{17}$$

As one can see, this frequency does not depend on the magnetic field. This is due to the fact that most of the energy of radiation is generated during reflection of the particle from the surface. In this process the particle reverses its transverse velocity in a very short period of time in comparison with the period of motion T. The duration of the braking process in the vicinity of crystal plane is of order of α^{-1}. Taking into account the Doppler effect, the characteristic frequency of the radiation is $\gamma^2\alpha$. For example, for parameters $eU_0 = 30$ eV, $a = 1$ Å, and $\gamma = 100$ the photon energy is equal to $\hbar\tilde{\omega}_{cr} \approx 15$ keV. The photon energy increases with the particle energy as $\gamma^{3/2}$.

If we devide expression (15) by $\hbar\tilde{\omega}$ where $\hbar$ is the Planck's constant, we get the number of photons N_f per unit frequence per unit solid angle

$$\frac{\mathrm{d}N_f}{\mathrm{d}\Omega\mathrm{d}(\hbar\tilde{\omega})} = \frac{e^2\gamma^2 L\sqrt{ay_m\eta}(1+\psi^2)^2}{\pi\hbar^2 c^2 R}(\rho_\sigma + \rho_\pi)F(z), \tag{18}$$

where $F(z) = S(z)/z$ can refered to as a normalized photonic spectrum. The function $F(z)$ is plotted in figure 7.

4. Conclusion

Under assumption that the magnetic field is much weaker than the average electric field of atoms in the surface layer, the main part of the particle trajectory lies exterior to the screening radius of the surface layer, and represents an arc of a circle, on which the particle moves in the magnetic field. The frequency of the first harmonic of radiation is determined by the length of the arc of the circle. The main part of radiation is generated during the sharp braking of the particles in the electric field of atoms forming the surface. The radiation frequency is inversely proportional to time spent by the particle within the range of screening radius and is much greater than the frequency of the first harmonic. The spectrum extends to the harmonic numbers of order y_0/a, where a is the screening radius, and $y_0 \gg a$ is the maximal distance between the particle and the crystal surface.

The frequency of radiation at the first harmonic and the number of harmonics forming the spectrum are dependent on the magnetic field. However, the spectral range in which radiation is generated by a parallel beam of particles does not depend on the magnetic field, and is determined only by the energy of the particles and the average potential of the surface layer of atoms. Angular distribution and polarization of radiation are typical for radiation of an ultra-relativistic particle moving along a flat trajectory and oscillating with a small amplitude: the radiation is concentrated in a narrow cone in the direction of the average velocity and polarized mainly in the plane of the trajectory.

Radiation effects caused by the surface channeling were also observed at grazing incidence of the electrons on the surface of silicon crystal placed into a betatron chamber [8]. However, the theory developed in this paper is not applicable in this case because interaction of negatively charged particles with the surface is more complicated.

Acknowledgments

This research has been supported by the grant for LRSS, project No. 88.2014.2.

References

[1] Baryshevsky V and Tikhomirov V 2013 *Nucl. Instrum. Methods Phys. Res., Sect. B* **309** 30–36
[2] Bagli E, Bandiera L, Bellucci V, Berra A, Camattari R, Salvador D D, Germogli G, Guidi V, Lanzoni L, Lietti D, Mazzolari A, Prest M, Tikhomirov V and Vallazza E 2014 *Eur. Phys. J. C* **14** 1–7
[3] Baryshevsky V, Dubovskaya I and Grubich A 1980 *Phys. Let. A* **77** 61–64 ISSN 0375-9601
[4] Kaplin V, Plotnikov S and Vorobiev S 1980 *Zh. Tekh. Fiz.* **50** 1079
[5] Vorobiev S, Kaplin V and Rozum E 1980 Patent SU 876044 A URL `http://www.findpatent.ru/patent/87/876044.html`
[6] Vorobiev S, Kaplin V and Rozum E 1982 Patent SU 1101050 A URL `http://www.findpatent.ru/patent/110/1101050.html`
[7] Vysotskii V, Kuzmin R and Maksyuta N 1995 *Laser Physics* **5** 396–400
[8] Kaplin V 2015 *Nucl. Iinstrum. Meth. B* **355** 257–260
[9] Bordovitsyn V (ed) 1999 *Synchrotron Radiation Theory and its Development* (World Scientific)
[10] Hofmann A 2004 *The Physics of Synchrotron Radiation* (Cambridge Univ. Press)

Polarization bremsstrahlung of fast electrons on metal nanospheres in a dielectric matrix in view of plasmon interference effects

V A Astapenko[1], S V Sakhno[1] and Yu A Krotov[2]

[1] Moscow Institute of Physics and Technology, 9 Institutsky Lane, Dolgoprudny 141700, Moscow Region, Russia
[2] "Polyus" R&D Institute named after M.F. Stelmakh, 1 bldg 3 Vvedensky St., Moscow 117342, Russia

E-mail: [1]astval@mail.ru, [2]kryuri@yandex.ru

Abstract. The paper is devoted to the theoretical analysis of polarization bremsstrahlung (PB) [1] due to scattering of fast electrons by a metal nanosphere embedded in a dielectric matrix in a frequency region in the vicinity of dipole and quadrupole plasmon resonances. Here we take into account plasmon interference effects that arise in the frequency-angular distribution of PB. These effects are a consequence of interrelation between contributions to PBs due to a plasmon on the sphere surface of different multipolarity. Our approach is based on the Fermi method of equivalent photons [2] and the Mie theory of radiation scattering by small metal particles [3]. It is shown that taking into account plasmon interference in the PB differential cross-section results in specific features in the spectral distribution of an emitted photon that depend strongly on the radiation angle and nanosphere radius.

1. Introduction

There are two components in bremsstrahlung (B) of a charged particle on a metal nanosphere in a dielectric matrix. One component is static bremsstrahlung (SB) of a projectile on a Coulomb force center [4]. The second component is radiation caused by the fact that a projectile polarizes electron shells (surface plasmons) of the nanosphere, and this polarization is also a source of radiation. This second component is called, for any targets, polarization bremsstrahlung (PB) [5].

2. Cross-section of virtual photon scattering by a metal nanosphere in a dielectric matrix

In case of metal nanospheres, the spectral-angular cross-section of scattering of monochromatic linearly polarized radiation is represented as

$$\sigma_{sc}(\omega,\theta,\varphi)=\pi\, r_s^2\, Q_{sc}(\omega,\theta,\varphi), \tag{1}$$

where r_s is the nanosphere radius, $Q_{sc}(\omega,\theta,\varphi)$ is the spectral-angular efficiency of scattering that is expressed in terms of the scattering function $F_{sc}(\omega,\theta,\varphi)$ as follows [6]:

$$Q_{sc}(\omega,\theta,\varphi)=\frac{4}{x^2}F_{sc}(\omega,\theta,\varphi), \tag{2}$$

Content from this work may be used under the terms of the Creative Commons Attribution 3.0 licence. Any further distribution of this work must maintain attribution to the author(s) and the title of the work, journal citation and DOI.
Published under licence by IOP Publishing Ltd

where $x = k r_s$, $k = \sqrt{\varepsilon_m}\, \omega/c$ is the propagation vector of radiation in a matrix with a dielectric permittivity ε_m, in which nanospheres with a dielectric permittivity $\varepsilon_m(\omega)$ are placed. The geometry of the process is shown in figure 1, where θ, φ are the polar and azimuth angles of the propagation vector of scattered radiation **k**.

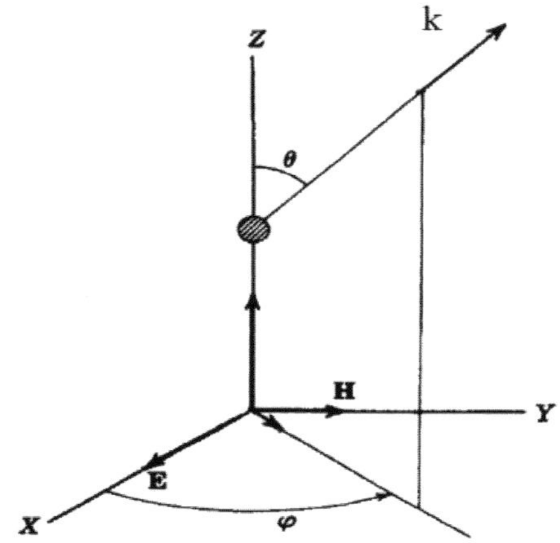

Figure 1. The geometry of the process.

For the function of scattering of monochromatic radiation at a frequency ω to a specified solid angle, we have [6]:

$$F_{sc}(\omega,\theta,\varphi) = |S_1(\omega,\theta)|^2 \sin^2\varphi + |S_2(\omega,\theta)|^2 \cos^2\varphi, \tag{3}$$

$$\text{where } S_1(\theta,\omega) = \sum_{n=1}^{\infty} \frac{2n+1}{n(n+1)} \{a_n(\omega)\pi_n(\theta) + b_n(\omega)\tau_n(\theta)\}, \tag{4}$$

$$S_2(\theta,\omega) = \sum_{n=1}^{\infty} \frac{2n+1}{n(n+1)} \{a_n(\omega)\tau_n(\theta) + b_n(\omega)\pi_n(\theta)\}, \tag{5}$$

$$\text{and } \pi_n(\theta) = \frac{P_n^1(\cos\theta)}{\sin\theta}, \tau_n(\theta) = \frac{dP_n^1(\cos\theta)}{d\theta}. \tag{6}$$

In our case of cylindrical symmetry, the cross section should be averaged over the azimuth angle φ. Then the scattering function loses the dependence on the azimuth angle, and the scattering cross-section depends only on the frequency and the polar angle.

The multipole expansion coefficients a_n, b_n are given within the framework of the Mie theory [3]:

$$a_n(x,y,m) = \frac{\psi_n'(y)\psi_n(x) - m\psi_n'(x)\psi_n(y)}{\psi_n'(y)\zeta_n(x) - m\zeta_n'(x)\psi_n(y)}, \tag{7}$$

$$b_n(x,y,m) = \frac{m\psi_n'(y)\psi_n(x) - \psi_n'(y)\psi_n'(x)}{m\psi_n'(y)\zeta_n(x) - \zeta_n'(x)\psi_n(y)}, \tag{8}$$

where $m = \sqrt{\varepsilon_s(\omega)/\varepsilon_m}$.

The coefficients $\psi_n(z) = z\, j_n(z)$ and $\zeta_n(z) = z\chi_n(z)$ are the functions introduced by Debye, where $j_n(z)$, $\chi_n(z)$ are the spherical Bessel and Hankel functions.

The dielectric permittivity of a metal sphere can be expressed in terms of the real $n_s(\omega)$ and imaginary $\kappa_s(\omega)$ parts of the refractive index of a metal:

$$\varepsilon_s(\omega)=\varepsilon_1(\omega)+i\,\varepsilon_2(\omega)=[n_s(\omega)]^2-[\kappa_s(\omega)]^2+2i\,n_s(\omega)\kappa_s(\omega). \quad (9)$$

For the functions $n_s(\omega)$ and $\kappa_s(\omega)$, we use experimental data obtained in the work [7].

We assume the dielectric permittivity of the matrix to be real and independent of radiation frequency:

$$\varepsilon_m = \text{const}.$$

Following the Fermi method of equivalent photons [2], the general formula for the PBs differential cross-section used in our calculations looks like

$$\frac{d\sigma^{(PB)}}{d\omega\, d\Omega}=\frac{c}{(2\pi)^2}\int_0^\infty \sigma_{sc}^{(ph)}(\omega,\theta)F^2\left(\sqrt{(\omega/\mathrm{v})^2+\rho^{-2}}\right)\frac{|E(\omega,\rho,\mathrm{v})|^2}{\hbar\omega}\rho\, d\rho, \quad (10)$$

where ρ is the impact parameter, v is the electron velocity, Ω, θ are the solid and polar angles of radiation, ω is the PBs frequency, $\sigma_{sc}^{(ph)}(\omega,\theta)$ is the cross-section of radiation scattering by a metal nanosphere, $E(\omega,\rho,\mathrm{v})$ is the Fourier transform of the scattered electron electric field strength for the specified impact parameter, $F(q)$ is the nanosphere form factor, $F(q)=3\frac{j_1(q r_s)}{q r_s}$ and $j_1(x)=\frac{\sin(x)}{x^2}-\frac{\cos(x)}{x}$. In the approximation of straight trajectories, for the normal component of the Fourier transform of the electric field strength in medium with the impact parameter ρ we have $E_n=\frac{2e\omega}{\mathrm{v}^2\tilde{\gamma}}K_1\left(\frac{\omega\rho}{\mathrm{v}\tilde{\gamma}}\right)$, where $K_1(z)$ is the Macdonald function of the first order and $\tilde{\gamma}=\left(1-\varepsilon_m \mathrm{v}^2/c^2\right)^{-1/2}$ is relativistic factor in medium with the dielectric permittivity ε_m. For parallel electric field strength we have: $E_\tau=\frac{2e\omega}{\mathrm{v}^2\tilde{\gamma}^2}K_0\left(\frac{\omega\rho}{\mathrm{v}\tilde{\gamma}}\right)$, where $K_0(z)$ is the Macdonald function of the zero order.

In the following calculations we consider the electron velocity v =90 at.u. For this velocity and glass dielectric permittivity $\varepsilon_m = 2.25$ we have $\tilde{\gamma}\cong 5.87$. Numerical analysis shows that for such parameters the contribution of parallel electric field into PB cross section is negligible. Thus we can substitute E_n in expression (10) instead of E in our calculations.

3. Results of calculations

In this section we present the results of calculations of electron PBs on silver and gold nanospheres in a glass matrix in different regions of PB frequencies in the vicinity of plasmon resonances for different radiation angles and nanosphere radii.

For metal nanospheres, values of radii about 50 nm were chosen, which corresponds to optical frequencies of corresponding plasmon resonances. The PB directional pattern was studied mainly for forward and backward directions; besides, integration and averaging over the PB solid angle were carried out. All calculations were done in the MathCAD system with the use of experimental data on dielectric permittivities of gold and silver in corresponding frequency ranges [7].

The comparison of the PB spectra in the vicinity of surface plasmon resonances for different radiation angles θ on a silver nanosphere in glass is given in figures 2 and 3. Solid line (forward PB) – θ = 0.1, dotted line – the calculation was performed with the scattering cross-section averaged over the solid angle of radiation, dashed line (backward PBs) – θ = 9π/10. The ordinate is plotted in nm^2/ (eV sr), the abscissa is plotted in eV.

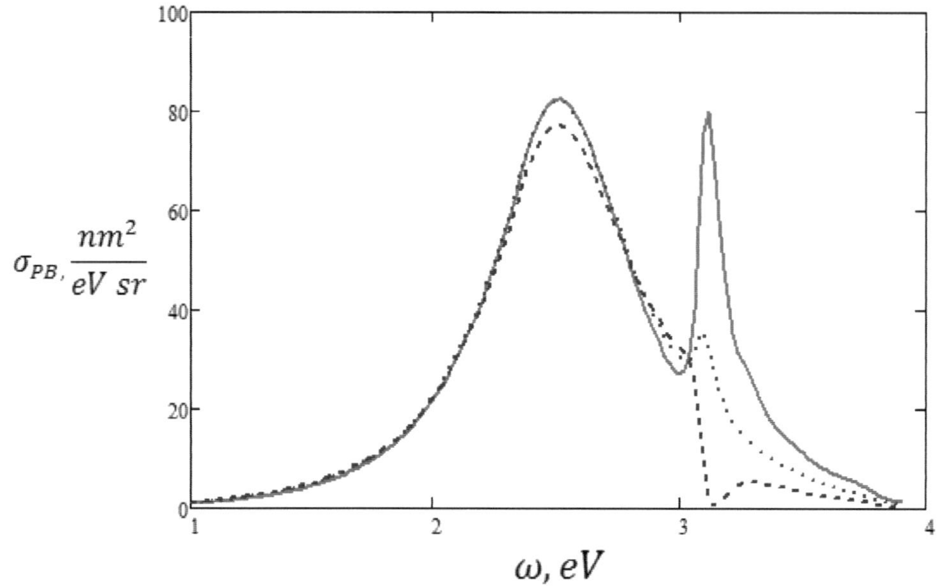

Figure 2. For a silver nanosphere in glass with the radius r_s = 40 nm, the electron velocity v = 90 atomic units: solid line – forward PB, dotted line – the calculation was performed with the scattering cross-section averaged over the solid angle of radiation, dashed line – backward PB.

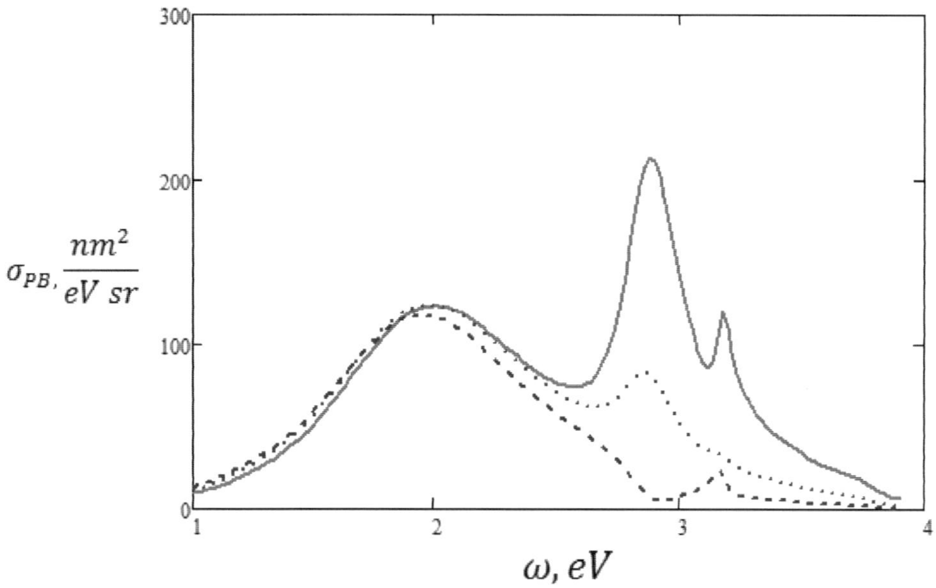

Figure 3. For a silver nanosphere in glass with the radius r_s = 60 nm, the electron velocity v = 90 atomic units: solid line – forward PB, dotted line – the calculation was performed with the scattering cross-section averaged over the solid angle of radiation, dashed line – backward PB.

The dependences of the PB cross-section on the nanosphere radius for two different photon energies are shown in figures 4, 5.

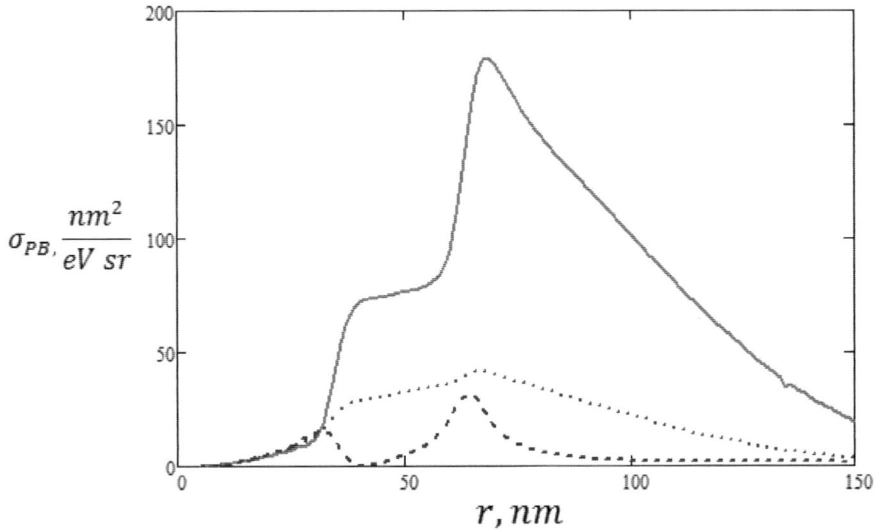

Figure 4. For a silver nanosphere in glass, $\hbar\omega = 3.14$ eV (in the vicinity of the spectral extremum), the electron velocity v = 90 atomic units: solid line – forward PB, dotted line – the calculation was performed with the scattering cross-section averaged over the solid angle of radiation, dashed line – backward PB.

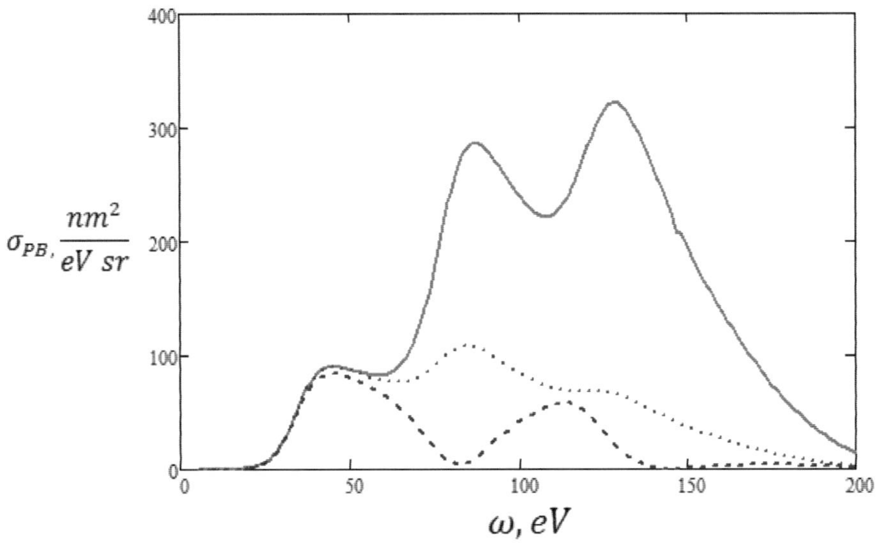

Figure 5. For a silver nanosphere in glass, $\hbar\omega = 2.48$ eV (in the vicinity of the dipole maximum), the electron velocity v = 90 atomic units: solid line – forward PB, dotted line – the calculation was performed with the scattering cross-section averaged over the solid angle of radiation, dashed line – backward PB.

The comparison of the PB spectra for different radiation angles θ on a gold nanosphere in glass with the radius $r_s = 40$ nm is given in figure 6.

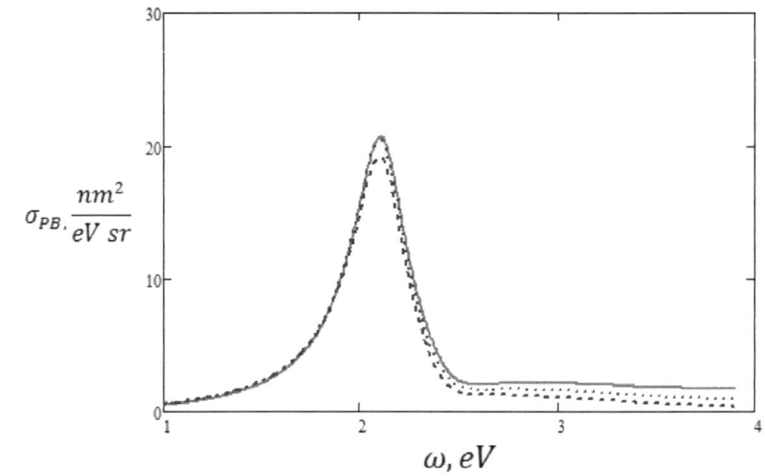

Figure 6. For a gold nanosphere in glass with the radius r_s = 40 nm, the electron velocity v = 90 atomic units: solid line – forward PB, dotted line – the calculation was performed with the scattering cross-section averaged over the solid angle of radiation, dashed line – backward PB.

The imaginary parts of dielectric permittivities of silver (solid line) and gold (dotted line) as functions of photon energy are demonstrated in figure 7.

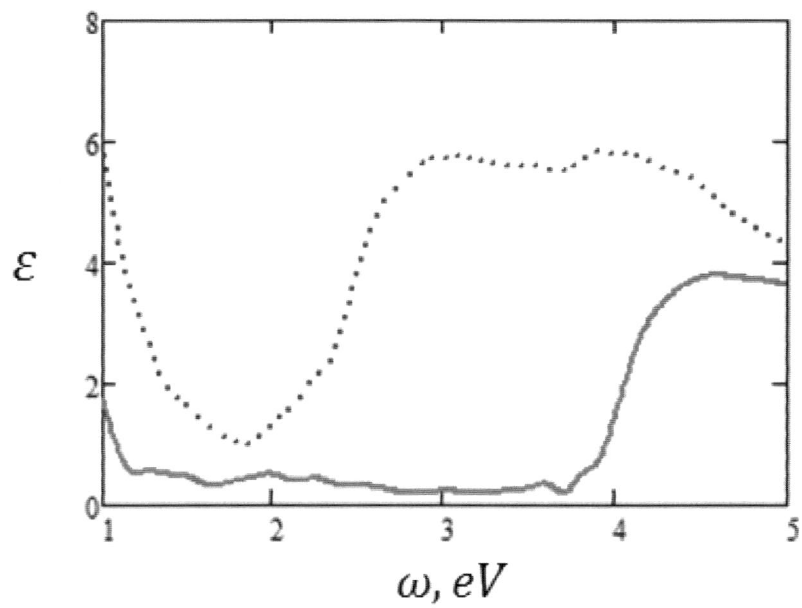

Figure 7. The imaginary parts of dielectric permittivities of silver (solid line) and gold (dotted line) as functions of photon energy are taken from [7].

The normalized angular dependence of PB cross-section for various frequencies is given in figure 8.

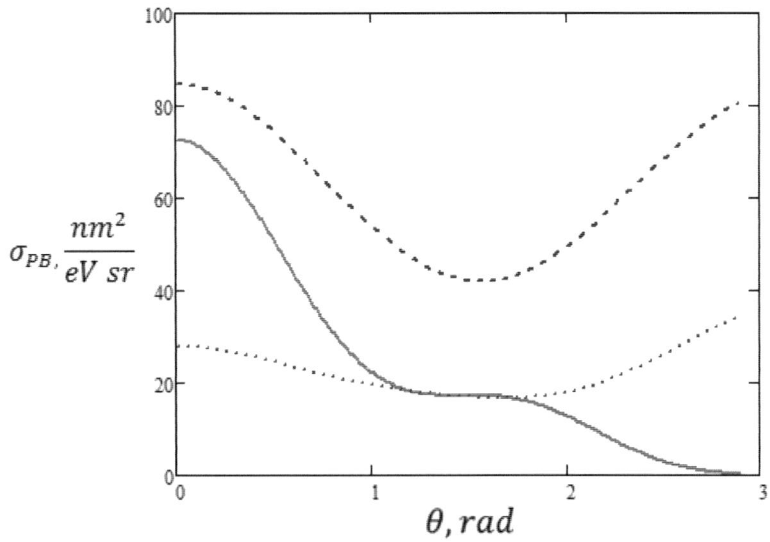

Figure 8. For a silver nanosphere in glass with the radius r_s = 40 nm, the electron velocity v = 90 atomic units: solid line – ħω=3.14 eV, dotted line – ħω=3.0 eV, dashed line – ħω=2.48 eV.

4. Conclusion

Polarization bremsstrahlung (PB) of fast electron scattered by a metal nanosphere embedded in a dielectric matrix in a frequency region in the vicinity of dipole and quadrupole plasmon resonances was studied theoretically. The numerical analysis of electron PB on a metal nanosphere in a dielectric was carried out in view of surface plasmon interference effects.

It was shown that the spectrum of forward PB on a silver nanosphere in glass has a maximum due to constructive interference between dipole and quadrupole plasmon contributions, while the spectrum of backward PBs has a minimum at the same photon energy.

The sharpness of spectral extrema depends on the nanosphere radius and the nanosphere metal.

The angular distribution of PB depends strongly on the photon energy and the nanosphere radius.

The calculation has shown that the role of plasmon interference effects is negligible for PB on a gold nanosphere because of interband transitions in gold, which results in the high value of the imaginary part of the dielectric permittivity of gold in the actual spectral range.

Acknowledgments

The authors are thankful to Professor A. Potylitsyn for useful discussions. The reported study was funded by RFBR according to the research project No. 16-37-00108 mol_a.

References

[1] Astapenko V 2013 *Polarization Bremsstrahlung on atoms, plasmas, nanostructures, and solids* (Springer Series on Atomic, Optical and Plasma Physics 72)
[2] Fermi E 1924 *Z. Phyzik.* **29** 315
[3] Mie G 1908 *Ann. Phys.* **330** (3) 377
[4] Heitler W 1954 *The Quantum Theory of Radiation* (Oxford: Clarendon Press)
[5] Amusia M *et al.* 1992 *Polarization Bremsstrahlung* (New York: Plenum Press)
[6] Van de Hulst H 1981 *Light scattering by small particles* (New York: Dover Publications)
[7] Johnson P and Chirsty R 1972 *Phys. Rev.* B **6** 4370

Coherent bremsstrahlung on a deformed graphene sheet

A R Mkrtchyan V V Parazian and A A Saharian

Institute of Applied Problems in Physics, 25 Nersessian Street, 0014 Yerevan, Armenia

E-mail: malpic@sci.am, vparazian@gmail.com, saharian@ysu.am

Abstract. We investigate coherent bremsstrahlung by electrons in a periodically deformed two-dimensional crystal. The formula for the corresponding spectral-angular density of the cross-section is derived for an arbitrary deformation field. Applications are given for a graphene sheet.

1. Introduction

The bremsstrahlung of high-energy electrons in crystals is one of the most effective methods for producing quasimonochromatic gamma-quanta. It has been extensively investigated either theoretically and experimentally over the last decade (see, for instance, Refs. [1]-[4]). The corresponding radiation has a number of remarkable properties and has found many important applications. Among them is the generation of intense positron beams. The basic source to creating positrons for high-energy electron-positron colliders is the electron-positron pair creation by hard bremsstrahlung photons produced when a powerful electron beam hits an amorphous target. One possible approach to increase the positron production efficiency is to use a crystal target as a positron emitter. When the crystal axis is aligned with the direction of the incident beam, intense photons are emitted through the coherent bremsstrahlung process and the channeling radiation process. These photons are then converted to electron-positron pairs.

From the point of view of controlling the parameters of the coherent bremsstrahlung it is of interest to investigate the influence of external fields, such as acoustic waves, a temperature gradient, etc., on corresponding characteristics. In Refs. [5, 6] it has been shown that the periodical deformations induced in a crystal by hypersonic waves can either enhance or reduce the bremsstrahlung cross-section (for the influence of the hypersonic waves on the coherent electron-positron pair production in crystals see [7, 8]). In the present paper we investigate the coherent bremsstrahlung by relativistic electrons on a two-dimensional deformed crystal. The general case of the profile function for the deformation will be discussed. As an application, we specify the general formula for the cross-section for the case of graphene lattice.

2. Cross-section of the bremsstrahlung on a two-dimensional deformed crystal

We start our consideration by the bremsstrahlung of electrons on a two-dimensional deformed crystal with a complex unit cell. We assume that the undeformed crystal is flat and coincides with the (y, z)-plane of the the cartesian coordinate system (x, y, z). The momentum and

Content from this work may be used under the terms of the Creative Commons Attribution 3.0 licence. Any further distribution of this work must maintain attribution to the author(s) and the title of the work, journal citation and DOI.
Published under licence by IOP Publishing Ltd

the energy of the radiated photon will be denoted by $\mathbf{k}$ and ω, respectively (throughout the paper we use the units with $\hbar = c = 1$). The atoms in the unit cell will be specified by the superscript j and $u_{\mathbf{q}}^{(j)}$ is the Fourier transform of the corresponding potential. Let us denote by $d\sigma_0^{(j)} = |u_{\mathbf{q}}^{(j)}|^2\sigma_0(\mathbf{q})d\theta_\gamma d\omega d\mathbf{q}$ the cross-section of the bremsstrahlung on an isolated jth atom as a function of the transfer momentum $\mathbf{q} = \mathbf{p}_1 - \mathbf{p}_2 - \mathbf{k}$, with $\mathbf{p}_1$ and $\mathbf{p}_2$ being the momenta for the electron in the initial and final states, θ_γ is the angle between the vectors $\mathbf{p}_1$, $\mathbf{k}$, and the factor $\sigma_0(\mathbf{q})$ does not depend on the type of the atom. For the transferred momenta $q \ll m_e$, with m_e being the electron mass, one has [9]

$$\sigma_0(\mathbf{q}) = \frac{e^2 q_\perp^2}{8\pi^4 E_1^2 q_\parallel^2}\left[1 + \frac{\omega^2}{2E_1E_2} - 4y^2(\mathbf{q})\frac{\delta}{q_\parallel}\left(1 - \frac{\delta}{q_\parallel}\right)\right]\frac{dy(\mathbf{q})/d\theta_\gamma}{\sqrt{1-y^2(\mathbf{q})}}, \tag{1}$$

where $\delta = 1/l_c$ is the minimum longitudinal momentum transfer and the function $y(\theta_\gamma)$ is given by the relation

$$y(\mathbf{q}) = \frac{m_e}{2q_\perp h(\mathbf{q})}\left[\left(\frac{E_1\theta_\gamma}{m_e}\right)^2 - h^2(\mathbf{q}) - \frac{q_\perp^2}{m_e^2}\right], \tag{2}$$

with the notation

$$h(\mathbf{q}) = \left(\frac{q_\parallel}{\delta} - 1 - \frac{q_\perp^2}{2E_1\delta}\right)^{\frac{1}{2}}. \tag{3}$$

In (1), e is the electron charge, E_1 is the energy of the initial electron, $q_\parallel$ and $\mathbf{q}_\perp$ are the parallel and perpendicular components of the transferred momentum with respect to $\mathbf{p}_1$. For the formation length of the bremsstrahlung we have the expression $l_c = 2E_1E_2/(\omega m_e^2)$, where E_2 is the energy of the electron after the radiation. The allowed values of the variables are restricted by the conditions

$$q_\parallel \geqslant \delta + \frac{q_\perp^2}{2E_1},\ q_\perp \geqslant 0,\ -1 \leqslant y(\mathbf{q}) \leqslant 1. \tag{4}$$

Now the differential cross-section per atom in a two-dimensional crystal with the atoms located at $\mathbf{r} = \mathbf{r}_n^{(j)}$ is written in the form

$$d\sigma = \sigma(\mathbf{q})\, d\omega d\theta_\gamma d\mathbf{q} = \frac{1}{N_0}\left|\sum_{n,j} u_{\mathbf{q}}^{(j)} e^{i\mathbf{q}\mathbf{r}_n}\right|^2 \sigma_0(\mathbf{q})\, d\omega d\theta_\gamma d\mathbf{q}, \tag{5}$$

where N_0 is the number of atoms in the crystal and the index n enumerates the elementary cells. In a way similar to that for a 3-dimensional crystal, after averaging over the thermal fluctuations, the cross-section is presented in the form

$$\sigma(\mathbf{q}) = \frac{1}{N_0}\left[N\sum_j \left|u_{\mathbf{q}}^{(j)}\right|^2\left(1 - e^{-q^2\overline{u_t^{(j)2}}}\right) + |U_{\mathbf{q}}|^2\right]\sigma_0(\mathbf{q}), \tag{6}$$

where N is the number of the elementary cells, $\overline{u_t^{(j)2}}$ is the temperature-dependent mean-squared amplitude of the thermal vibrations for the jth atom. In (6) we have introduced the notation

$$U_{\mathbf{q}} = \sum_{n,j} u_{\mathbf{q}}^{(j)} e^{i\mathbf{q}\mathbf{r}_{n0}^{(j)}} e^{-\frac{1}{2}q^2\overline{u_t^{(j)2}}}, \tag{7}$$

where $\mathbf{r}_{n0}^{(j)}$ determines the equilibrium positions of the atoms in deformed crystal and $e^{-\frac{1}{2}q^2\overline{u_t^{(j)2}}}$ is the Debye-Waller factor.

The part in the cross-section with the first term in the square brackets of (6) does not depend on the orientation of the crystal axis with respect to the momentum of the electron and corresponds to the incoherent effects. The coherent effects are described by the part

$$\sigma_c\left(\mathbf{q}\right) = \frac{1}{N_0}\left|U_{\mathbf{q}}\right|^2\sigma_0\left(\mathbf{q}\right). \tag{8}$$

For a deformed crystal the vectors determining the positions of the atoms are presented in the form $\mathbf{r}_{n0}^{(j)} = \mathbf{r}_{ne}^{(j)} + \mathbf{u}_n^{(j)}$, where $\mathbf{r}_{ne}^{(j)}$ corresponds to the equilibrium position in the undeformed crystal and $\mathbf{u}_n^{(j)}$ is the atom displacement induced by the deformation. We assume that the undeformed two-dimensional crystal is planar and, hence, the vector $\mathbf{r}_{ne}^{(j)}$ determines the points on the (y, z)-plane. We will consider periodic deformation fields of the form

$$\mathbf{u}_n^{(j)} = \mathbf{u}_0 f(\mathbf{k}_s\mathbf{r}_{ne}^{(j)}), \tag{9}$$

where $f(x)$ is a periodic function with the period equal 2π and $\max|f(x)| = 1$, $\mathbf{u}_0$ is the amplitude of the deformation. For the vector $\mathbf{k}_s$ one has $\mathbf{k}_s = (0, k_{sy}, k_{sz})$ and $k_s = |\mathbf{k}_s| = 2\pi/\lambda_s$ with λ_s being the period of the deformation.

The coordinates of the atoms can be presented in the form $\mathbf{r}_{ne}^{(j)} = \mathbf{R}_n + \rho^{(j)}$, where $\mathbf{R}_n$ corresponds to the positions of the atoms of one of the primitive lattices, and $\rho^{(j)}$ are the equilibrium positions for the other atoms inside nth elementary cell with respect to $\mathbf{R}_n$. Expanding the function $e^{i\mathbf{q}\mathbf{u}_n^{(j)}}$ over the Fourier series, the sum over n in (7) is written in the form

$$U_{\mathbf{q}} = \sum_{m=-\infty}^{+\infty} F_m\left(\mathbf{q}\mathbf{u}_0\right) S(\mathbf{q}, \mathbf{q}{+}m\mathbf{k}_s) \sum_n e^{i(\mathbf{q}+m\mathbf{k_s})\mathbf{R}_n}, \tag{10}$$

where the factor

$$S(\mathbf{q}, \mathbf{q}') = \sum_j u_{\mathbf{q}}^{(j)} e^{i\mathbf{q}'\rho^{(j)}} e^{-\frac{1}{2}q^2\overline{u_t^{(j)2}}}, \tag{11}$$

is determined by the structure of the elementary cell. In (10), $F_m\left(y\right)$ is the Fourier transform of the function $e^{iyf(x)}$:

$$F_m\left(y\right) = \frac{1}{2\pi}\int_{-\pi}^{+\pi} e^{iyf(x)-imx}dx. \tag{12}$$

For an infinite two-dimensional crystal one has

$$\sum_n e^{i(\mathbf{q}+m\mathbf{k_s})\mathbf{R}_n} = \frac{(2\pi)^2}{\Delta}\sum_{\mathbf{g}}\delta(\mathbf{q}^{\parallel} - \mathbf{g}_m), \tag{13}$$

where $\mathbf{q}^{\parallel} = (q_y, q_z)$ is the component of the vector $\mathbf{q}$ parallel to the (y, z)-plane (undeformed crystal), Δ is the surface area of the unit cell, $\mathbf{g} = (g_y, g_z)$ is the reciprocal lattice vector and

$$\mathbf{g}_m = \mathbf{g} - m\mathbf{k}_s. \tag{14}$$

In this case the coherent part of the cross-section is written in the form

$$\sigma_c\left(\mathbf{q}\right) = \frac{N}{N_0}\frac{(2\pi)^2}{\Delta}\sum_{m=-\infty}^{+\infty}|F_m\left(\mathbf{q}\mathbf{u}_0\right)|^2|S(\mathbf{q}, \mathbf{q}{+}m\mathbf{k}_s)|^2\sum_{\mathbf{g}}\delta(\mathbf{q}^{\parallel} - \mathbf{g}_m)\sigma_0\left(\mathbf{q}\right). \tag{15}$$

Substituting this into $d\sigma_c = \sigma_c(\mathbf{q})\, d\omega d\theta_\gamma d\mathbf{q}$, after the integration over the components of the transferred momentum parallel to the undeformed sheet we get

$$\begin{aligned} d\sigma_c &= \frac{e^2 N}{2\pi^2 N_0 E_1^2 \Delta} \sum_{m,\mathbf{g}} |F_m(\mathbf{q}_m \mathbf{u}_0)|^2 \frac{|S(\mathbf{q}_m, \mathbf{q})|^2}{\sqrt{1-y^2(\mathbf{q}_m)}} \frac{q_{m\perp}^2}{q_{m\parallel}^2} \\ &\times \left[1 + \frac{\omega^2}{2E_1E_2} - 4y^2(\mathbf{q}_m)\frac{\delta}{q_{m\parallel}}\left(1-\frac{\delta}{q_{m\parallel}}\right)\right] d\omega dq_x dy(\mathbf{q}_m), \end{aligned} \tag{16}$$

where

$$\mathbf{q} = (q_x, g_y, g_z), \ \mathbf{q}_m = (q_x, g_{my}, g_{mz}), \tag{17}$$

and the summation goes under the conditions

$$q_{m\parallel} \geqslant \delta + \frac{q_{m\perp}^2}{2E_1}, \ -1 \leqslant y(\mathbf{q}_m) \leqslant 1. \tag{18}$$

The momentum conservation is written as $\mathbf{q}_m = \mathbf{p}_1 - \mathbf{p}_2 - \mathbf{k}$.

For a flat crystal, when the deformation is absent one has $F_m(\mathbf{q}_m \mathbf{u}_0) = F_m(0) = \delta_{m0}$ and the general formula is reduced to

$$\begin{aligned} d\sigma_c &= \frac{e^2 N}{2\pi^2 N_0 E_1^2 \Delta} \sum_{\mathbf{g}} \frac{|S(\mathbf{q}, \mathbf{q})|^2}{\sqrt{1-y^2(\mathbf{q})}} \frac{q_{\perp}^2}{q_{\parallel}^2} \\ &\times \left[1 + \frac{\omega^2}{2E_1E_2} - 4y^2(\mathbf{q})\frac{\delta}{q_{\parallel}}\left(1-\frac{\delta}{q_{\parallel}}\right)\right] d\omega dq_x dy(\mathbf{q}), \end{aligned} \tag{19}$$

with $\mathbf{q} = (q_x, g_y, g_z)$. The corresponding constraints are obtained from (18) taking $m = 0$. As is seen, the periodic deformation leads to an additional summation over m in the expression for the coherent bremsstrahlung cross-section. This corresponds to the generation of a superlattice with the period λ_s. The separate terms in the summation over m are weighted by the factor $|F_m(\mathbf{q}_m \mathbf{u}_0)|^2$ which is determined by the profile function of the deformation. In particular, for the sinusoidal deformation field $f(x) = \sin(x + \varphi_0)$ we have

$$F_m(y) = e^{im\varphi_0} J_m(y). \tag{20}$$

From (16), integrating over y and q_x, for the spectral dependence of the cross-section we obtain

$$\begin{aligned} \frac{d\sigma_c}{d\omega} &= \frac{e^2 N}{2\pi N_0 E_1^2 \Delta} \sum_{m,\mathbf{g}} \int dq_x |F_m(\mathbf{q}_m \mathbf{u}_0)|^2 |S(\mathbf{q}_m, \mathbf{q})|^2 \\ &\times \frac{q_{m\perp}^2}{q_{m\parallel}^2}\left[1 + \frac{\omega^2}{2E_1E_2} - \frac{2\delta}{q_{m\parallel}}\left(1-\frac{\delta}{q_{m\parallel}}\right)\right], \end{aligned} \tag{21}$$

with the constraint $q_{m\parallel} \geqslant \delta + q_{m\perp}^2/(2E_1)$. Note that one has $q_{m\perp}^2/(2E_1\delta) = (E_2/\omega)(q_{m\perp}/m_e)^2$ and, hence, for the photons with the energies $\omega > E_2(q_{m\perp}/m_e)^2$ the second term in the right-hand side of this constraint is subdominant. We recall that our consideration is valid under the condition $q_{m\perp}/m_e \ll 1$. For a simple crystal and for the exponential screening of the atomic potential, ignoring the thermal effects, one has

$$|S(\mathbf{q}_m, \mathbf{q})|^2 = |u_{\mathbf{q}_m}|^2 = \frac{16\pi^2 Z^2 e^4}{(\mathbf{q}_m^2 + R^{-2})^2}, \tag{22}$$

with R being the screening radius and Z is the number of electrons in an atom.

3. Special cases and applications to graphene

Let us denote by θ_e and α_e the spherical angles for the vector $\mathbf{p}_1$ in the coordinate system connected with the undeformed crystal:

$$\mathbf{p}_1 = p_1(\cos\theta_e, \sin\theta_e \cos\alpha_e, \sin\theta_e \sin\alpha_e). \tag{23}$$

Here θ_e is the angle between the vector $\mathbf{p}_1$ and x-axis and α_e is the angle between the projection of the vector $\mathbf{p}_1$ on the (y, z) plane and y-axis. Then one gets

$$\begin{aligned} q_{m\|} &= q_x \cos\theta_e + g_m^{(yz)}, \\ q_{m\perp}^2 &= q_x^2 + g_{my}^2 + g_{mz}^2 - q_{m\|}^2, \end{aligned} \tag{24}$$

with

$$g_m^{(yz)} = \sin\theta_e \left(g_{my} \cos\alpha_e + g_{mz} \sin\alpha_e\right). \tag{25}$$

For fixed m and $\mathbf{g}$, the condition $q_{m\|} \geqslant \delta + q_{m\perp}^2/(2E_1)$ determines the limits of the integration over q_x in (21). The formulas (19) and (21), in combination with (24) and (25), give the dependence of the cross-section on the angles θ_e and α_e. The coherence effects induced by the crystal are important when the electron moves close to the crystal plane that corresponds to the angles θ_e close to $\pi/2$. Relatively simple results are obtained for the motion parallel to the undeformed crystal, $\theta_e = \pi/2$. In this case, one has $q_{m\|} = g_m^{(yz)} = g_{my} \cos\alpha_e + g_{mz} \sin\alpha_e$ and $q_{m\|}$ does not depend on q_x.

Let us specify the general results, given in the previous section, for the case of a graphene sheet. Graphene is a carbon monolayer with honeycomb lattice having the interatomic distance $a \approx 0.142$ nm. The unit cell contains two atoms each of which generates a triangular sublattice (see figure 1). To have an orthogonal lattice, as an elementary cell we take the one containing four atoms of carbon (see figure 1). For the radius-vectors $\rho^{(j)} = (\rho_y^{(j)}, \rho_z^{(j)})$ of these atoms one has

$$\begin{aligned} \rho^{(0)} &= (0,0)\,,\ \rho^{(1)} = \frac{a}{2}\left(-1, \sqrt{3}\right), \\ \rho^{(2)} &= \frac{a}{2}\left(3, \sqrt{3}\right),\ \rho^{(3)} = (a, 0)\,. \end{aligned} \tag{26}$$

For the corresponding reciprocal lattice vector one has

$$\mathbf{g} = (g_y, g_z) = \frac{2\pi}{3a}\left(n_y, \sqrt{3}n_z\right), \tag{27}$$

with $n_y, n_z = 0, \pm 1, \pm 2, \ldots$. For graphene one has one sort of atoms and, hence,

$$S(\mathbf{q}_m, \mathbf{q}) = u_{\mathbf{q}_m} e^{-\frac{1}{2}q_m^2 \overline{u_t^2}} \sum_j e^{i\mathbf{q}\cdot\rho^{(j)}}. \tag{28}$$

Now, by taking into account that the vectors $\rho^{(j)}$ lie in the (y, z)-plane and, hence, $\mathbf{q}\cdot\rho^{(j)} = \mathbf{g}\cdot\rho^{(j)}$, for an elementary cell under consideration one gets

$$\Big|\sum_j e^{i\mathbf{q}\cdot\rho^{(j)}}\Big|^2 = 4\left[1 + (-1)^{n_y+n_z}\right]\left[1 + (-1)^{n_y}\cos(\pi n_y/3)\right]. \tag{29}$$

In addition, one has $N/N_0 = 1/4$.

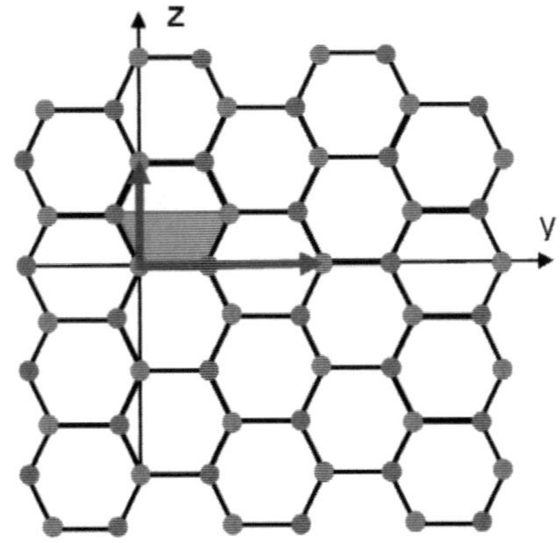

Figure 1. Graphene hexagonal lattice with two triangular sublattices. In order to deal with an orthogonal lattice, we take an elementary cell containing four carbon atoms (shaded region). The corresponding lattice vectors are depicted by arrows.

4. Conclusion

We have investigated the coherent bremsstrahlung of electrons on a deformed two-dimensional crystal with a complex base. For the deformation field we have considered general profile function. The cross-section for the spectral-angular distribution of the radiated photons is given by the expression (16), where the function $y(\mathbf{q})$ is related to the angle θ_γ between the momentum of the initial electron and the wave vector of the photon by (2). In a special case, when the deformation is absent we obtain the cross-section on a flat crystal. In the presence of periodical deformation, the expression for the cross-section contains an additional summation over the reciprocal lattice vector of the corresponding one dimensional superlattice. After the integration over the angle θ_γ, we obtain the spectral density of the cross-section, given by the expression (21). The general results are specified for the graphene sheet. The numerical analysis of the cross-section is in progress and will be presented elsewhere.

References

[1] Ter-Mikaelian M L 1972 *High Energy Electromagnetic Processes in Condensed Media* (New York: Wiley Interscience)
[2] Akhiezer A I and Shulga N F 1996 *High Energy Electrodynamics in Matter* (Amsterdam: Gordon and Breach)
[3] Baier V N, Katkov V M and Strakhovenko V M 1998 *Electromagnetic Processes at High Energies in Oriented Single Crystals* (Singapore: World Scientiic)
[4] Rullhusen R, Artru X and Dhez P 1998 *Sources Using Relativistic Electrons* (Singapore: World Scientific)
[5] Saharian A A, Mkrtchyan A R Parazian V V and Grigoryan L Sh 2004 *Mod. Phys. Lett.* **A19** 99
[6] Mkrtchyan A R, Saharian A A and Parazian V V 2009 *Mod. Phys. Lett.* **B23** 2573
[7] Mkrtchyan A R, Saharian A A, Grigoryan L Sh and Khachatryan B V 2002 *Mod. Phys. Lett.* **A17** 2571
[8] Mkrtchyan A R, Saharian A A and Parazian V V 2006 *Mod. Phys. Lett.* **B20** 1617
[9] Akhiezer A I, Boldyshev V F and Shulga N F 1975 *J. Nucl. Phys.* **22** 1185

The measurement results of carbon ion beam structure extracted by bent crystal from U-70 accelerator

A G Afonin[1], E V Barnov[1], G I Britvich[1], Yu A Chesnokov[1], P N Chirkov[1], A A Durum[1], M Yu Kostin[1], V A Maisheev[1], V I Pitalev[1], S F Reshetnikov[1], A A Yanovich[1], R M Nazhmudinov[2], A S Kubankin[2] and A V Shchagin[2,3]

[1]Institute for High Energy Physics in National Research Centre "Kurchatov Institute", Protvino 142281, Russia;
[2]Belgorod State University, Belgorod 308015, Russia;
[3]Kharkov Institute of Physics and Technology, Kharkov 61108, Ukraine.

E-mail: chesnokov@ihep.ru

Abstract. The carbon ion ^{+6}C beam with energy 25 GeV/nucleon was extracted by bent crystal from the U-70 ring. The bent angle of silicon crystal was 85 mrad. About 2×10^5 particles for 10^9 circulated ions in the ring were observed in beam line 4a after bent crystal. Geometrical parameters, time structure and ion beam structure were measured. The ability of the bent monocrystal to extract and generate ion beam with necessary parameters for regular usage in physical experiments is shown in the first time.

The bent monocrystals are successfully used long time for proton beam extraction in circular accelerators included large modern colliders [1, 2, 3]. The usage of monocrystals to extract the part of a proton beam with energy 50 – 70 GeV in various beam lines is usual practice in IHEP [4, 5]. The carbon ions instead of protons were recently accelerated in the U-70 synchrotron [6]. The mode of extraction of six-charged carbon ions with energy 25 GeV on a nucleon in the beam line 4a was investigated in spring-2015 operating run of U-70. It is necessary to note that the opportunity of deflection and extraction of ion beams by the crystals has been demonstrated about twenty years ago [7, 8] and collimation of circulated heavy ions was investigated at the RHIC and the SPS in later experiments [9, 10]. The main purpose of our work is a creation and study of stable ion beam extracted by a silicon crystal to apply in regular U-70 beam runs.

The scheme of extraction of carbon ions by the crystal deflector and beam transportation in beam line 4a is shown in figure 1.

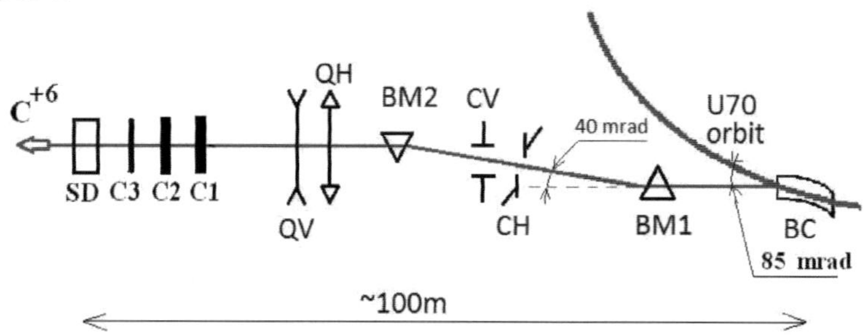

Figure 1. The scheme of crystal extraction of ions in the beam line 4a. BC is the bent crystal, BM1 and BM2 are the dipole magnets, QH and QV are the quadrupole lenses, CH and CV are the beam collimators, C1 - C3 are the scintillation counters, SD is semiconductor detector.

Content from this work may be used under the terms of the Creative Commons Attribution 3.0 licence. Any further distribution of this work must maintain attribution to the author(s) and the title of the work, journal citation and DOI.
Published under licence by IOP Publishing Ltd

The crystal deflector BC has been installed in vacuum goniometer inside the 27th magnetic block of U-70. The lengthened silicon crystal plate with orientation (111) was bent by 85 mrad angle by the metallic holder (see figure 2) and had dimensions 0.3×3×60 mm (thickness, height, length along a beam). Such significant angle of the crystal bend is determined by the specificity of the beam line 4a. Nominally, this beam line is used for negative particles production, which is formed on internal amorphous target in U-70. Negative particles are pushed out by a magnetic field of the accelerator in the external side of a ring, but the protons or ions deflected by a crystal are turned inside the ring. Circulating beam is tuned on the crystal by slow increasing field in bump-magnets. The ion beam from bent crystal was directed on an axis of 4a beam line by two dipole magnets BM1 and BM2 and then formed by two quadrupole lenses QH and QV. The copper collimators CH and CV have been opened completely. A couple of big scintillation counters C1 and C2 measured the beam intensity.

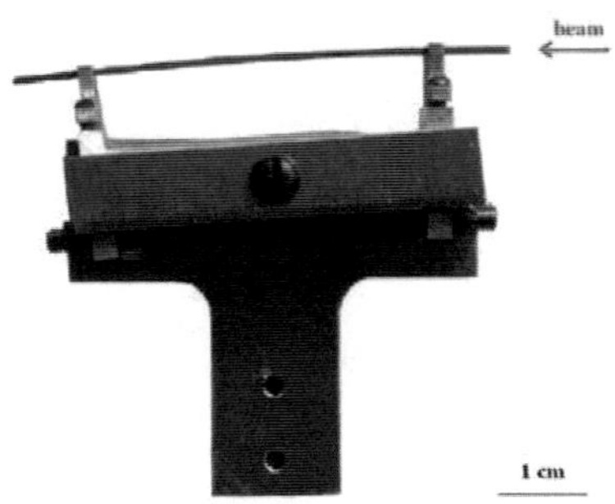

Figure 2. The crystal device with 85 mrad bends.

There is orientation curve, which describes a dependence of particle beam intensity in the beam line versus crystal alignment angle with respect to direction of accelerator orbit. It is a direct proof of beam channeling in crystal when the experimental points are fitted by this curve. The results of angular scanning are shown in figure 3.

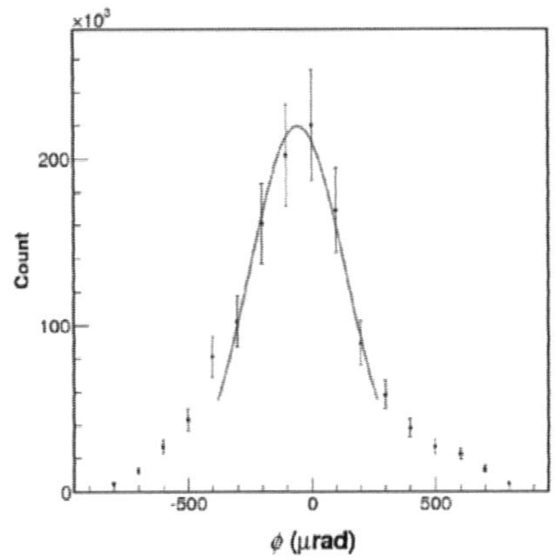

Figure 3. Dependence of particles quantity in the beam line versus crystal alignment angle ϕ with respect to direction of beam orbit in accelerator.

Maximum of (2.2 ± 0.3) $\times 10^5$ particles in the beam line 4a were observed at 10^9 ions on crystal deflector. The curve width is 0.2 mrad. It corresponds to angular divergence of a beam at the accelerator. The obtained results for efficiency of carbon ion extraction correspond to theoretical expectations, as shown below.

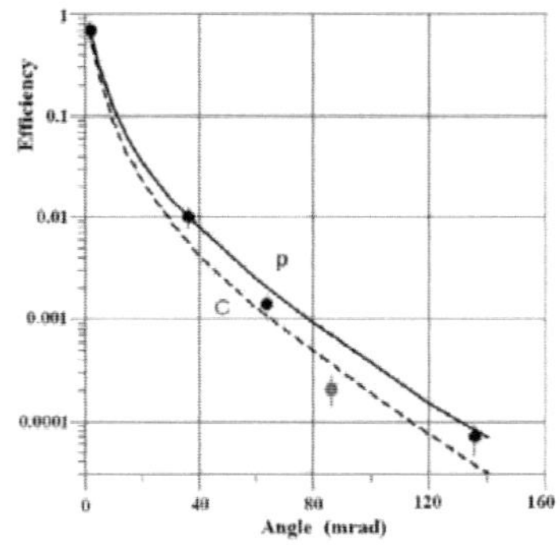

Figure 4. Efficiency of crystal extraction for a circulating beam versus crystal bend angle.

The experimental point (red color in figure 4) for carbon falls on a theoretical curve [11]. The data for a proton beam (a theoretical curve and experimental points [11]) are shown also. The calculations of crystal efficiency for beam extraction were done by the program SCRAPER [12]. This program described the processes of particle channeling and dechanneling in crystal. And also the nuclear interactions during particles penetration in a crystal material and multi-turn movement in accelerator ring were taken into account. It is necessary to note, the extraction efficiency of carbon ions is below in few times for corresponding value of a proton beam, although the channeling properties of crystals are identical to protons and ions with identical magnetic rigidity (p/z), which is the ratio of particle momentum to its charge [7,8]. The crystal extraction efficiency for ions is much smaller than for protons, which was described early [8, 9, 10]. It is well explained by a more intensive nuclear interaction of ions with a crystal target compare to protons during the particles multi-turn movement at accelerator.

The signals from scintillation monitors near the point of beam prompting on a crystal are shown in figure 5. It is visible, that uniformity of beam tuning and time structure is satisfactory in spite of the small intensity of circulating particles.

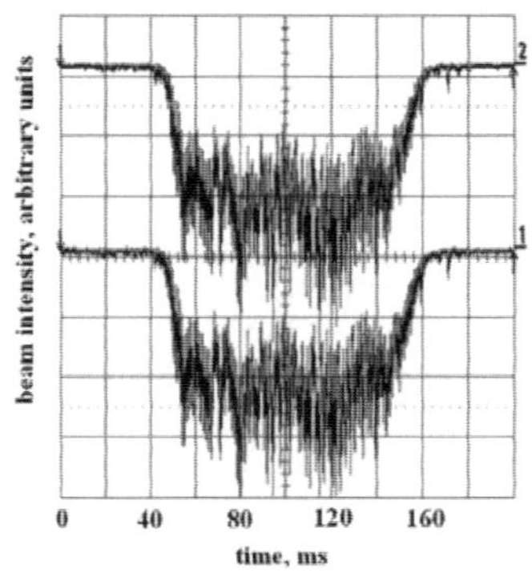

Figure 5. The time structure of ion beam directed on a crystal, on monitors 1 and 2 (the picture with the oscilloscope).

The image of the extracted beam is shown in figure 6. This image was obtained by dosimetric film EBT3, which placed at the detectors position. The size of carbon ions beam (FWHM) is equal ~10 mm in horizontal and vertical planes. It corresponds to the size of proton beam, which was extracted earlier.

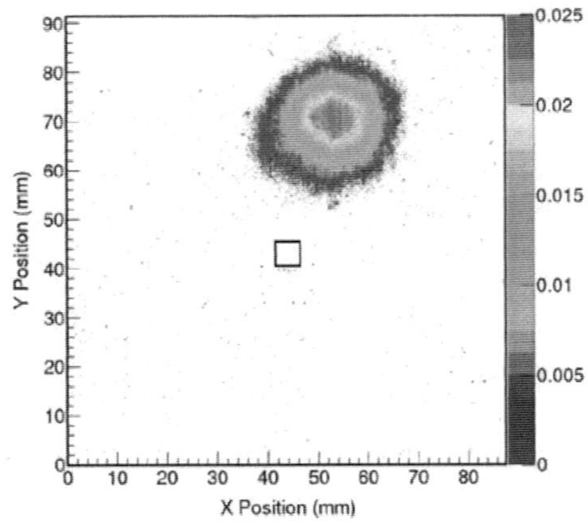

Figure 6. The image of particles density in the extracted ions beam.

The ionization energy losses in the thin scintillator C3 were used to determinate of beam composition in the center of beam spot. The results are shown in figure 7. The ions identification with a various charge was done by square-law dependence of ionization losses. It is visible, that the one peak dominates (approximately 80%). It matches carbon ion having charge 6.

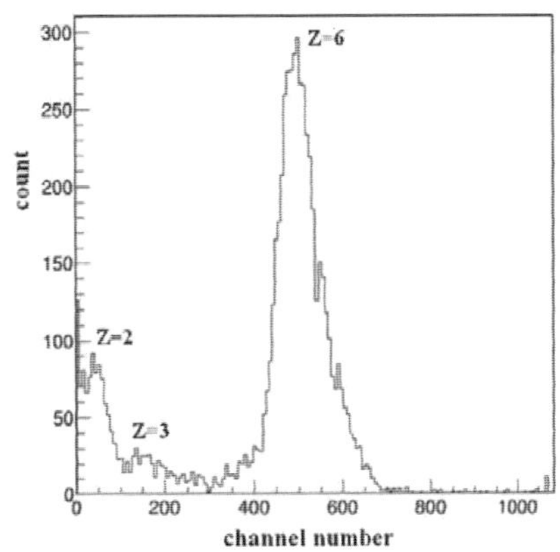

Figure 7. Ionization energy loss spectra of beam particles in the scintillator C3.

The semiconductor surface-barrier detector SD was used to measure the beam composition on periphery of the beam spot at 1% basic intensity level. The position of the detector with respect to the center of the beam is shown in figure 6 by a black square. Spectra of ionization losses in the detector are shown in figure 8. About 30% of carbon and the large quantity of fragments were there. The main reason of fragments is interaction of the extracted carbon beam with vacuum partition foils and other substance in the beam line 4a (about 5 g/cm^2).

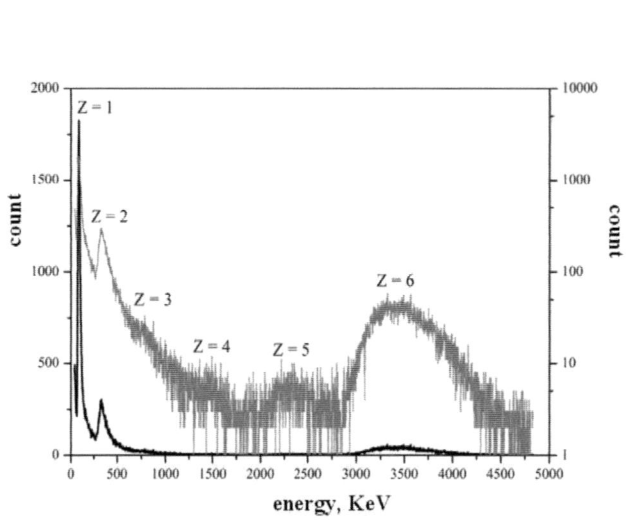

Figure 8. Ionization energy loss spectra of beam particles in the semiconductor detector (a black line – linear scale, a grey line – logarithmic scale).

Summary

Our investigation shown the possibility of bent monocrystal to extract the carbon ion beam from the accelerator with reasonable parameters (intensity, time structure and beam composition) for some physical experiments planned in regular U-70 runs.

References

[1] Afonin A G, Baranov V T, Biryukov V M et al 2001 *Phys. Rev. Lett.* **87** 094802
[2] Scandale W, Arduini G, Assmann R et al 2010 *Phys. Lett.* B **692** 78
[3] Shiltsev V, Annala G, Drozhdin A et al 2010 FERMILAB-CONF-10-127-APC May 2010
[4] Asseev A A, Bavizhev M D, Lyudmirsky E A et al 1991 *Nucl.Instr.Meth.* A **309** 1
[5] Chesnokov Yu A, Afonin A G, Baranov V T et al 2013 *Nucl.Instr.Meth.* B **309** 105
[6] Ivanov S V 2012 Advances of light-ion acceleration program in the U70 2012 *Proceedings of RUPAC* Saint-Petersburg Russia p 120
[7] Bel'zer L I, Bodyagin V A, Vardanyan I N et al 1988 *JETP Lett.* **46** 381
[8] Arduini G, Biino C, Clement M et al 1997 *Phys.Rev.Letters* **79** 4183
[9] Fliller R P, Drees A, Gassner D et al 2006 *Phys.Rev. ST Accel. Beams* **9** 013501
[10] Scandale W, Arduini G, Assmann R et al 2011 *Phys.Lett.* B **703** 547
[11] Arkhipenko A A, Afonin A G, Baranov V T et al 2008 *JETP Letters* **88** 229
[12] Degtyarev I, Liashenko O A and Yazynin I A 2001 *Proc. of PAC2001* 2506

Wave functions of channeling electrons in regular and chaotic cases

N F Shul'ga[1,2], V V Syshchenko[3], A I Tarnovsky[3] and A Yu Isupov[4]

[1] Akhiezer Institute for Theoretical Physics of the NSC "KIPT", Akademicheskaya Street, 1, Kharkov 61108, Ukraine
[2] V.N. Karazin National University, Svobody Square, 4, Kharkov 61022, Ukraine
[3] Belgorod State University, Pobedy Street, 85, Belgorod 308015, Russian Federation
[4] Laboratory of High Energy Physics (LHEP), Joint Institute for Nuclear Research (JINR), Dubna 141980, Russian Federation

E-mail: syshch@bsu.edu.ru, syshch@yandex.ru

Abstract. The spectral method applicability to the problem of the charged particles axial channeling is studied. The quantum state characteristics for the electron motion near the atomic string [110] direction of the silicon crystal are obtained using this method of Schrödinger equation numerical solution. The full set of eigenenergies and eigenfunctions for the electron with longitudinal energy of 5 MeV is computed. The results obtained illustrate the difference in the wave function morphology for regular and chaotic cases.

1. Introduction

The fast charged particles incident onto the crystal under a small angle to any crystallographic axis densely packed with atoms can perform the finite motion in the transverse plane; such motion is known as the axial channeling [1]–[3]. The particle motion in the axial channeling mode could be described with a good accuracy as the one in continuous potential of the atomic string, i.e. in the potential of atoms averaged along the string axis. During motion in this potential the longitudinal particle momentum $p_{\|}$ is conserved, so the motion description is reduced to two-dimensional problem of motion in the transversal plane. This transverse motion could be substantially quantum [1].

From the viewpoint of the dynamical systems theory, the channeling problem is interesting because the particle's motion could be both regular and chaotic. The wave functions of the stationary quantum states in the situations of regular and chaotic motion in the classical limit are computed in the present paper. The qualitative difference between them illustrates the basic statements of the quantum chaos theory [4]–[8].

2. Method

The electron transversal motion in the atomic string continuous potential is described by the two-dimensional Schrödinger equation

$$\hat{H}\Psi(x, y, t) = i\hbar \frac{\partial}{\partial t}\Psi(x, y, t) \tag{1}$$

Content from this work may be used under the terms of the Creative Commons Attribution 3.0 licence. Any further distribution of this work must maintain attribution to the author(s) and the title of the work, journal citation and DOI.
Published under licence by IOP Publishing Ltd

with Hamiltonian

$$\hat{H} = -\frac{\hbar}{2E_{\parallel}/c^2}\left[\frac{\partial^2}{\partial x^2} + \frac{\partial^2}{\partial y^2}\right] + U(x,y) \tag{2}$$

and the value $E_{\parallel}/c^2$ (here $E_{\parallel} = (m^2c^4 + p_{\parallel}^2c^2)^{1/2}$) instead of the particle mass [1].

The Hamiltonian eigenfunctions have been found numerically as well as the transverse energy eigenvalues using the so-called spectral method [9]. For the channeling problem it has been applied for the first time in [10].

The spectral method idea is the following. Every solution of the time-dependent Schrödinger equation could be expressed as the superposition

$$\Psi(x,y,t) = \sum_{n,j} A_{n,j}\psi_{n,j}(x,y)\exp(-iE_nt/\hbar) \tag{3}$$

of the Hamiltonian eigenfunctions $\psi_{n,j}(x,y)$,

$$\hat{H}\psi_{n,j}(x,y) = E_n\psi_{n,j}(x,y) \quad ,$$

where the j index is used to distinguish the degenerate states corresponding to the energy E_n. Computation of the correlation function

$$P(t) = \int_{-\infty}^{\infty}\int_{-\infty}^{\infty} \Psi^*(x,y,0)\Psi(x,y,t)\,dxdy \tag{4}$$

for the wave function of the form (3) gives

$$P(t) = \sum_{n,n',j,j'} \exp(-iE_{n'}t/\hbar)A^*_{n,j}A_{n',j'}\int_{-\infty}^{\infty}\int_{-\infty}^{\infty} u^*_{n,j}(x,y)\psi_{n',j'}(x,y)dxdy =$$

$$= \sum_{n,n',j,j'} \exp(-iE_{n'}t/\hbar)A^*_{n,j}A_{n',j'}\delta_{nn'}\delta_{jj'} = \sum_{n,j}|A_{n,j}|^2\exp(-iE_nt/\hbar) \quad . \tag{5}$$

Fourier transformation of (5) leads to the expression

$$P_E = \int_{-\infty}^{\infty} P(t)\exp(iEt/\hbar)\,dt = 2\pi\hbar\sum_{n,j}|A_{n,j}|^2\,\delta(E-E_n) \quad . \tag{6}$$

We see the Fourier transform of the correlation function looks like a series of δ-form peaks, whose positions indicate the energy eigenvalues. So, the method is based on the numerical simulation of the time evolution of initial wave function $\Psi(x,y,0)$ according to time-dependent Schrödinger equation.

Of course, the numerical integration of (1) could be performed for the finite time interval T only, so the Fourier transform of the correlation function takes the form

$$P(E) = \int_0^T P(t)\exp\left(\frac{i}{\hbar}Et\right)dt = \sum_{n,j}|A_{n,j}|^2\frac{\sin\left[(E_n-E'_n)T/2\hbar\right]}{(E_n-E'_n)/2\hbar}\exp\left[\frac{i}{2\hbar}(E-E_n)T\right] \quad . \tag{7}$$

The $\sin(xT)/x$ function in the right part of (7) has the main maximum at $x = 0$, whose magnitude is increased and width is narrowed with T interval increasing. Its other maxima and minima at both sides from the main maximum have the lower magnitudes. So, instead of δ-like peaks set we have the superposition of maxima, whose widths are inverse proportional to

T. The required resolution of the computation dictates the T value, while the expected minimal distance ΔE between the neighbor energy levels dictates the resolution, and condition for T is

$$T > 2\pi\hbar/\Delta E \ . \tag{8}$$

To avoid the superposition of side maxima of $\sin(xT)/x$ function from neighbor energy levels this requirement should be enhanced. In our calculations we use

$$T = 16\pi\hbar/\Delta E \ . \tag{9}$$

The numerical procedure based on these ideas has been developed and successfully used for transverse energy levels search with high precision in [11]–[13].

The stationary wave functions could be obtained with the spectral method also [9, 10]. The superposition (3) multiplied by $\exp(iE_n t/\hbar)$ and integrated over the time interval T provides the eigenfunctions $\psi_{n,j}(x,y)$ of the eigenvalue E_n:

$$\int_0^T \Psi(x,y,t)\exp\left(\frac{i}{\hbar}E_n t\right)dt = \sum_{n',j} A_{n',j}\,\psi_{n',j}(x,y)\int_0^T \exp\left[\frac{i}{\hbar}(E_n - E'_n)t\right]dt = \tag{10}$$

$$= \sum_{n',j} A_{n'}\,\psi_{n',j}(x,y)\,\frac{\sin\left[(E_n - E'_n)T/2\hbar\right]}{(E_n - E'_n)/2\hbar}\exp\left[\frac{i}{2\hbar}(E_n - E'_n)T\right] \ .$$

For the large enough T the left part integral in (10) is proportional to either eigenfunction $\psi_n(x,y)$ (without state degeneration), or $\psi_{n,j}(x,y)$ superposition (if n-th energy level is degenerated) with a good accuracy.

3. Wave functions in the regular case

The field of single atomic string has the axial symmetry $U(x,y) = U(r)$, if the neighbor strings influence is neglected. This symmetry allows us to classify the electron quantum states. In the polar coordinates

$$\hat{H} = -\frac{\hbar}{2E_{\parallel}/c^2}\left[\frac{1}{r}\frac{\partial}{\partial r}\left(r\frac{\partial}{\partial r}\right) + \frac{1}{r^2}\frac{\partial^2}{\partial\varphi^2}\right] + U(r) \ . \tag{11}$$

The equation for the eigenfunctions and eigenvalues of the operator (11)

$$-\frac{\hbar}{2E_{\parallel}/c^2}\left[\frac{1}{r}\frac{\partial}{\partial r}\left(r\frac{\partial}{\partial r}\right) + \frac{1}{r^2}\frac{\partial^2}{\partial\varphi^2}\right]\psi(r,\varphi) + U(r)\psi(r,\varphi) = E_{\perp}\psi(r,\varphi) \tag{12}$$

allows us to separate the variables, and their solutions are

$$\psi_{n_r,m}(r,\varphi) = \frac{1}{\sqrt{2\pi}}e^{im\varphi}\rho_{n_r,|m|}(r) \ , \tag{13}$$

where the $\rho_{n_r,|m|}(r)$ function is the solution of

$$\frac{1}{r}\frac{\partial}{\partial r}\left(r\frac{\partial}{\partial r}\right)\rho_{n_r,|m|}(r) + \frac{2E_{\parallel}/c^2}{\hbar}\left(E_{\perp} - U(r) - \frac{\hbar}{2E_{\parallel}/c^2}\frac{m^2}{r^2}\right)\rho_{n_r,|m|}(r) = 0 \ . \tag{14}$$

The particle quantum states in the potential $U(r)$ with axial symmetry are characterized by two quantum numbers: the radial one n_r and projection m of the orbital momentum to the symmetry axis of the field. The n_r is equal to the number of zeroes of the function $\rho_{n_r,|m|}(r)$ at finite r

Journal of Physics: Conference Series **732** (2016) 012028 doi:10.1088/1742-6596/732/1/012028

(except the possible zero at $r = 0$). The states with $m = 0$ are non-degenerated, while ones with non-zero m are doubly degenerated by the signum of m. The eigenvalues of Hamiltonian (11), i.e. the eigenvalues of the transversal motion energy of the channeling electron, depend in general on both quantum numbers: $E_\perp = E_{n_r,|m|}$.

First of all let's consider the low energy ($E_\parallel = 5$ MeV) electron's motion near direction of the atomic string [110] of the Si crystal. The continuous potential could be represented by the modified Lindhard potential [1]

$$U_1(x, y) = -U_0 \ln\left(1 + \frac{\beta R^2}{x^2 + y^2 + \alpha R^2}\right) \quad , \tag{15}$$

where $U_0 = 60$ eV, $\alpha = 0.37$, $\beta = 3.5$, $R = 0.194$ Å (Thomas–Fermi radius).

In the upper left plot in figure 1 the wave function of the ground state is shown. It has single maximum in center ($r = 0$) and has not zeroes at all (except the asymptotic zero at $r = \infty$).

The other wave functions are represented by black (for $\psi(x, y) < 0$) and white (for $\psi(x, y) > 0$) regions on the plane (x, y) (see figure 1). The irregularity on the periphery of some plots is due to numerical uncertainties. This representation allows us to classify the wave functions using the quantum numbers n_r and m easily, by the wave function zeroes counting.

Indeed, at the $m = 0$ (upper row in figure 1) the wave functions $\psi_{n_r,0}(r, \varphi)$ depend on radial coordinate r only, so the $\psi(x, y) = 0$ zero lines are concentric circles.

Each next solution $\rho_{n_r,0}(r)$ of the equation (14) at $m = 0$ in the solution sequence, where the corresponding eigenenergy is increased, has more zeroes than previous by exactly one. So we can enumerate the corresponding quantum eigenstates by the quantum number n_r which is equal to the number of zeroes of the radial wave function $\rho_{n_r,0}(r)$. Practically n_r is equal to number of borders between the concentric regions.

For the $m \neq 0$ case the situation is analogous. The degeneration by signum of m means the our method provides for each $E_\perp$ eigenvalue the superposition of $\psi_{n_r,m}(r, \varphi)$ and $\psi_{n_r,-m}(r, \varphi)$ functions with equal weights instead of eigenfunction (13). This superposition looks as follows:

$$\rho_{n_r,|m|}(r) \cos\left[|m|\varphi + \alpha_m\right] \quad . \tag{16}$$

So, the m value for calculated wave function equals to number of white (or black) sectors on the plot in figure 1. The energy levels of the considered system classified by quantum numbers n_r and m are presented in figure 2.

The character feature of the calculated stationary wave functions is the crossing of $\psi(x, y) = 0$ lines, which forms the chessboard-like picture. The counting of these zero lines allows us to find the quantum number values easily. The crossings presence is due to the considered system integrability. Indeed, this two-dimensional system has two degrees of freedom and two integrals of motion — transversal energy $E_\perp$ and projection m of orbital momentum to the field symmetry axis. So, the variables separation is possible in the motion equation, which appears integrable in quadratures. The presence of the zero lines (surfaces for dimensions number more than two) crossings (or thin quasicrossings) appears a general property of quantum systems, whose classical counterparts demonstrate the regular dynamic [4]–[7].

4. Wave functions in the chaotic case

The qualitatively different picture is observed for the non-integrable systems, for example, the electron motion in the field of two neighbor atomic strings [110] of Si crystal. This field (the contribution from other atomic string pairs could be neglected) could be written as

$$U(x, y) = U_1(x, y + a/8) + U_1(x, y - a/8) \quad , \tag{17}$$

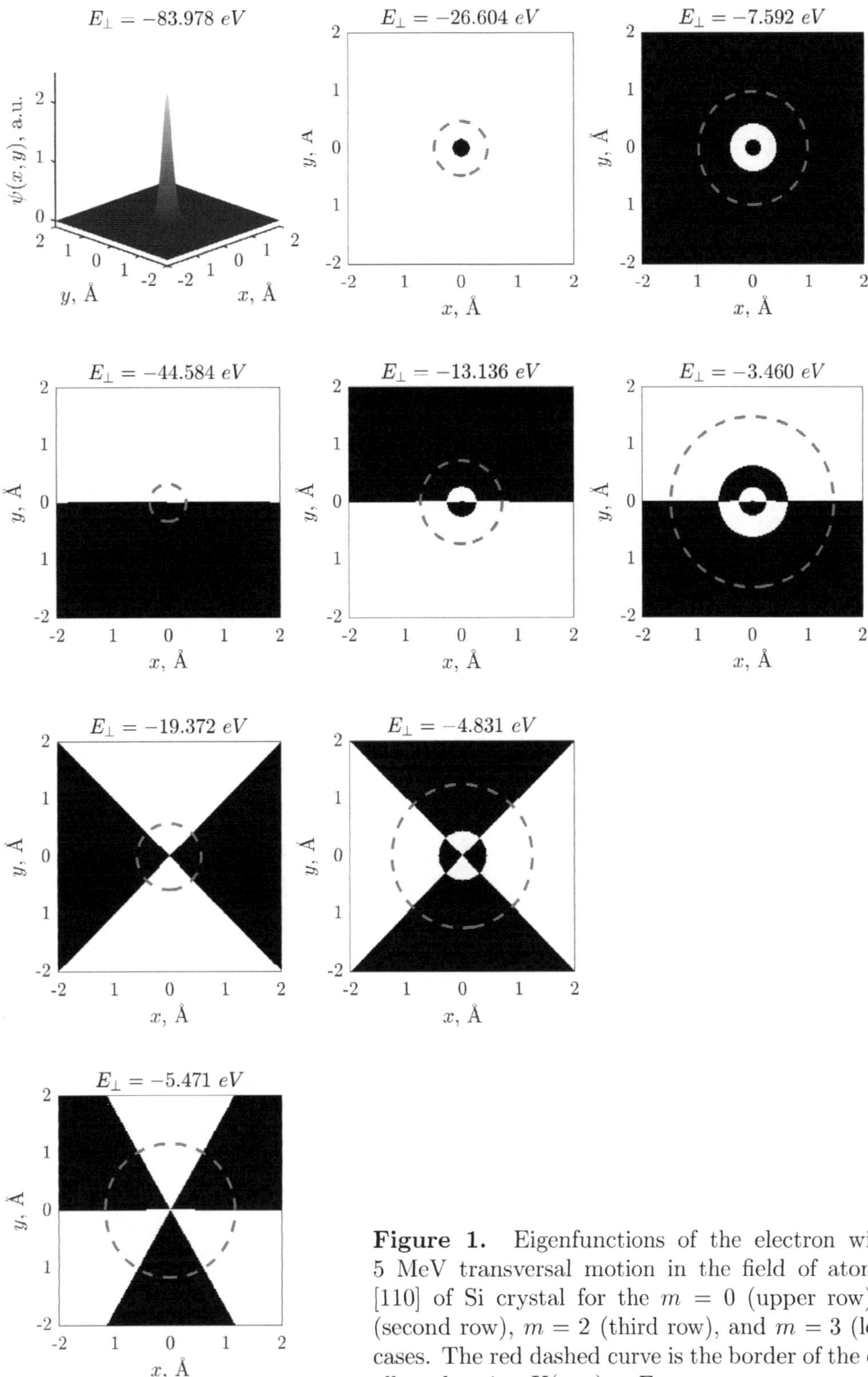

Figure 1. Eigenfunctions of the electron with $E_\parallel$ = 5 MeV transversal motion in the field of atomic string [110] of Si crystal for the m = 0 (upper row), m = 1 (second row), m = 2 (third row), and m = 3 (lower row) cases. The red dashed curve is the border of the classically allowed region $U(x, y) < E_\perp$.

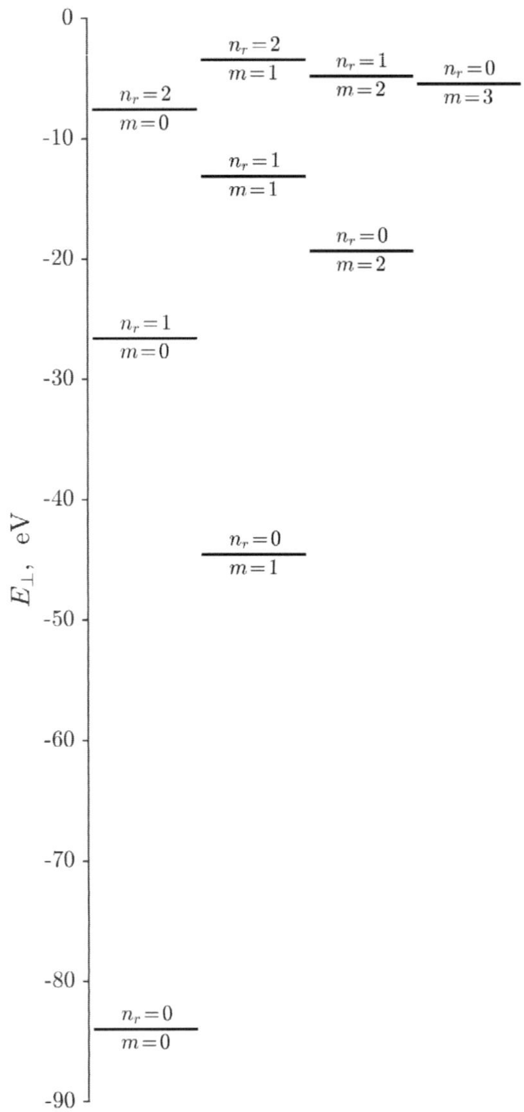

Figure 2. The energy levels classification of the electron with $E_\parallel = 5$ MeV transversal motion in the field of atomic string [110] of Si crystal.

$E_\perp$, eV	n_r	m
−83.978	0	0
−44.5838	0	1
−26.604	1	0
−19.3717	0	2
−13.136	1	1
−7.59175	2	0
−5.4715	0	3
−4.83075	1	2
−3.46025	2	1

where $a = 5.431$ Å is the lattice period. Some calculated eigenfunctions of the electron with longitudinal energy $E_\parallel = 5$ MeV channeling in this field are shown in figure 3, while all the eigenvalues — in table 1. Without the field axial symmetry the single integral of motion (energy $E_\perp$) still exists for two-dimensional system. This leads to dramatical change of the wave functions nature: the zero lines $\psi(x, y) = 0$ have not crossings. However the low energy $E_\parallel = 5$ MeV case not allows us to see this change clearly, so some wave functions of the electron channeling with $E_\parallel = 500$ MeV are presented in figure 4. The latter energy is high enough to ensure big total number of energy levels and small inter-level distances [12]. Namely this situation, similar to one in the classical mechanics with its continuous energy spectra, is needed for the quantum chaos manifestation studies [5, 6, 8]. We can see bizarre picture of black and white "islands" (see figure 4) instead of chessboard-like one for the axially symmetrical field. This behavior is general for the non-integrable systems [4]–[7].

In the classical mechanics the integrability (or not) of the system is closely related with regularity (or chaoticity) of motion of this system [5, 6]. The dynamical chaos in the classical case is the essential sensitivity of the system motion to the initial conditions, which leads to

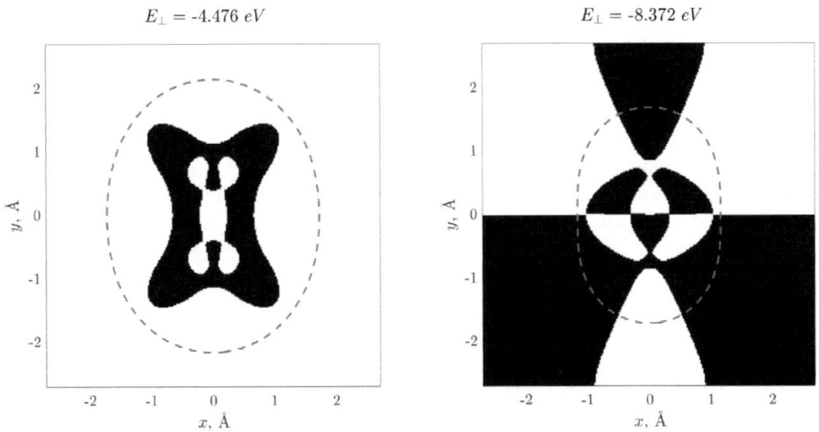

Figure 3. The eigenfunction samples for the electron with longitudinal energy of $E_{\|} = 5$ MeV channeling in the field of two neighbor atomic strings [110] of Si crystal.

Table 1. Full set of eigenenergies $E_\perp$ (eV) of the electron with $E_{\|} = 5$ MeV in the field of two neighbor atomic strings for each of four symmetry classes of the eigenfunction.

$\Psi(-x,y) = \Psi(x,y)$ $\Psi(x,-y) = \Psi(x,y)$	$\Psi(-x,y) = \Psi(x,y)$ $\Psi(x,-y) = -\Psi(x,y)$	$\Psi(-x,y) = -\Psi(x,y)$ $\Psi(x,-y) = \Psi(x,y)$	$\Psi(-x,y) = -\Psi(x,y)$ $\Psi(x,-y) = -\Psi(x,y)$
−49.0303	−48.7077	−48.711	−48.982
−31.6285	−23.7033	−24.3553	−30.8612
−24.894	−16.7748	−17.8317	−23.529
−19.98	−8.5155	−12.2515	−16.036
−12.3253	−6.3875	−9.101	−9.967
−11.214	−4.696	−6.39	−8.37175
−7.95275		−3.583	−6.071
−5.254		−3.22625	−3.36075
−4.4755			
−3.25275			
−2.84325			

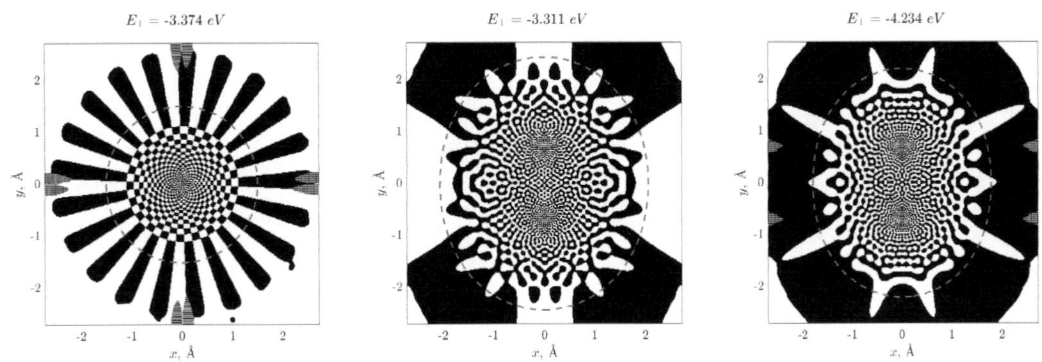

Figure 4. The eigenfunction samples for the electron with longitudinal energy of $E_{\|} = 500$ MeV channeling in the field of one (left plot) and two neighbor (center and right plots) atomic strings [110] of Si crystal.

exponential divergence of the initially close trajectories. The trajectories still deterministic (because the noise or accidental forces are assumed absent in the system) however appear very like to random ones.

Table 2. The spectral method algorithm parameters.

longitudinal energy $E_{\parallel}$, MeV	space lattice	time step value dt, eV^{-1}	number of time steps N_t	energy step value dE, eV
5	256 × 256	9.2×10^{-5}	27324375	2.5×10^{-4}
500	384 × 384	9.2×10^{-5}	136621875	5.0×10^{-5}

In the quantum mechanics case the exponential dependence on the initial conditions is absent. However the behavior of systems, which are regular and chaotic in the classical limit, is qualitatively different by the number of features. The search and investigation of these differences is the quantum chaos topic contents [8]. The observed difference of the wave functions structure, namely, zero lines $\psi(x, y) = 0$ picture of these functions is the one of manifestations of the quantum chaos.

5. Conclusion

The quantum mechanical problem of the fast electron motion in the axially-symmetric field of single atomic string [110] and the field of two neighbor ones of Si crystal is considered. The spectral method suitability for numerical solving, i.e. obtaining the eigenfunctions and energy eigenvalues, of the Schrödinger equation for this system in the axial channeling mode is demonstrated. This approach could be used for investigation of the quantum chaos manifestations in the electron axial channeling through the crystal.

Acknowledgments

The work was supported by the Ministry of Education and Science of the Russian Federation (the State assignment N 3.500.2014/K).

Appendix

First of all the spectral method algorithm was implemented on the MATLAB for debugging. After the code quality was proven the algorithm was implemented on C language using FFTW (for fast Fourier transforms) and GSL (for complex values calculations) libraries, because the MATLAB FFT performance and memory constraints appear unsatisfactory. The production calculations were performed under UNIX–like operating systems on computers of `amd64` multi-processor architecture. The same algorithm with minor changes allows us to calculate both energy levels and (in the separate run) eigenfunctions of the considered system. The algorithm parameters for present data calculations at $E_{\parallel} = 5$ and 500 MeV are summarized in table 2. For field of both single and two neighbor atomic strings the full set of eigenenergies and eigenfunctions at $E_{\parallel} = 5$ MeV was computed.

References

[1] Akhiezer A I and Shul'ga N F 1996 *High-Energy Electrodynamics in Matter* (Amsterdam: Gordon and Breach)
[2] Akhiezer A I, Shul'ga N F, Truten' V I, Grinenko A A and Syshchenko V V 1995 *Physics-Uspekhi* **38** 1119
[3] Uggerhøj U I 2005 *Rev. Mod. Phys.* **77** 1131
[4] Stratt R M, Handy N C and Miller W H 1979 *J. Chem. Phys.* **71** 3311
[5] Gutzwiller M C 1990 *Chaos in Classical and Quantum Mechanics* (New York: Springer-Verlag)
[6] Schuster H G 2005 *Deterministic Chaos. An Introduction* (Weinheim: WILEY-VCH Verlag)
[7] Berezovoj V P, Bolotin Yu L and Cherkaskiy V A 2004 *Phys. Lett.* A **323** 218
[8] Berry M V 1987 *Proc. R. Soc. Lond.* A **413** 183
[9] Feit M D, Fleck J A Jr and Steiger A 1982 *J. Comput. Phys.* **47** 412
[10] Dabagov S and Ognev L I 1988 *Nucl. Instrum. and Methods in Phys. Res.* B **30** 185

[11] Shul'ga N F, Syshchenko V V and Neryabova V S 2013 *Nucl. Instrum. and Methods in Phys. Res.* B **309** 153
[12] Shul'ga N F, Syshchenko V V and Isupov A Yu 2014 *Problems Atom. Sci. Technol.* **63** 5 (93) 120
[13] Shul'ga N F, Syshchenko V V, Neryabova V S and Isupov A Yu 2014 *J. Phys.: Conf. Series* **517** 012030

Characterization of multi-strip crystal deflector for high energy proton beams by synchrotron radiation topography with angular scanning

A A Kaloyan[1], S A Tikhomirov[1], K M Podurets[1], Yu A Chesnokov[2], Yu E Sandomirskiy[2]

[1]National Research Centre "Kurchatov Institute", Moscow 123182, Russia;
[2]Institute for High Energy Physics in National Research Centre "Kurchatov Institute", Protvino 142281, Russia.

E-mail: chesnokov@ihep.ru

Abstract. Currently, for extraction and collimation of proton beam at the large accelerators bent silicon single crystals are used. Recently the new device for multiple deflection of the proton beam was developed, it consists of several bent strips of silicon in the reflection mode. In the device the successive bending of silicon strips at the surface of a thick plate is achieved due to internal stresses in the material of the crystal due to the Twyman effect as a result of applying mechanical grooves. Method of X-ray topography at synchrotron radiation with angular scanning was applied for measurement of bending of the individual strips of the deflector and the crystal as a whole. The results of the measurement are compared with the results obtained in the proton beam.

1. Introduction

The bent crystals, due to the high values of the internal electric field, are increasingly used at large circular accelerators in the systems for the extraction and collimation of high-energy proton beams [1–6]. Such crystals have been used mainly in a mode of planar channeling. Recently, a new physical phenomenon has been experimentally discovered – reflection of the beam of high-energy protons by the bent atomic planes of the silicon crystal, and the use of this phenomenon at the accelerators has begun [7–10].

Volume reflection is caused by the interaction of the incident proton with the potential of the bent crystal lattice and occurs at a small length in the region, tangential to the curved atomic plane. The deflection angle of the particles at the reflection from crystallographic planes is limited to 1.5 θ_c, where $\theta_c = (2U_c/pv)^{1/2}$ – critical angle of channeling, U_c ~20 eV – the potential barrier of the planar channel (111) in silicon, p, v – momentum and velocity of the incoming particle. Protons with an energy of $E = 400$ GeV are deflected by the single strip in the direction, opposite to the bend, at an angle of 15 μrad [10]. Probability of a single reflection is high and at energies around 100 GeV approaches unity. To increase the angle of reflection and to make the use of the volume reflection practical for extraction and collimation of the proton beams it is necessary to increase the value of the deflection angle of the charged particles.

In experiments with accelerators of CERN and IHEP it has been shown that proton beams with energies of 400 and 70 GeV are effectively deflected by the reflection in the devices consisting of several aligned crystals due to multiple enhancement of the deflection angle [11–14]. In these devices separate silicon strips are installed in a metal holder which creates a bending moment for each crystal.

A disadvantage of the multi-crystal structures with successively arranged strips with bending by the mechanical holder, is a misalignment of the individual strips by about 50 μrad [13]. Therefore, for applications such as at the Large Hadron Collider (LHC), the device goes beyond the requirements for the system of localization of the LHC, which requires the crystals disoriented to no more than few μrad from each other, with the total bending of a few tens of μrad with a bending radius about 100 meters [15].

Content from this work may be used under the terms of the Creative Commons Attribution 3.0 licence. Any further distribution of this work must maintain attribution to the author(s) and the title of the work, journal citation and DOI.
Published under licence by IOP Publishing Ltd

Thus, for ultrahigh-energy particle beams a chain of successive crystals with very similar parameters of bending of each of them is required. To create such a device there was recently proposed a new scheme of bending [16], which is based not on the external force exerted by the holder but caused by internal stresses created by the grooves mechanically applied on the surface of a thick crystal (figure 1). Bending of the crystal strips between the grooves is caused by deformation of their surface layers caused by the Twyman effect known in optics [17].

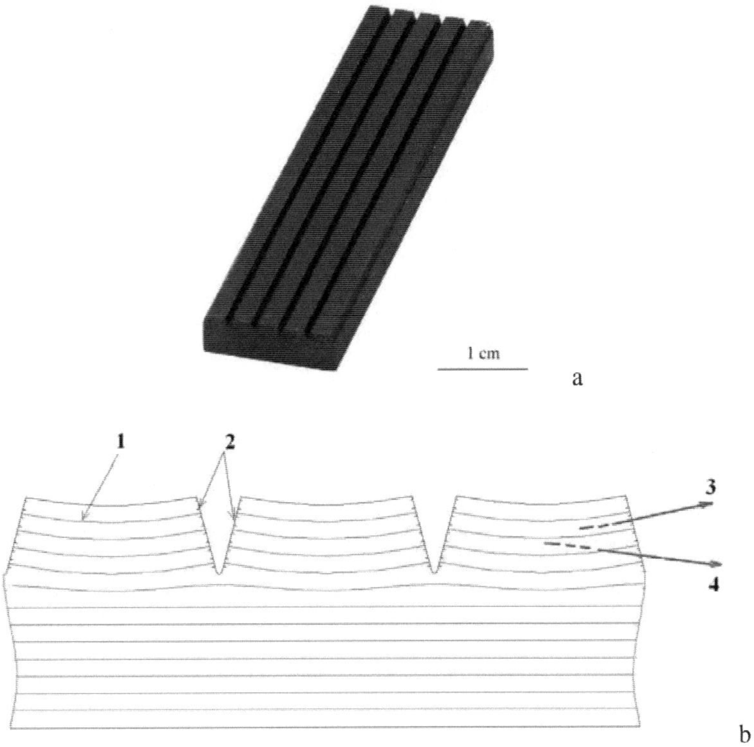

Figure 1. Silicon plate with grooves (a) and the supposed layout of the bent crystal planes due to the Twyman effect (b). 1 – bent crystallographic plane, 2 – rough surfaces of grooves, 3 – beam deflected by channeling, 4 – reflected beam of particles.

A recent experiment [18] at the SPS accelerator of CERN has shown that by using this crystal device it is possible to deflect the beam of protons with energy of 400 GeV. However, for use of such device at LHC with beam energy of 7 TeV detailed information on the size and shape of the bent crystal strips is needed to be known. Calculation of the deformation caused by surface treatment is difficult, so that data on the deflector can only be obtained experimentally. The aim of this work is to study the silicon deflector using diffraction of synchrotron radiation, which is sensitive to small local bending of the crystal.

2. Method

Method of the investigation of the deflector must meet several requirements. Firstly, it should be local, since the deflector is bent unequally in different points of the crystal. Secondly, the method should characterize the bent crystal quantitatively, i.e. rocking curves of the crystal should be measured in its different regions. Method of X-ray topography at synchrotron radiation with angular scanning, which satisfies these requirements, was proposed in [19]. The method consists in to obtaining a series of digital topographic images of the crystal at different angles of incidence of radiation on the investigated crystal, which allows one to restore local rocking curve at each point of the image. In the present study, this method has been implemented at the "Mediana" station of the Kurchatov synchrotron radiation source [20]. The sample was a silicon plate with the surface treated with chemical mechanical polishing, (111) orientation, thickness of 5 mm, size 71 × 16 mm, with 4 triangular shaped grooves of about 1 mm depth made along its long side. The beam incident on the sample was formed by an asymmetric silicon monochromator in (511) reflection. The sample was mounted in a symmetric Bragg reflection (333), so the non-dispersive plane-wave conditions were realized (figure 2). The wavelength was 0.72 Å, the Bragg angle was 20°, the depth of X-ray absorption of 100 microns. Estimated rocking curve of the sample had a width at half maximum (FWHM) 3 μrad. The sample was oriented so that the grooves lie perpendicular to the diffraction plane, so bending in the plane perpendicular to the grooves was registered. Images were recorded by area detector of 4008 × 2672 dimensions with a pixel size of 8.9 microns. The crystal was rotated with piezo actuator by steps of 0.7 μrad in the direction shown in figure 2, an image of the crystal was recorded at each position.

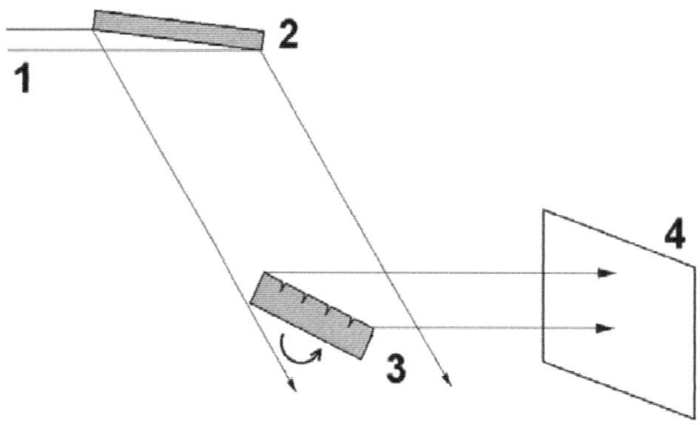

Figure 2. Scheme of the experiment: 1– radiation beam from the source, 2– monochromator crystal, 3– sample, 4– area detector.

3. Results

Figure 3 shows the topographs of the crystal obtained at different values of the angle of the sample. It can be seen that in the studied angular range of 200 μrad, grooves were always in reflecting position, indicating a strong broadening of the rocking curve of silicon in the grooves. When the crystal was rotated counterclockwise (figure 2), different regions of the crystal consecutively entered the reflecting position. In the topographs obtained at the increasing angle of rotation, the crystal region which reflected the X-ray beam, appeared at the lower end of each strip and moved to its top end. The uneven appearance of reflections on different strips is also observed, movement of the most bright parts of the crystal from the top down is visually traced.

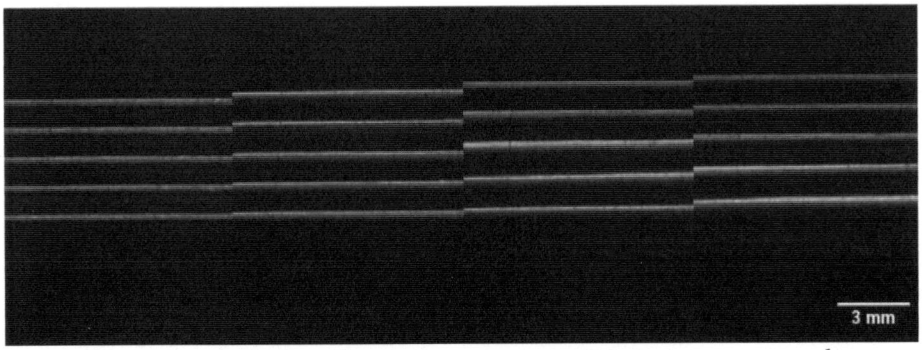

Figure 3. Topographs of the sample obtained at the position of the crystal of 0 (a), 81.2 μrad (b), 97.3 μrad (c), 105 μrad (d).

In order to quantitatively characterize the bending of the crystal the integration region of the size of 3618 × 12 pixels was taken in the topographs, at that the width of the image of the single strip was about 90 pixels. Integration region was shifted in the direction across the strips, so the intensity of crystal reflection of the chosen region was obtained as the dependence from its position on the crystal and from the angle of rotation of the crystal. Thus the rocking curves were reconstructed for different parts of the crystal. The rocking curves for a single strip are shown in figure 4. Attention is drawn to the fact that the rocking curves corresponding to the areas at the edges of the strips that are near the grooves, are greatly broadened with respect to the estimated width of the rocking curve (3 μrad). This means that the material close to the grooves has a strongly disturbed crystal structure. As the distance from the groove grows, the width of the rocking curve becomes closer to that expected for a perfect crystal structure. Peak point position of the curves is a quantitative measure of bending of the crystal, so the angular distance between the extreme peaks of one strip in the average equal to 66 μrad. With strip width of 2.4 mm this corresponds to average curvature radius of ~40 m, and the crystal planes parallel to the surface are concave as shown in figure 1. Figure 5 shows the dependence of the local maxima of the rocking curves on the coordinate across the crystal. It can be seen that the strips are slightly turned relative to each other, the overall bending of the crystal is 15 μrad. The bending is directed in the opposite to the bending of each strip, i.e. the entire crystal is convex.

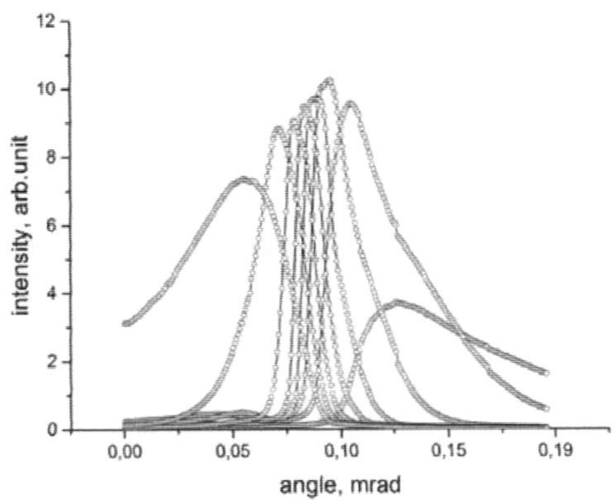

Figure 4. The rocking curves of the crystal within the same strip.

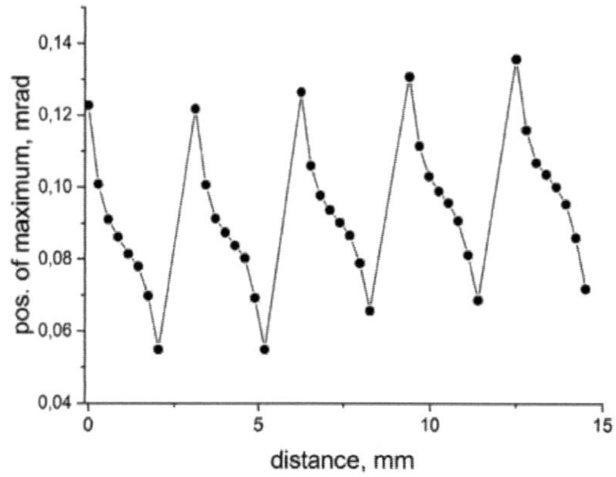

Figure 5. Dependence of the position of local maxima of the rocking curves on the coordinate across the crystal.

Using the data of figure 5 it is possible to receive the local values of radius of curvature of a separate bent crystal strip $R(l)= dl/d\alpha$. Changes of radius versus the coordinate are shown in figure 6. It is visible, that at length l ~ 1 mm (it is just a characteristic length of process of reflection of 7 TeV protons $R{\cdot}\theta_c$ ~ 1 mm) the radius is constant and equal to 80 m that is close to requirements of the LHC at 7 TeV energy. The fact of turn of the separate bent strips relative to each other by an angle of 3.7 μrad is also positive because this angle is equal approximately to reflection angle of 7 TeV protons. In this case, orientation offset of consecutive stripes will match the trajectory of the deflected particles, and operation of the crystal device will be improved.

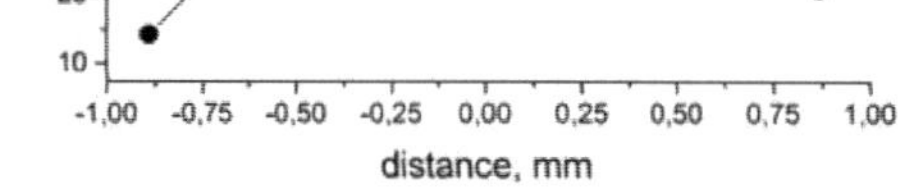

Figure 6. The curvature radius of separate crystal strip versus the coordinate.

4. Summary

Thus, with the help of synchrotron radiation topography distortion of the crystal lattice in a multi-strip deflector was observed. These distortions lead to the appearance of macroscopic stress, causing bending of the crystal. Bending of the crystal was observed at the scale of each strip and the crystal as a whole. The measured parameters of deformed crystalline strips well explain the data of experiment on deflection of the 400 GeV proton beam [18], where the beam is effectively deflected by successive reflections on five strips. The angular region of the proton deflection complies with the bending angle of the stripes, measured here with the help of x-rays.

Thus, the method of X-ray topography with angular scanning performed at synchrotron radiation makes it possible to characterize the deflector prior to its use at the high-energy beams of protons at large accelerators and can be used in the development of manufacturing technology of the deflector. We hope, that using a method of X-ray topography one can optimize the multi-crystal deflectors and for higher energy of protons which it is planned to reach by the future machines, such as FCC in CERN and SPPC in China with the energy up to 100 TeV.

Acknowledgments
In this work the facility "Kurchatov synchrotron radiation source", funded by the Ministry of Education and Science (project ID RFMEFI61914X0002) was used, also work was supported by grant RFBR № 14-22-01001 ofi-m.

References
[1] Afonin A G, Biryukov V M, Gavrilushkin V A *et al* 1998 *JETP Lett.* **67** 781
[2] Afonin A G, Arkhipenko A A, Baranov V I *et al* 1998 *Phys.Lett.* **B435** 240
[3] Afonin A G, Baranov V T, Biryukov V M *et al* 2001 *Phys.Rev.Lett.* **87** 094802
[4] Afonin A G, Baranov V T, Biryukov V M *et al* 2005 *Phys.Part.Nucl.* **36** 21
[5] Scandale W, Arduini G, Assmann R *et al* 2010 *Phys.Lett.* **B692** 78
[6] Mokhov N V, Annala G E, Apyan A *et al* 2009 *Proc. of PAC 09, Vancouver, BC, Canada* 1836
[7] Ivanov Yu M, Petrunin A A, Skorobogatov V V *et al* 2006 *Phys. Rev. Lett.* **97** 144801
[8] Scandale W, Still D, Carnera A *et al* 2007 *Phys. Rev. Lett.* **98**, 154801
[9] Taratin A M and Vorobiev S A 1987 *Phys. Lett.* **A119** 425
[10] Maisheev V A 2007 *Phys.Rev.ST Accel.Beams* **10** 084701
[11] Scandale W *et al* 2008 *Phys.Lett.* **B658** 109
[12] Scandale W *et al* 2009 *Phys. Rev. Lett.* **102** 084801
[13] Scandale W *et al* 2010 *Phys. Lett.* **B688** 284
[14] Afonin A G *et al* 2009 *Atomic Energy* **106** 409
[15] Assmann R, Redaelli S, Previtali V, Yazynin I 2009 *Proc. of PAC 09, Vancouver, BC, Canada* 1823
[16] Afonin A G *et al* 2013 *Instrum. Exp. Tech.* **56** 617
[17] Lambropoulos J C, Xu S , Fang T, Golini D 1996 *Appl. Opt.* **35** 5704
[18] Scandale W *et al* 2014 *Nucl.Instrum.Meth.* **B338** 108
[19] Lübbert D, Baumbach T, Härtwig J, Boller E, Pernot E 2000 *Nucl. Instrum. Meth.* **B160** 521
[20] Manushkin A A, Pogoreliy D K, Podurets K M, Vazina A A, Lagoda T S, V.A. Somenkov V A 2007 *Nucl. Instrum. Meth.* **A575** 225

Thermal damage at short electron bunches passage through a thin target

A A Babaev and A S Gogolev

Tomsk Polytechnic University, Lenin Ave 30, Tomsk 634050, Russia

E-mail: krass58ad@mail.ru (A Babaev)

Abstract. The thin target could be used for beam diagnostics by means the radiation that is induced by interaction of beam particles with target matter. The electron beams used in modern applications (as, for example, modern FELs) have very large brightness, small emittance as well as very short bunch length. For example, the bunch length of XFEL is about of 25 um at bunch charge order of 1 nC and with electrons energy of 17.5 GeV. The passage of this powerful short bunches could damage the target or even completely destroy it. In the presented work the train of such bunches passages through the target is investigated. It is shown the target works in extreme regime close to phase transition temperature.

1. Introduction

Intense ultrashort soft and hard X-ray pulses will give rise to new opportunities for studies in time-resolved spectroscopy, diffraction, and imaging. These pulses can be generated from free-electron lasers (FELs) such as FLASH (DESY, Germany) [1], LCLS (SLAC, USA) [2], SACLA (Spring-8, Japan) [3], or the European XFEL (DESY, Germany) [4]. A key requirement in realizing an accelerator like an x-ray FELs is a high brightness electron beam that possesses a high peak current, i.e. a high bunch charge (typically few nC) compressed to an extremely small longitudinal extension (few tens um), and a very low transverse emittance. The diagnostics of the electron bunch length which typically is the order of few tens of fs is of crucial importance. The information on the bunch shape could be obtained by the analysis of the radiation that is generated in the solid target placed in the beamline [5,6].

The applicability of beam diagnostics schemes that use the radiation in solid targets at FELs parameters demands the additional investigations. The interaction of the short powerful bunches with a solid medium leads to fast and significant local heating in the small volume of target. From one hand, the target could be overheated because the train typically consists of few thousands of bunches. Hence, one should be sure the local temperature in the interaction area does not exceed the critical temperature (for example, the phase transition temperature). Moreover, the radiation properties also could change because the medium parameters depends on the temperature. From the other hand, the fast heating in the small area generates the mechanical stress waves spreading over the target. The stress oscillations could also damage the target even if there is not overheating.

This work is devoted to the investigation of the target heating at FELs electron beam parameters. Namely, the passage of short electron bunches through the thin diamond target is considered. The beam parameters assumed below are: electron energy $E = 17.5$ GeV, bunch

Content from this work may be used under the terms of the Creative Commons Attribution 3.0 licence. Any further distribution of this work must maintain attribution to the author(s) and the title of the work, journal citation and DOI.
Published under licence by IOP Publishing Ltd

charge $q_0 = 4$ nC, bunch length RMS $l = 25$ um, bunch transverse dimension RMS $\sigma = 10$ um, the time between bunch passages is $\Delta t = 100$ ns [4]. The angular divergence is negligible for the task. The target is the thin round plate with thickness $h = 10$ um and radius $R = 5.6$ mm. In simulations the critical temperature is $T_{cr} = 2273$ K when the phase transition takes place, and the target is broken due to the overheating. The beam heats the center of the target along its axis, see in figure 1.

In the work firstly the applicability of heat conductivity theory is verified at conditions of the task. Than the simulations are carried out to estimate possible thermal damage of the target as well as to find the safe regime of the beam passage.

2. Theoretical model

When the train of electron bunches passages through the solid target the particles interact with the medium, and they loss their energy. The energy can be transferred to the medium (ionization loss) or can be radiated (radiation loss). In this work the radiation absorption in the medium is neglected so the target heating is caused by the ionization loss. Initially the projectiles transfer their energy to electrons of the medium. After that this energy spreads to the ionic lattice of target while the temperature of electronic and ionic components of the medium do not became equal. The characteristic time of relaxation is $t_r \sim 0.1$ ns. Hence, the heat spreading over the medium can not be considered by using the usual conduction theory at the time range $t \leq t_r$.

Assuming the Gaussian longitudinal density of bunch electrons, the estimation of bunch length is $6l$. Hence, the bunch passage time, when energy is transferred to the target, is $t \approx (h + 6l)/c = 0.53$ ps (c is the speed of light, it is approximately equal to the velocity of beam's electron). So, $t \ll t_r \ll \Delta t$. Hence, the beam interaction with the target could be considered at two steps. At the first step the energy transfer could be considered instantaneous, whereas at the second step the heat spreading between bunch passages should be considered based on the heat conductivity theory.

The ionization loss $|-dE/dx| = 4.76$ MeV/cm for one 17.5 GeV electron was evaluated by using the GEANT4 toolkit [7]. The bunch contains q_0/e electrons. Hence, at the bunch passage the target gets the heat $Q_b = |-dE/dx|q_0h/e = 19.04$ uJ (e is the electron charge). For the Gaussian beam considered here the electron flux at the distance r from the target axis depends on the time t as

$$n(r,t) = n_0 \exp\left[-r^2/(2\sigma^2)\right] \exp\left[-(ct)^2/(2l^2)\right].$$

The electron number that passage through the thin ring layer with the radius r, thickness dr, and coaxial with the target is

$$dn(r) = \int_{-\infty}^{\infty} n(r,t)dt \cdot 2\pi r dr = \frac{(2\pi)^{3/2} n_0}{c} lr \exp\left[-r^2/(2\sigma^2)\right] dr, \quad (1)$$

where $2\pi r dr$ is the thin layer cross-section. The constant n_0 is defined by the condition on the total number of electrons in the bunch:

$$\int_0^{\infty} dn(r) = \frac{q_0}{e}, \text{ therefore, } n_0 = \frac{q_0 c}{(2\pi)^{3/2} e l \sigma^2}. \quad (2)$$

Substitution of equation (2) in equation (1) gives

$$dn(r) = \frac{q_0}{e\sigma^2} r \exp\left[-r^2/(2\sigma^2)\right] dr.$$

During the bunch passage the thin layer gets heat

$$dQ = Q_b \frac{e}{q_0} dn(r) = \frac{Q_b}{\sigma^2} r \exp\left[-r^2/(2\sigma^2)\right] dr.$$

The thin layer volume is $dV = 2\pi rhdr$. Therefore, the instantaneous jump of temperature depends on the distance r to target axis and is defined by the expression

$$\Delta T(r) = \frac{dQ}{\rho c_p dV} = \frac{Q_b}{2\pi\rho c_p h\sigma^2} \exp\left[-r^2/(2\sigma^2)\right],$$

where ρ is the target density, and c_p is the target heat capacity. In particular, the maximal temperature jump (evaluated at initial temperature $T_0 = 300$ K) is $\Delta T(0) = 1724$ K and it takes place at the target axis. It should be noticed here the simulations are carried out within the wide temperature range. So, both the density and the heat capacity should be considered as temperature dependent functions at the moment when the bunch hits the target.

The temperature spreading over the target between bunches (during the time Δt) is described by the equation:

$$\frac{\partial T}{\partial t} = a\left[\frac{1}{r}\frac{\partial}{\partial r}\left(r\frac{\partial T}{\partial r}\right) + \frac{\partial^2 T}{\partial x^2}\right],$$

where a is the thermal diffusivity of the target material. The target is considered to be placed in the vacuum and anchored at lateral cylindrical surface (see in figure 1). So, the boundary conditions for both beam entrance and exit round surfaces were black body radiation with the ambient temperature T_0:

$$\left(\frac{\partial T}{\partial x}\right)_{\text{entrance, } x=-h/2} = -\epsilon\sigma_{SB}\left(T^4(-h/2, r, t) - T_0^4\right),$$

$$\left(\frac{\partial T}{\partial x}\right)_{\text{exit, } x=h/2} = \epsilon\sigma_{SB}\left(T^4(h/2, r, t) - T_0^4\right),$$

where ϵ is the surface emissivity and σ_{SB} is the Stefan Boltzmann constant. The temperature of the lateral surface was fixed at T_0. The same temperature $T_0 = 300$ K is the initial temperature for whole task.

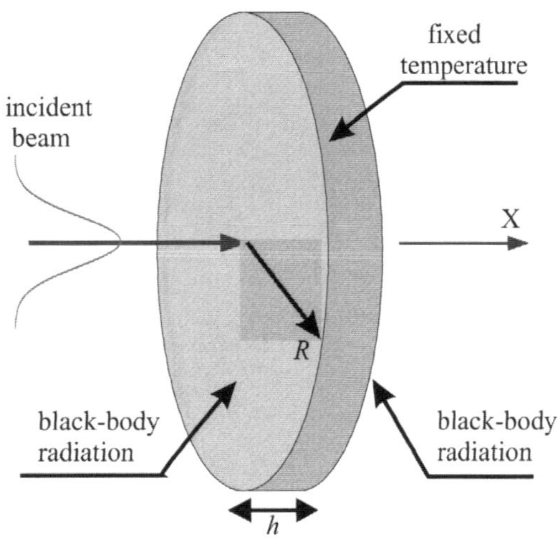

Figure 1. The geometry of the model and boundary conditions. The beam hits the target along target axis denoted by X.

3. COMSOL simulations and discussion

The model developed above was simulated by COMSOL Multifysics software [8]. Target parameters were taken from COMSOL database (namely, C diamond [solid,thin film] was used

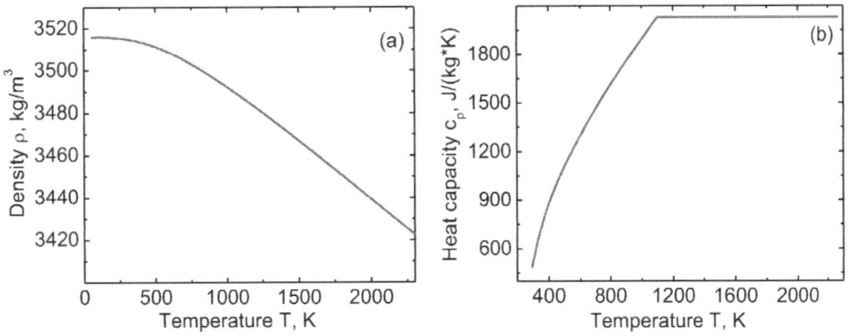

Figure 2. COMSOL data. Dependencies on the temperature T: (a) the material density $\rho(T)$, (b) the material heat capacity $c_p(T)$.

as material) taking into account their temperature dependencies (see in figure 2). The diamond thermal conductivity $k = 2000$ W/(m·K) ($a^2 = k/(\rho c_p)$) as well as surface emissivity $\epsilon = 1$ were introduced manually. The maximal temperature during the train passage is suspected at the target axis. The corresponding temperature evolution is shown in figure 3. Here one can see both the temperature jumps when a bunch hits the target and the relaxation between bunches. It should be noticed the first bunch at $t = 0$ produces the maximal temperature jump. After that jumps are drastic reduced due to the increasing of heat capacity as shown in figure 2(b). These jumps have close values because the material parameters change slightly in the temperature range where simulations are carried out. The relaxation stages between bunches produce close values of temperature decreasing, about of 400 K, excepting the fist relaxation stage when the temperature falls down at 600 K. The detailed data are presented in table1. As the matter of fact the temperature decreasing during relaxation stages become bigger with every successive stage (starting from the second relaxation stage) due to the spreading of the heat from the central area of the target to its periphery. In principle, the temperature decreasing at the relaxation would be equal to the temperature jump for the large enough bunch numbers. But the simulations show the target will be broken at 11-th bunch passage. Usually train consists of several thousands of bunches. So, the application of thin target at FELs beam diagnostics is obstructed.

To make the heat spreading over the target more clear several samples of the temperature distribution along the target radius are shown in figure 4. The area shown in the figure includes the range of $0 \leq r \leq 8\sigma$ where the beam hits the crystal and where the heat spreads while the target keeps undamaged. The difference between curves 1 and 2 demonstrates the temperature jump when the second bunch hits the target (see in figure 3). One can see the jump depends on the distance from the crystal axis. The curve 3 demonstrates the temperature at the finish of the second relaxation stage. The comparison of curves 2 and 3 illustrates the heat flux in the target during one relaxation stage. Heat transfers from the central area of the target so the temperature increases at the periphery of the shown area whereas the temperature decreases near the axis. The curve 4 shows the temperature in the middle of the last relaxation stage. One can see the temperature at the distance $r > 8\sigma$ remains practically equal to initial temperature T_0. This means the additional cooling of the lateral surface can not protect the target because $R \gg \sigma$, and the target is overheated in its central area before the heat flux can reach the target lateral edge.

The simulations show the target is overheated at modern FELs electron beams with the bunch charge order of 1 nC. In principle, there would be possible the stable regime when

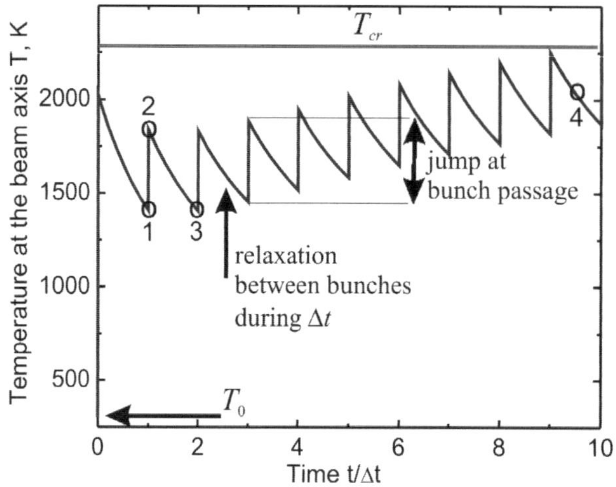

Figure 3. Temperature at the target axis when train of electron bunches passages through the target. The initial target temperature is $T_0 = 300$ K, the critical temperature of phase transition is $T_{cr} = 2273$ K. The target is overheated after 11-th bunch. Open circles denote corresponding curves in figure 4.

Table 1. Data on the beam passage through the target which is shown in figure 3.

Bunch number	Temperature jump at the bunch passage, K	Initial temperature at relaxation, K	Temperature decreasing at relaxation, K	Final temperature at relaxation, K
1	1724	2024	616	1408
2	435	1838	435	1403
3	430	1833	380	1453
4	431	1884	367	1517
5	431	1948	364	1584
6	431	2015	367	1648
7	431	2079	371	1708
8	432	2140	375	1765
9	433	2198	379	1819
10	433	2252	382	1870
11	433	2303	-	-

the temperature jump at the bunch passage equals to temperature decreasing at successive relaxation. More realistic, at the large time limit the difference between target temperatures at the finish of two successive bunch passages should tend to zero.

The drastic way to find the safe regime for target is to decrease the bunch charge q_0. The simulation of the whole train by using the COMSOL is not possible due to technical restrictions. To estimate the target temperature after the 1000 bunches the next method was used. The passage of 19 bunches with the charge q_0 was simulated. After that the difference of temperatures δT after 18-th and 19-th bunches at the target axis was evaluated. The corresponding value for

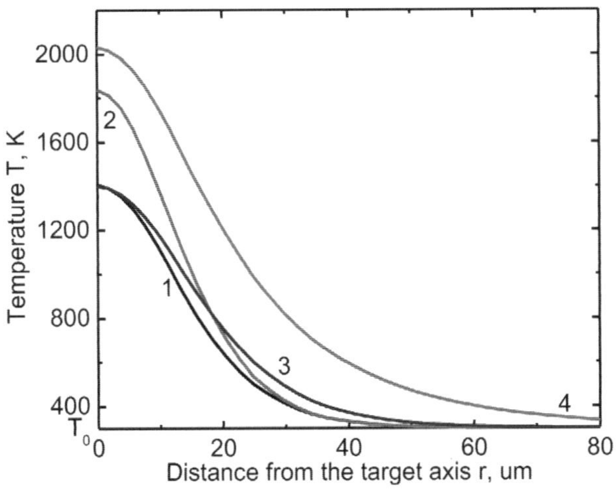

Figure 4. Dependencies of temperature on the distance from the target axis at different time moments: 1 — the finish of the first relaxation stage, 2 — the finish of the second bunch passage, 3 — the finish of the second relaxation stage, 4 — the middle moment of the tenth relaxation stage (see in figure 3).

Table 2. The top estimation $T_{\rm est}$ of the target temperature after the train passage in dependence on bunch charge q_0.

q_0, nC	T_{est}, K	Comparison with T_{cr}	Result
1.0	8300	$\gg T_{\rm cr}$	Crash
0.5	5300	$\gg T_{\rm cr}$	Crash
0.25	2300	$\sim T_{\rm cr}$	Danger
0.1	1370	$< T_{\rm cr}$	Safe

next bunches (after 19-th bunch) should be less than δT so the value $T_{\rm est} = T_0 + 1000\delta T$ gives the rough top estimation of the maximal target temperature after the whole train passage. If this estimation gives $T_{\rm est} \gg T_{\rm cr}$ one could surely say the target will be broken, if $T_{\rm est} \sim T_{\rm cr}$ there is the danger of overheating, if $T_{\rm est} \ll T_{\rm cr}$ the regime of the train passage is safe. These estimations for different values of q_0 is given in table 2. One can see the target could not be damaged if the bunch charge would be about of 0.1 nC, i.e. one order less than the FELs bunch charge that we are interested here.

4. Conclusion

In this paper the heating of thin target was investigated when it is irradiated by the train of short powerful electron bunches. The electron beam properties, in general, corresponds to parameters of modern FELs facilities. The overheating of the target was demonstrated clearly. It was shown the bunch charge should be significantly reduced to avoid the thermal damage of the target.

Finally, it should be underlined the target heating depends strongly on both bunch charge and material properties that define the temperature jumps at bunch passages. Moreover, the large temperature jumps concentrated in the small central area of the target should lead to the appearance of mechanical stress waves in the target. This stress can be additional cause of the target both damage and breakage. This problem needs additional investigation in view of solid target application for the beam diagnostics.

Acknowledgments
The work was supported by the grant of Ministry of Education and Science of the Russian Federation within Nauka Program project #3761.

References
[1] Ackermann W, Asova G, Ayvazyan V, Azima A, Babaoi N, Bähr J, Balandin V, Beutner B, Brandt A, Bolzmann A *et al* 2007 *Nature Photonics* **1** 336
[2] Pile D 2011 *Nature Photonics* **5** 456
[3] Ishikawa T, Aoyagi H, Asaka T, Asano Y, Azumi N, Bizen T, Ego H, Fukami K, Fukui T, Furukawa Y *et al* 2012 *Nature Photonics* **6** 540
[4] Altarelli M, Brinkmann R, Chergui M, Decking W, Dobson B, Düsterer S, Grübel G, Graeff W, Graafsma H, Hajdu J, Marangos J *et al* (eds) 2007 The European X-Ray Free-Electron Laser *DESY Technical Design Report* DESY 06-097
[5] Gogolev A, Potylitsyn A, Kube G 2012 *J. Phys.: Conf. Series* **357** 012018
[6] Takabayashi Y 2012 *Phys. Lett.* A **376** 2408
[7] Geant4 Collaboration, CERN/LHCC 98-44, Geant4: An Object-Oriented Toolkit for Simulation in HEP; http://geant4.web.cern.ch/geant4/
[8] COMSOL Multiphysics: www.comsol.com

Comparison of the quantum and classical calculations of flux density of (220) channeled positrons in Si crystal

K B Korotchenko[1], TATukhfatullin[1], Yu L Pivovarov[2] and Yu L Eikhorn[1]

[1]National Research Tomsk Polytechnic University, 30 Lenin Avenue, 634050 Tomsk, Russia
[2]V.E. Zuev Institute of Atmospheric Optics SB RAS, 1 Zuev Square. 634034 Tomsk, Russia

E-mail: aihornyurij@mail.ru

Abstract. Simulation of flux-peaking effect of the 255 MeV positrons channeled in (220) Si crystals is performed in the frame of classical and quantum mechanics. Comparison of the results obtained using both approaches shows relatively good agreement.

1. Introduction
The flux-peaking effect (FPE) is one of the known orientation effects at channeling of positively charged particles. A detailed review of the FPE was given in [1].Theoretical investigation of the FPE has been carried out on the basis of kinetic equations for the particle distribution functions or numerical solution of classical equation of motion [2]. In this work we perform analysis of FPE using both classical and quantum methods as in work [3]. We use computer code "Basic Channeling with Mathematica©" BCM–1.0 [4], which was successfully applied to describe recent theoretical [5-10] and experimental results [11-15], obtained at SAGA Light Source. Calculations were performed for 255 MeV positrons channeled along (220) planes in a thin Si crystal.

2. Classical description
For the calculation of every individual charged particles trajectory it is necessary to solve classical equation of motion. Under the channeling condition the longitudinal component of relativistic particles velocity $\upsilon_{\|}$ is nearly equal to the speed of light c. The transversal component of velocity is much less than the longitudinal one: $\upsilon_{\perp} \equiv \dot{x} << \upsilon_{\|}$; $\upsilon_{\perp}/c << 1/\gamma$ (γ denotes the relativistic factor).Thus, equations of motion are reduced to the non-relativistic equation with the replacement of the particle mass m to the relativistic mass γm:

$$\gamma m\ddot{x} = -\frac{\partial U(x)}{\partial x}, \quad \gamma m\ddot{z} = 0, \tag{1}$$

Where x is the transversal coordinates, z the longitudinal coordinate, $U(x)$ here describes the potential energy of the particles in the field of periodically arranged crystal planes.

The initial conditions for this system contain the positron entrance point into the crystal $x(0) \equiv x_0$ and transversal components of initial velocity:

Content from this work may be used under the terms of the Creative Commons Attribution 3.0 licence. Any further distribution of this work must maintain attribution to the author(s) and the title of the work, journal citation and DOI.
Published under licence by IOP Publishing Ltd

$$\upsilon_x(0) = c\sqrt{1-\frac{1}{\gamma^2}}\sin(\theta), \tag{2}$$

where θ is the incidence angle i.e. the angle between positron momentum and channeling plane. The point of entry of particles were uniformly distributed within a period of change the function $U(x)$(within one potential well). For the numerical simulation of the equation (1) we use computer code "Basic Channeling with Mathematica©" BCM–1.0 [4]. This code computes numerical solutions of the classical equations of relativistic particles using the Doyle-Turner potential for the crystallographic planes [9]. Simulated trajectories of 255-MeV positrons in 5μm Si crystal and potential energy of positrons in the (220) crystal planes electric field are shown in figure 1. Only 50 trajectories are presented. The incidence angles for all positrons are the same.

To simulate the angular divergence of positron beam the procedure described in Ref. [5] is used. For each point of entry, 20 values of the incident angle θ are generated randomly with a normal distribution. The standard deviation of the distribution had the value 0.1 mrad≈$\theta_c/4$ (θ_c is the critical channeling angle). Thus, 10 000 trajectories are calculated for different entry points within one period of the planar channeling potential (figure 1(a)). Then, using these trajectories we obtain 2D spatial distribution of the positrons inside the crystal.

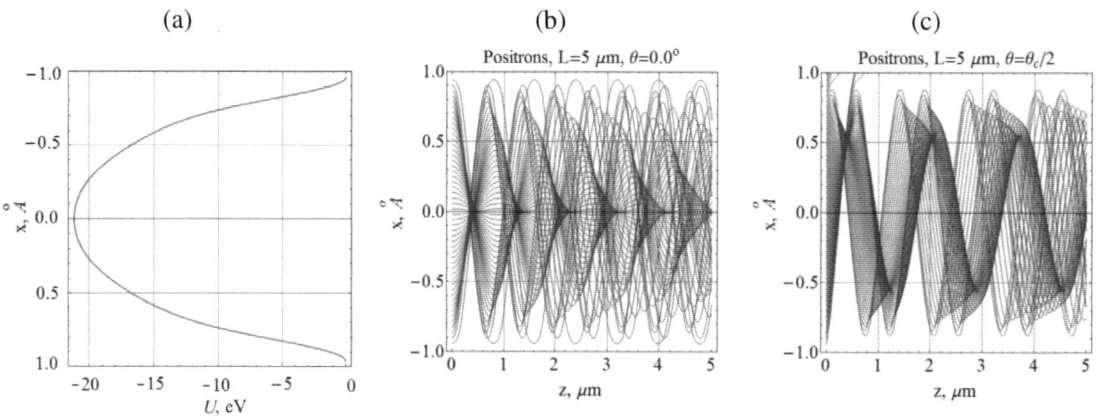

Figure 1. Simulated trajectories of 255MeV positrons in 5μm Si crystal and potential energy of positrons in the electric field of the crystal planes. (a) (220) interplanar potential energy, (b) 50 simulated trajectories for incidence angle θ=0.0, (c) 50 simulated trajectories for incidence angle $\theta=\theta_c/2$.

3. Quantum description

The channeled particle wave function $\Phi_i(x)$ was calculated using computer code "Basic Channeling with Mathematica©" BCM–1.0 [4]. This code computes numerical solutions of the Schrödinger equations with relativistic mass [10] for relativistic particles using the Doyle-Turner approximation for the crystallographic planes potential [9].

The conventional wave packet $\Psi(x,t)$ based on the wave functions $\Phi_i(x)$ we have chosen as

$$\Psi(x,t,\theta) = \sum_i P_i(\theta)\exp\left(-\mathrm{i}\frac{\varepsilon_i}{\hbar}t\right)\Phi_i(x), \tag{3}$$

where $P_i(\theta)$ is the initial population of the transverse sub-barrier energy levels:

$$P_i(\theta) = \frac{1}{d}\left|\int_{-d/2}^{d/2} \exp(ik_x\theta \cdot x)\Phi_i(x)dx\right|^2 . \tag{4}$$

Here θ is the positrons initial angle of incidence, k_x the transverse wave vector component, d the interplanar distance, ε_i the eigenvalues of Schrödinger equation.

In addition, the probability density of positrons wave packet was obtained by analogy with the free particles Gaussian wave packet:

$$\Psi_G(x,t,\theta) = \sum_i \sum_m P_i(\theta)\tilde{\Phi}_{i,m} \exp\left(\mathrm{i}\, m\, g\, x - \mathrm{i}\frac{\varepsilon_i}{\hbar} t\right), \tag{5}$$

where $\tilde{\Phi}_{i,m}$ is the Fourier transform of wave function $\Phi_i(x)$, g reciprocal lattice vector.

4. Results of simulation

The spatial distribution of the 255 MeV positrons channeled in a 5 μm thick Si crystal along (220) planes are shown in figure 2 obtained using classical approach. Left column represents spatial distribution for parallel positron beam. In the right column spatial distributions of positrons inside the crystal for beam with angular divergences are shown. It is clearly seen that number of maximum of spatial distribution equal 6 for crystal thickness 5 μm and angle of incidence θ=0° (see, in figure 2(a)).

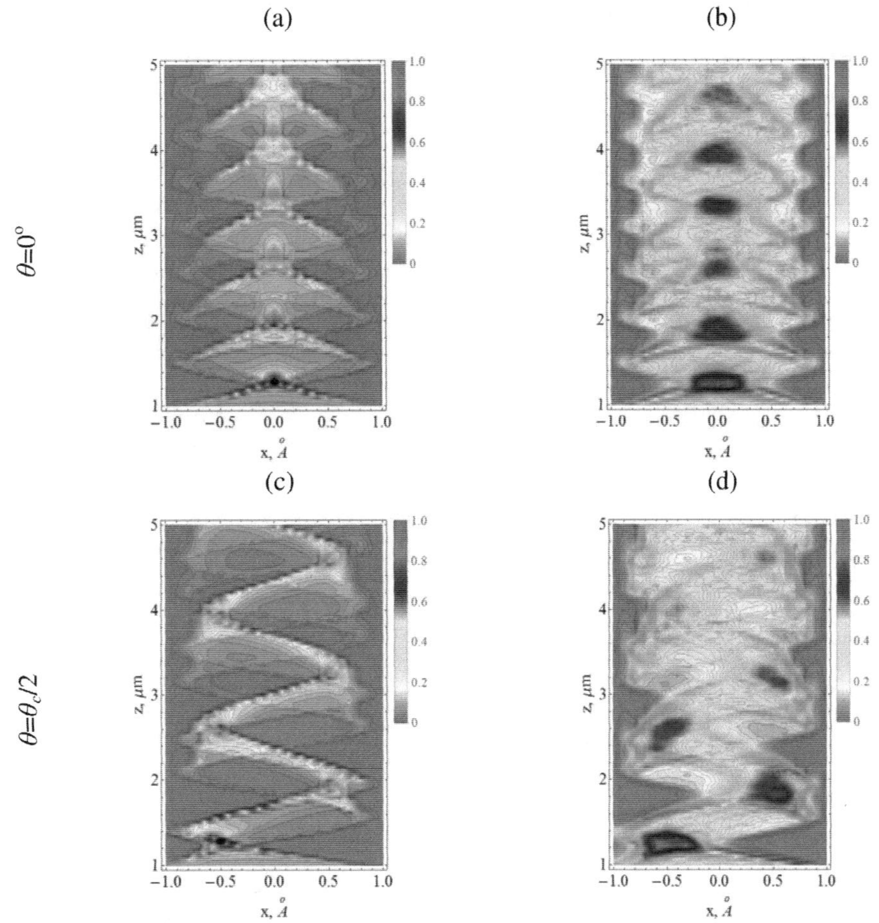

Figure 2. Spatial distribution of the 255 MeV positrons channeled in (220) Si crystal for incident angle θ=0°and $\theta=\theta_c/2$=0.012°=0.2 mrad. (a, c) parallel positron beam (no angular divergence), (b, d) beam with angular divergences equals 0.1 mrad≈$\theta_c/4$.

For beam with angular divergence the quantity of maximum does not change but it became wider (see, in figure 2(b)). Spatial distributions of positrons for incident angle $\theta=\theta_c/2$ are shown in figure 2 (c,d). In this case the spatial distribution has specific "zigzag" shape. This is due to the form of positron trajectories which are determined interplanar potential which is close to parabolic (see, in figure 1).

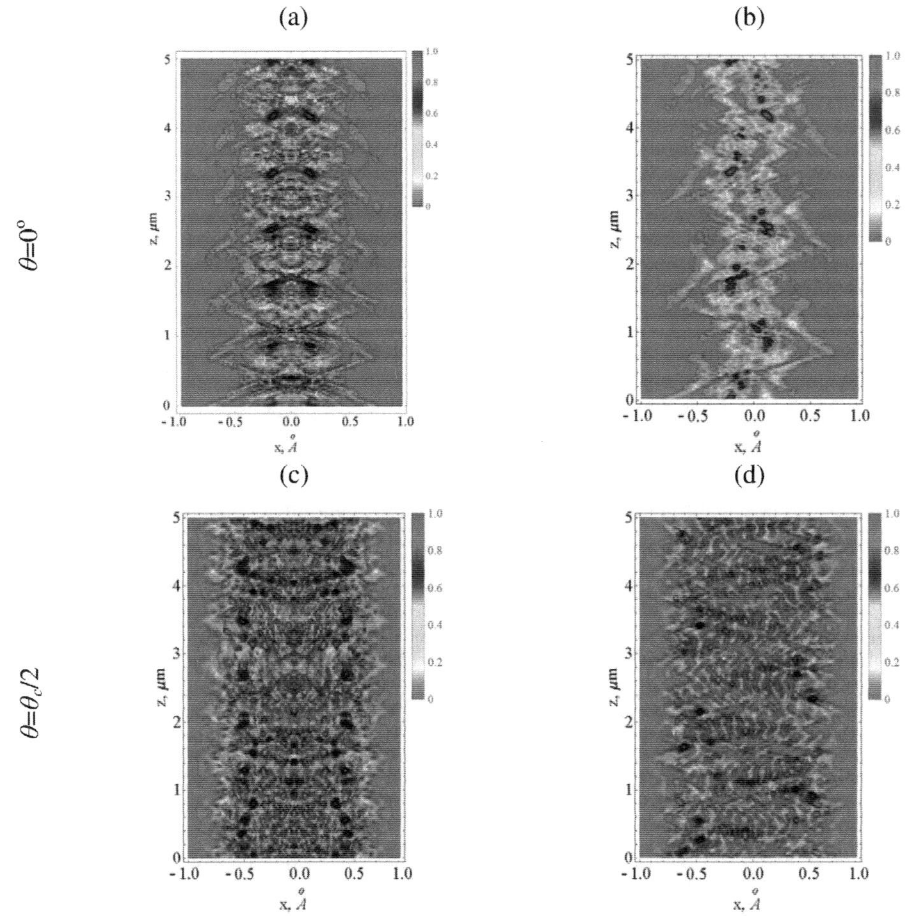

Figure 3. Wave packet probability density of the 255 MeV positrons channeled in (220) Si crystal for incident angles θ=0° and $\theta=\theta_c/2$. (a, c) The conventional wave packet $\Psi(x,t,\theta)$; (b, d) the Gaussian wave packet $\Psi_G(x,t,\theta)$.

Probability density of positrons wave packet calculated using formula (4) are shown in figure 3(a,c). The shape of the probability density is more particular than in the case of classical approach but the number of maxima is the same as for classical spatial distribution (see, in figure 2(a)). The probability density of positrons calculated using Gaussian wave packet are sown in figure 3(b, d). It is seen that the form of probability density resembles the classical spatial distribution calculated for incident angle $\theta = \theta_c/2$.

5. Conclusions

Simulation of flux-peaking effect (FPE) of the 255 MeV positrons channeled in (220) Si crystals is performed in the frame of classical and quantum mechanics using computer code "Basic Channeling with Mathematica©" BCM–1.0. Comparison of the calculation results of the positron beam spatial distribution and the wave packet probability density shows relatively good agreement. This is, mainly, confirms the classical model validity.

However, there are some features: the positron beam "zigzag" shape spatial distribution occurs if angle of incidence is not zero, while a similar distribution for the wave packet probability density occurs only for a Gaussian wave packet. In our opinion, such result means, that the Gaussian wave packet is the most appropriate model of the classical spatial distribution if incident angle of positrons is comparable to the critical angle of channeling.

On the contrary, a good coincidence of classical and quantum results on the flux-peaking effect simulation allows us to conclude that the choice of method is determined by the description of the problem. If we need to estimate the numerical values of the beam parameters, you should use the classic approach. If you need to obtain detailed description of beam spatial distribution, then use the quantum approach.

6. Acknowledgements

The reported study was funded by RFBR according to the research project No. 16-32-00464 мол_а.

References

1. Kumakhov M A and Shirmer G 1980*Atomnie stolknovenija v kristallakh (Atomic Collisions in Crystals(in Russian))* (Moscow: Nauka)
2. Dedkov G V, Kumakhov A M and Sokhov M Z 1983 *Radiation Effects* **71** 261
3. Bogdanov O V, Korotchenko K B, Pivovarov Yu L and Tukhfatullin T A 2008 *Nucl. Instrum. & Methods* B**266** 3858
4. Bogdanov O V, Fiks E I, Korotchenko K B, Pivovarov Yu L and Tukhfatullin T A 2010 *J. Phys.: Conf. Ser.***236** 012029
5. Bogdanov O V, Korotchenko K B, Pivovarov Yu L 2008 *J. Surf. Investigat.* **2, 2,** 290
6. Bogdanov O V, Korotchenko K B, Pivovarov Yu L 2008 *J. Phys.* **B**: *Atomic, Molecular and Optical Physics* **41**, 055004
7. Korotchenko K B 2010 *J. Surf. Investigat.* **4, 4,** 599
8. Korotchenko K B, Pivovarov Yu L, Tukhfatullin T A 2011 *Nucl. Instrum. & Methods* B**269** 2840
9. Korotchenko K B, Pivovarov Yu L 2014 *J. Phys.: Conf. Ser.* **517** 012019
10. Korotchenko K B and Pivovarov Yu L 2015 *JETP Letters* **101** 719
11. Bogdanov O V, Pivovarov Yu L, Takabayashi, Tukhfatullin TA 2012 *J. Phys.: Conf. Ser.***357** 012030
12. Takabayashi Y, Korotchenko, K B, Pivovarov Yu L, Tukhfatullin T A 2013 *Nucl. Instrum. & Methods* B**315** 105
13. Korotchenko K B, Pivovarov Y L and Takabayashi Y 2013 *Nucl. Instrum. & Methods* B**309** 25
14. Korotchenko K B, Pivovarov Yu L Takabayashi Y 2013 *Nucl. Instrum. & Methods* B**309** 25
15. Takabayashi Y, Pivovarov Yu L and Tukhfatullin T A 2014 *Phys. Lett.* A**378** 1520
16. Doyle P A and Turner P S 1968 *Acta Crystallogr.* A**24** 390
17. Kimball J C and Cue N1985 *Physics Reports (Review Section of Physics Letters)* **125** 69

Estimation of the sensitivity in dual wave X-ray absorptiometry

A Gogolev[1], R Rezaev[1,2], Yu Cherepennikov[1], A Vukolov[1], T Gogoleva[1]

[1] National Research Tomsk Polytechnic University, Tomsk, Russia
[2] National Research Nuclear University MePhI, Moscow, Russia

E-mail: gogolev@tpu.ru

Abstract. Dual wave X-ray absorptiometry is considered theoretically and the application of suggested technique extends to the multiphase flow analysis. Proposed method allows for specifying dynamically the percentage of fluid components with the resolution as high as 0.25% (according to the mathematical simulating). The accuracy of this measurement is one order higher by magnitude than that provided by the state of the art flow analyzing devices.

A non-separative real time measurement is one of the most relevant technical issues discussed during a few last decades. The determination of the amount and parameters of extracted well production without usage movable details and manual process control is a very high challenge and still needs to be addressed. There is a wide variety of different approaches to measure multiphase fluid parameters. The transmission radiometric methods based on the analysis of radiation such as X-ray, gamma or neutron passed through the studied object, are the most perspective ones which are currently known [1]. Particularly, X-ray absorptiometry, which is widely used, for example in medicine and biology, is very promising [2, 3].

The state of the art for non-separative flow metering is Vx technology in which ^{133}Ba radioactive source with an activity equals to 10 mCi is used. Radiation of two different energies (in the discussed case 32 and 81 keV) is absorbed, during the passing through the multiphase medium, in varying degrees in according to the composition of medium. Analysis of the absorption degree allows determining the component composition of the three-phase medium [4]. However, radioactive isotope based devices have a few limitations which can be crucial for multiphase flow analysis. The most significant of them are relative low intensity of produced radiation, which limits minimal time of a measurement and leads to increase statistical error, and environmental hazards. In order to avoid these disadvantages, X-ray based radiation sources, which are able to provide much higher radiation flux, can be used, that promises good prospectives in application of aforementioned devices [1]. On the other hand, continuous spectrum of X-ray tubes causes essential difficulties on the stage of radiation detection and data processing. Thus, the main challenge in this area is to produce linear X-ray spectrum using continuous one.

Research and development in that direction resulted in the device implementation «X-ray based densitometer for multiphase flow measurement» [5] which is one of the most advanced devices up today realizing the radiation-based method of measurement. In this device it is suggested to generate X-ray beam having a linear spectrum of the secondary fluorescence (see for example, FluorX described in Ref. [6, 7]).

Content from this work may be used under the terms of the Creative Commons Attribution 3.0 licence. Any further distribution of this work must maintain attribution to the author(s) and the title of the work, journal citation and DOI.
Published under licence by IOP Publishing Ltd

As it previously shown [1], there are a few drawbacks of proposed technology. Such, intensity of characteristic lines is significantly decreased due to the reradiation resulting in about 3 order of intensity loss compared to the primary fluorescence [8]. Also, the presence of intensive background radiation, which is comparable with the intensity of useful K_α line, increases errors. Therefore although the proposed technology helps to improve the radiation safety it does not provide a significant increase of the accuracy and rate of measurement compared to the technology based on Vx.

An alternative method based on considering a wave dispersive analysis of the radiation [9] passed through a multiphase medium was suggested by the authors of the present work. The basic idea of the method consists of the following: generation of radiation based on X-ray tube as a source of radiation, passing of complex in spectral composition radiation through the multiphase flow, extraction two narrow monochromatic lines of radiation (first and second order of diffraction), detection of these lines and analysis of their degree of absorption.

The realization of the described approach is implemented in the device described in the patent [10]. Lines of X-ray radiation with such energies as 20 and 40 keV are considered as preferable for the analysis procedure. Suggested method provides a number of advantages. Firstly, the elimination of the stage of reradiation (generation of secondary fluorescence) results in the increasing the X-ray radiation intensity what improves the accuracy and the rate of analysis. The loss of intensity due to the monochromatization in this case is determined by the reflection coefficient of radiation in the diffraction directions. This coefficient constitutes the value more than 50% that gives the loss of intensity less than the half from the initial level and yet this is still much lower than in commercially available sources, for example, Fluor'X. Besides, the proposed method ensures a much lower level of the background radiation.

To check the possibilities of the suggested method a simulation of the radiation spectrum on the different stages was performed. The tube with the argentum anode working at the cathode-anode voltage 60 kV was considered as the radiation source. Simulating was completed using the software GEANT4. In figure 1 the radiation spectra generated by the tube before and after the passing through the studied object are presented. Characteristic X-ray lines of Ag with the energy 22.1 keV was chosen as the first working line for further evaluations and the line with double energy was chosen as the second one.

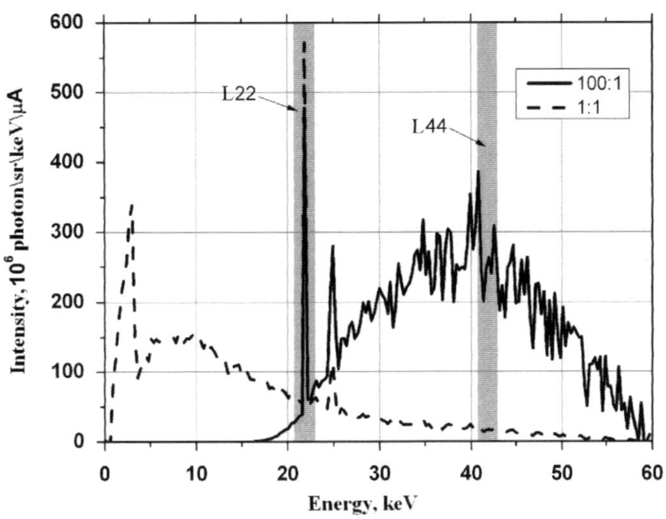

Figure 1. X-ray spectrum generated by the X-ray tube with Ag anode; the dashed line denotes the radiation before the studied object and the solid line (the magnification is by 100 times) – the radiation after the studied object.

In figure 2 X-ray spectrum passed without interaction through 73 mm of water (dashed line) and oil (solid line) after reflection from the crystal are presented. Shown spectrum are perfectly illustrating the difference in the degree of radiation absorption by different components of multiphase flow.

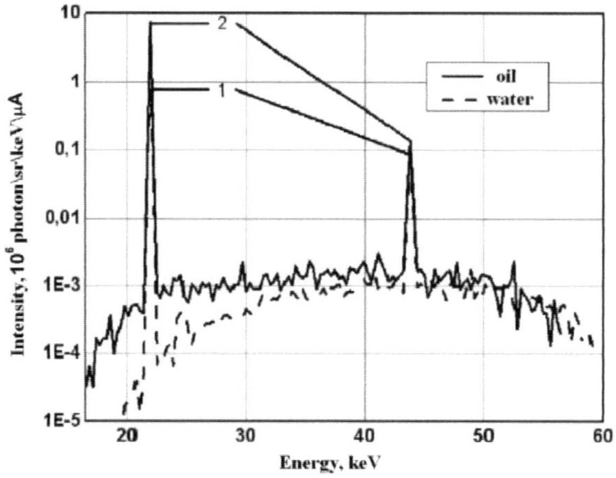

Figure 2. X-ray radiation spectrum passed without interaction through 73 mm of water (the dashed line) and oil (the solid line) after reflection from the crystal. By numbers 1 and 2 the maximal levels for water and oil are denoted, correspondingly.

As it shown in [1] and also figure 3 the ratio of number of monochromatic lines photons to the number of background radiation photons for FluorX source is close to unit, meantime by using the diffraction method to extract the analytical lines this ratio has constituted the value as low as 0.03 that points to the advantage of the discussed approach of monochromatization. Figure 3 shown here in order to provide convenience for the readers.

a)

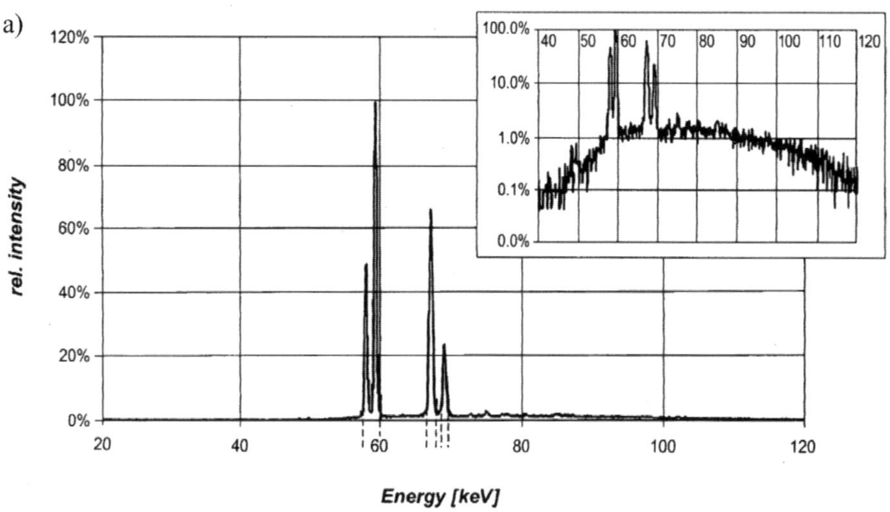

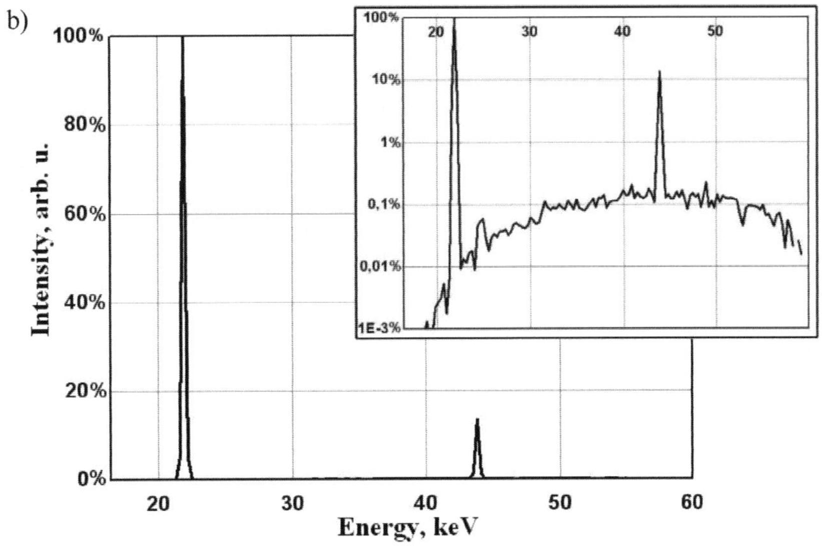

Figure 3. Radiation spectrum from X-ray radiation source are shown: a) – Fluor'X [7], b) – the proposed source.

From the other side in order to receive the statistical uncertainty on the level of stability of modern X-ray sources, it is about 0.1 % photons per second, we estimate the required time for accumulation the statistics. Under the assumption that the process of X-ray photons interaction with detector material obeys to Poisson distribution the expression for time determination reads

$$t = \frac{1}{\bar{N}\delta^2},$$

where δ is a relative statistical uncertainty, $\bar{N}$ is the average number of photons incident on the detector per second.

Thereby the required time for one measurement is $\max\{t_l, t_h\}$ in the case of source FluorX. Here l, h are indices corresponding to the line at high and low quantum energy.

$$t_l = \frac{1}{(0.0035 \cdot 10^6 \cdot 10^{-6})} = 286 \text{ sec},$$

$$t_h = \frac{1}{(0.0025 \cdot 10^6 \cdot 10^{-6})} = 400 \text{ sec}.$$

The time required for a measurement with the claimed statistical uncertainty in proposed method:

$$t_l = \frac{1}{(0.75 \cdot 10^6 \cdot 10^{-6})} = 1 \text{ sec},$$

$$t_h = \frac{1}{(0.1 \cdot 10^6 \cdot 10^{-6})} = 10 \text{ sec}.$$

In fact the device principle is based on measurement the average density (or thickness). Thereby in the first approximation the limit of accuracy should be a function of minimal its deviation which can be detected. The possibility to distinguish the specific characteristics is defined by SNR parameter describing the signal to noise ratio. The minimal value of this parameter at which the difference in the signal still can be extracted, is equal to 1. In the case of monochromatic radiation the signal to noise ratio reads

$$SNR = \frac{N_2 - N_1}{\sqrt{N_1}},$$

where N_1 is a number of photons, passed through the homogeneous medium of the thickness H (e.g. 100% water) and N_2 is a number of photons, passed through the same medium containing oil and gas. Now we evaluate the minimal linear size δ of the inhomogeneity with the absorption coefficient μ_2. The inhomogeneity is located in a homogeneous medium with the absorption coefficient μ_1. Let N_0 be a number of photons incident on the medium, N_1 and N_2 are the numbers of photons passed through the homogeneous medium with the absorption coefficient μ_1 and μ_2, correspondingly:

$$N_1 = N_0 \exp(-\mu_1 H);\ N_2 = N_0 \exp(-\mu_1(H-\delta) - \mu_1\delta).$$

Substituting these expressions into the equation for SNR results in

$$SNR = \frac{N_0 \exp(-\mu_1(H-\delta) - \mu_2\delta) - N_0 \exp(-\mu_1 H)}{\sqrt{N_0}\exp(-\frac{\mu_1 H}{2})} =$$

$$= \sqrt{N_0}[\exp(-\mu_1(0.5H-\delta) - \mu_2\delta) - \exp(-0.5\mu_1 H)].$$

Equating SNR to 1 we obtain the following expression

$$\exp((\mu_1 - \mu_2)\delta) - 1 = \frac{\exp(0.5\mu_1 H)}{\sqrt{N_0}}.$$

Expressing the exponent in the left part of the last equation gives:

$$(\mu_1 - \mu_2)\delta = \ln\left(\frac{\exp(0.5\mu_1 H)}{\sqrt{N_0}} + 1\right).$$

Finally

$$\delta = \frac{1}{\mu_1 - \mu_2}\ln\left(\frac{\exp(0.5\mu_1 H)}{\sqrt{N_0}} + 1\right) \approx \frac{\exp(0.5\mu_1 H)}{(\mu_1 - \mu_2)\sqrt{N_0}} = \frac{1}{(\mu_1 - \mu_2)\sqrt{N_1}}.$$

Absorption coefficients at radiation energy 44 keV for water and oil are $\mu_1 = 0.248\ \mathrm{cm}^{-1}$, $\mu_2 = 0.197\ \mathrm{cm}^{-1}$. These coefficients determine the minimal linear size of the oil drop which is detectable in water. Results of simulation give the number of photons passing into the solid angle 10^{-4} sr per 10 sec about $N_1 = 950000$. Thus for oil

$$\delta_{\mathrm{oil}} = 0.02\ \mathrm{cm}$$

and similarly for a methane bubble with $\mu_2 = 0.000234\ \mathrm{cm}^{-1}$ at the same energy

$$\delta_{\mathrm{met}} = 0.0041\ \mathrm{cm}.$$

Absorption coefficients for water, oil and methane are $\mu_1 = 0.655937\ \mathrm{cm}^{-1}$, $\mu_2 = 0.337753\ \mathrm{cm}^{-1}$ $\mu_2 = 0.000374\ \mathrm{cm}^{-1}$ at radiation energy 22 keV that gives the number of photons $N_1 = 7500000$. Corresponding sizes are

$$\delta_{\mathrm{oil}} = 0.0011\ \mathrm{cm}$$

for oil drop in water and

$$\delta_{\mathrm{met}} \approx 0.0007\ \mathrm{cm}$$

for methane bubble.

Finally the minimal dimensions for methane bubble in water which are detectable equal to 41 µm and 200 µm for oil at detector loading level 10^6 counts per sec and the tube diameter by 73 mm.

Taking into account the average density 0.9 g/cm^3 for oil and 0.001 g/cm^3 for methane and expressing the minimal sizes by mass fraction one can receive

$$m_{\text{oil}} = \frac{x}{H\langle\rho\rangle}\rho_{\text{oil}} = 0.0025 = 0,25\% \text{ for oil,}$$

$$m_{\text{met}} = \frac{x}{H\langle\rho\rangle}\rho_{\text{met}} = 5.6\cdot10^{-7} = 5.6\cdot10^{-5}\% \text{ for methane bubble.}$$

Interesting to note that the threshold of the present-to-day commercially available devices is as low as 2.5% by the liquid phase and 5% by the gas phase which at least one order lower than our technology provides.

Conclusion

An approach to perform the compositional analysis of multiphase flow based on dual wave X-ray absortiometry is suggested and shown to provide the higher sensitivity in comparison with present analogues. Numerical simulations are performed and the spectrum of X-ray radiation on different stages of analysis is determined that allows for sensitivity evaluation of suggested approach. According to these evaluations a drop of oil with the linear size of 0.011 mm in a layer of water with the thickness 73 mm can be detected. Corresponding size of a gas bubble in a water layers is 0.007 mm.

Acknowledgement

This work was partially supported by The Ministry of education and science of the Russian Federation within program “Nauka”.

References

[1] Gogolev A, Cherepennikov Yu, Rezaev R *et al* 2015 *Adv. Mater. Res.* **1084** 252
[2] Stuchebrov S, Verigin D, Lukyanenko Y *et al* 2014 *Adv. Mater. Res.* **880** 168
[3] Stuchebrov S, Batranin A, Verigin D *et al* 2014 *Adv. Mater. Res.* **880** 53
[4] Scheers A M, Slijkerman W F J 1996 *SPE publication* 36593
[5] Patent US № 20120087467, 12.04.2012
[6] Patent US № 4903287, 20.02.1990
[7] Patent US № 5157704, 20.10.1992
[8] Griaznov A Y 2004 Development of hardware and methodical ways to improve analytical parameters of energy dispersive XRF analyzer, Dissertation of Candidate of Technical Science (PhD), St. Petersburg, (in Russian)
[9] Cherepennikov Yu, Gogolev A 2014 *Journal of Physics: Conference Series* **517** 012037
[10] Patent RU № 2559119, 10.08.2015 (in Russian)

Numerical Simulation of the Medical Linear Accelerator Electron Beams Absorption by ABS-Plastic doped with Metal

S G Stuchebrov, I A Miloichikova, A A Krasnykh
Tomsk Polytechnic University, Tomsk, Russia

E-mail: stuchebrov@tpu.ru

Abstract. In this paper the numerical simulation results of the dose spatial distribution of the medical electron beams in ABS-plastic doped with different concentrations of lead and zinc are shown. The dependences of the test material density on the lead and zinc mass concentrations are illustrated. The depth dose distributions of the medical electron beams in the modified ABS-plastic for three energies 6 MeV, 12 MeV and 20 MeV are tested. The electron beam shapes in the transverse plane in ABS-plastic doped with different concentrations of lead and zinc are presented.

1. Introduction

Radiation therapy uses a wide range of radiation for neoplasms treatment: gamma rays, X-rays, protons, neutrons and electrons [1–8]. Nowadays, one of the most interesting directions of the electron beam medical application is an intraoperative therapy [3]. For these purposes it is necessary to have the possibility of beam profile modulation. For example, new methods of the collimators and filters production allow to increase the accuracy of dose delivery to the target.

The intraoperative therapy procedure requires careful planning for the determination of the geometry and operating modes of the treatment accelerator [3–4]. The dosimetric treatment planning methods is mainly based on computer simulations [4]. The presence of the materials with different densities allows to trigger the interaction of the electron beam with human tissue. It enables phantoms for experimental methods of the medical procedures planning to develop. For these purposes it is necessary to have the method of tissue-equivalent materials for the human body phantom production. Such requirements are met by rapid prototyping technology [9].

One solution for the electron beam filters and the human body phantoms production can be the creation of filaments with the particular densities for 3D-print methods [9–10].

In this investigation one of the most common types of the clinical linear accelerator "Elekta Synergy System" is taken as electron-emitting source. The electron beam energies were selected according to the accelerator operating mode – 6 MeV, 12 MeV, or 20 MeV. For the test material the ABS-plastic doped with different concentrations of lead and zinc was chosen.

The characteristics of the electron beam interaction with the test material were calculated by Monte Carlo method.

Content from this work may be used under the terms of the Creative Commons Attribution 3.0 licence. Any further distribution of this work must maintain attribution to the author(s) and the title of the work, journal citation and DOI.
Published under licence by IOP Publishing Ltd

2. Materials and methods

2.1. Test materials

For the numerical simulation the test material densities were selected according to the real anatomical structures density. For example, bone tissue density approximately equal to 1.75 g/cm^3, skin density approximately equal to 1.1 g/cm^3, muscular tissue density approximately equal to 1.06 g/cm^3 [12, 13].

In this investigation, as test materials we were used the ABS-plastic doped with different concentrations of lead and zinc. The estimation of the test materials density was carried out by the following equation:

$$\rho = \frac{\rho_{\mathrm{ABS}} \cdot \rho_{\mathrm{Pb/Zn}}}{k(Me) \cdot \rho_{\mathrm{ABS}} + (100 - k(Me)) \cdot \rho_{\mathrm{Pb/Zn}}}, \quad (1)$$

were ρ_{ABS} – the ABS-plastic density measured in this investigation (1.06 g/cm^3); $\rho_{Pb/Zn}$ – the lead density (11.34 g/cm^3) or zinc density (7.13 g/cm^3); $k(Me)$,% – the mass concentration of the metal in the test material [14].

Figure 1 shows the dependence of the test material density on the lead and zinc mass concentration obtained with the help of equation 1.

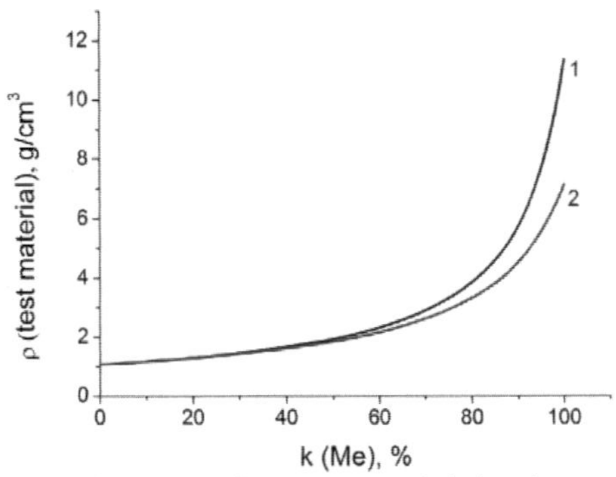

Figure 1. The dependence between the test material density and the concentration of: 1 – lead and 2 – zinc.

2.2. Emitting source

In this paper for the bases of the simulation the linear accelerator "Elekta Synergy System" was chosen as electron source. "Elekta Synergy Platform" is a radiation delivery system that integrates Image-Guided Radiation Therapy (IGRT). This accelerator allows to use leading technology in 3D conformal techniques, through static and dynamic Intensity Modulated Radiation Therapy (IMRT) to Volumetric Arc Therapy VMAT [15].

2.3. Simulation program

For the medical electron beams model, the possibility of the program "Computer laboratory (PCLab)" version 9.6 was analyzed. This software package uses the Monte Carlo method. The dose spatial distribution was calculated in the program operating mode "PHANTOM", which was created specially for the radiotherapy purposes. It allows to present energy and dose distributions in the matter for the three types of the initial particles: photons, electrons and protons. This program are modelled in the simulation the different types of physical processes, for example scattering, ionization, pair production and photoelectric effect [16].

2.4. Experiment geometry

In the simulation procedure the initial electron beam size was chosen conventionally, because the collimation system of the Elekta Synergy accelerator allows to obtain any radiation fields according to the task. The material-radiation interaction analysis was carried out in the ABS-plastic doped with different concentrations of lead and zinc. For convenience, the following parameters of the emitting source were used: the initial beam size – 2.0 mm, divergent electron beam (0.1 rad); electron energies were chosen according to the accelerator operating mode – 6 MeV, 12 MeV and 20 MeV.

3. Results and discussions

In this investigation the electron beam interaction with the ABS-plastic doped with different concentrations of lead and zinc was analyzed. For this purpose the depth dose distribution of the electron beam in the ABS-plastic doped with different concentrations of lead and zinc was calculated. The data was obtained for the three electron beam energies – 6 MeV, 12 MeV and 20 MeV. The curves are shown in Figure 2 for the test material densities equal to 1.1 g/cm^3, 1.5 g/cm^3, 2.0 g/cm^3 and 2.5 g/cm^3.

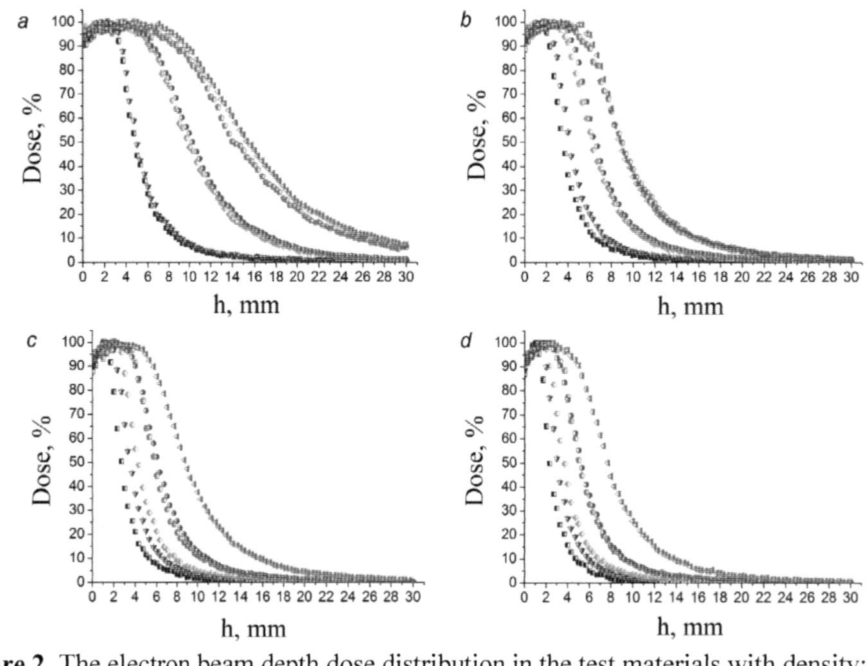

Figure 2. The electron beam depth dose distribution in the test materials with density: a – 1.1 g/cm^3; b – 1.5 g/cm^3; c – 2.0 g/cm^3; d – 2.5 g/cm^3.
The ABS-plastic modified by lead, the electron beam energy equal to: ◧ – 6 MeV; ◈ – 12 MeV; ⬟ – 20 MeV.
The ABS-plastic modified by zinc, the electron beam energy equal to: ▼ – 6 MeV; ◑ – 12 MeV; ◀ – 20 MeV.

Figure 2 shows that minor change of the lead and zinc concentration in the ABS-plastic significantly affects on the character electron beam interaction with the test material. It is shown that, at the equal values of the test material density, the difference between the depth dose distribution curves obtained in the ABS-plastic doped with the lead and zinc increase with the growth of the electron beam energy.

Figures 3 and 4 show examples of the electron beam shapes in the transverse plane at the depth 3.7 mm in the ABS-plastic modified by lead and zinc (the test material densities – 1.1 and 2.5 g/cm^3)

for the energies equal to 6 and 20 MeV. The dose results were averaged and normalized to the maximum simulation dose in all volume of the test material.

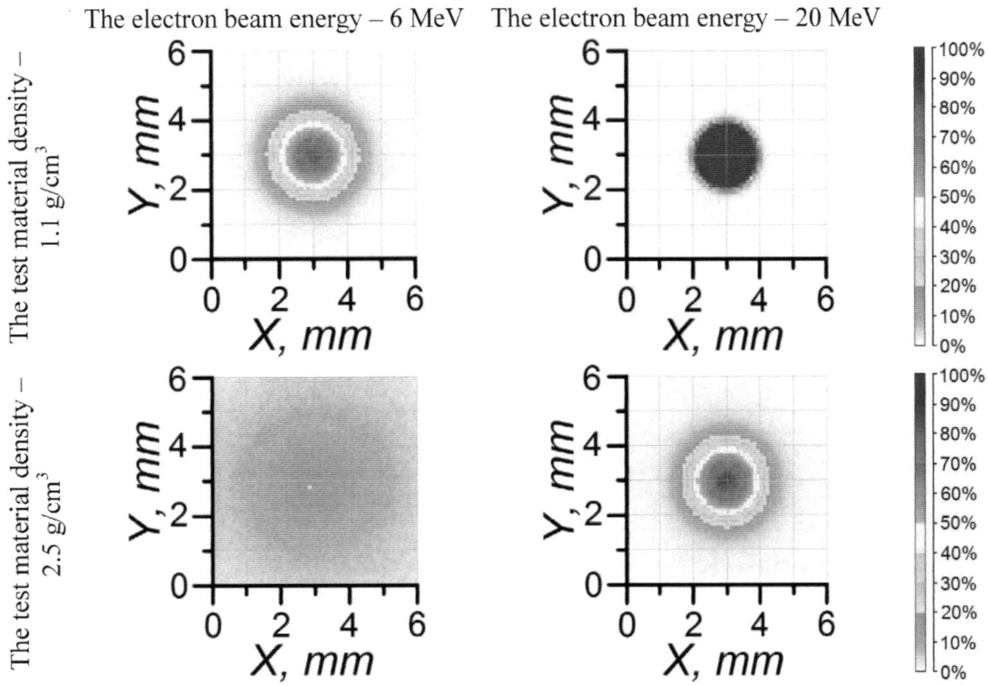

Figure 3. The electron beam shape in the transverse plane at the depth 3.67 mm in the ABS-plastic modified by lead.

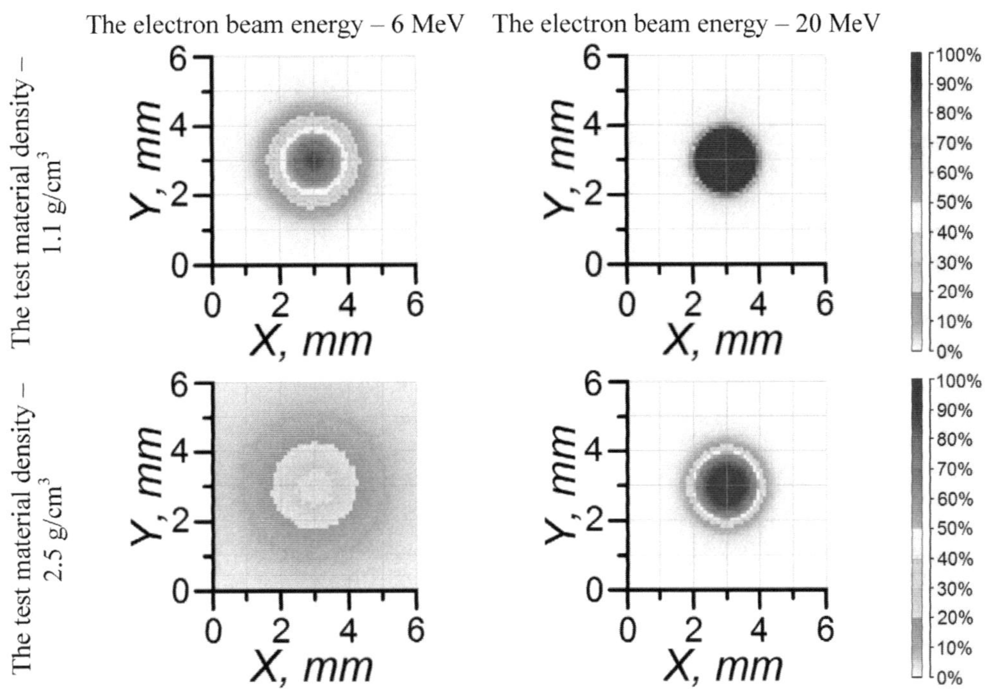

Figure 4. The electron beam shape in the transverse plane at the depth 3.67 mm in the ABS-plastic modified by zinc.

Figure 5 shows examples of the electron beam shapes in the transverse plane at the depth 3.67 mm in the ABS-plastic modified by lead and zinc (the test material density – 2.0 g/cm3) for the energy equal to 12 MeV. The dose results were averaged and normalized to the maximum simulation dose per test object.

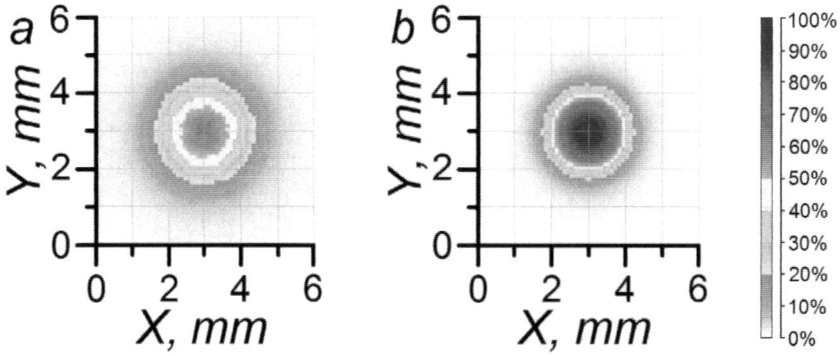

Figure 5. The electron beam shape in the transverse plane at the depth 3.67 mm in the ABS-plastic modified by a – lead and b – zinc (the test material density – 2.0 g/cm^3; the electron beam energy – 12 MeV).

The graphic data present that there is a difference in dose distribution with changing of the electron beam energy and the concentration of metal contamination in plastic. These effects are more expressed for the high electron energy and material density. It is shown that at equal density values of the test material there is a difference in the electron beam interaction with ABS-plastic modified by lead and zinc, it is conditioned by the different values of the interaction cross section.

4. Summary

The paper presents the numerical simulation results of the dose spatial distribution of the medical electron beams in ABS-plastic doped with different concentrations of lead and zinc. It is shown that the different metal contamination in ABS-plastic have an effect on the electron interaction cross section, at the equal test material density, therefore, for the human body phantom creation, it is preferred to use the Hounsfield scale for the test material instead of the mass densities.

On account of the Hounsfield's units are calculated for the X-ray interaction, further analysis of the electron beam distribution in the test materials needs additional experimental research. The purpose of a future investigation will be to modify ABS-plastic with lead and zinc rather than to obtain the Hounsfield's units and the real density of the materials.

Acknowledgements

This work was financially supported by the Ministry of Education and Science of the Russian Federation in part of the science program (N 5.1485.2015).

References

[1] Mayles P, Nahum A and Rosenwald J C 2007 *Handbook of radiotherapy physics: theory and practice* (USA: CRC Press) 1393

[2] Melnikov A A, Vasilyev S A, Musabaeva L I, Velikaya V V, Gribova O V, Startseva Zh A, Urazova L N, Lebedev I N and Choynzonov E L 2012 *Experimental oncology* **34(4)** 354-357

[3] Choinzonov E L and Musabaeva L I 2006 *The intraoperative electron and gamma-ray teletherapy of the malignant neoplasms* (Tomsk: NTL) 216 [in Russian]

[4] Barrett A, Dobbs J, Morris St and Roques T 2009 *Practical Radiotherapy Planning* (UK: Hodder Arnold) 468

[5] Musabaeva L I, Lisin V A and Velikaia V V 2014 *Radiatsionnaia biologiia, radioecologiia* **54(5)** 474-478 [in Russian]
[6] Amichetti M, Lombardi A, Algranati K and Schwarz M 2006 *Cancer Therapy* **4** 231-240
[7] Gribova O V, Musabaeva L I, Choinzonov E L and Mukhamedov M R 2012 *Vestnik otorinolaringologii* **5** 91-92 [in Russian]
[8] Hogstrom K R and Almond P R 2006 *Phys. Med. Biol.* **51** R455
[9] Christ S, Schnabel M, Vorndran E, Groll J, Gbureck U 2015 *Materials Letters* **139** 165–168
[10] Information on http://www.ptcommunity.com/system/files/pdf/ptj3910704.pdf
[11] Information on http://www.appliancedesign.com/ext/resources/AM/Home/Files/PDFs/themoplastics.pdf
[12] Information on http://www.scrollseek.com/training/densitiesofdifferentbodymatter.html
[13] Information on https://www.diva-portal.org/smash/get/diva2:328143/FULLTEXT01.pdf
[14] Information on http://www.coolmagnetman.com/magconda.htm
[15] Information on https://www.elekta.com/radiotherapy/treatment-delivery-systems/elekta-synergy/
[16] Bespalov V I 2015 *Computer laboratory* (Version 9.6) (TPU: Tomsk) 115 [in Russian]

Wave functions of channeling electrons in regular and chaotic cases

N F Shul'ga[1,2], V V Syshchenko[3], A I Tarnovsky[3] and A Yu Isupov[4]

[1] Akhiezer Institute for Theoretical Physics of the NSC "KIPT", Akademicheskaya Street, 1, Kharkov 61108, Ukraine
[2] V.N. Karazin National University, Svobody Square, 4, Kharkov 61022, Ukraine
[3] Belgorod State University, Pobedy Street, 85, Belgorod 308015, Russian Federation
[4] Laboratory of High Energy Physics (LHEP), Joint Institute for Nuclear Research (JINR), Dubna 141980, Russian Federation

E-mail: syshch@bsu.edu.ru, syshch@yandex.ru

Abstract. The spectral method applicability to the problem of the charged particles axial channeling is studied. The quantum state characteristics for the electron motion near the atomic string [110] direction of the silicon crystal are obtained using this method of Schrödinger equation numerical solution. The full set of eigenenergies and eigenfunctions for the electron with longitudinal energy of 5 MeV is computed. The results obtained illustrate the difference in the wave function morphology for regular and chaotic cases.

1. Introduction

The fast charged particles incident onto the crystal under a small angle to any crystallographic axis densely packed with atoms can perform the finite motion in the transverse plane; such motion is known as the axial channeling [1]–[3]. The particle motion in the axial channeling mode could be described with a good accuracy as the one in continuous potential of the atomic string, i.e. in the potential of atoms averaged along the string axis. During motion in this potential the longitudinal particle momentum $p_{\|}$ is conserved, so the motion description is reduced to two-dimensional problem of motion in the transversal plane. This transverse motion could be substantially quantum [1].

From the viewpoint of the dynamical systems theory, the channeling problem is interesting because the particle's motion could be both regular and chaotic. The wave functions of the stationary quantum states in the situations of regular and chaotic motion in the classical limit are computed in the present paper. The qualitative difference between them illustrates the basic statements of the quantum chaos theory [4]–[8].

2. Method

The electron transversal motion in the atomic string continuous potential is described by the two-dimensional Schrödinger equation

$$\hat{H}\Psi(x, y, t) = i\hbar \frac{\partial}{\partial t}\Psi(x, y, t) \tag{1}$$

Content from this work may be used under the terms of the Creative Commons Attribution 3.0 licence. Any further distribution of this work must maintain attribution to the author(s) and the title of the work, journal citation and DOI.
Published under licence by IOP Publishing Ltd

with Hamiltonian

$$\hat{H} = -\frac{\hbar}{2E_{\|}/c^2}\left[\frac{\partial^2}{\partial x^2} + \frac{\partial^2}{\partial y^2}\right] + U(x,y) \tag{2}$$

and the value $E_{\|}/c^2$ (here $E_{\|} = (m^2c^4 + p_{\|}^2c^2)^{1/2}$) instead of the particle mass [1].

The Hamiltonian eigenfunctions have been found numerically as well as the transverse energy eigenvalues using the so-called spectral method [9]. For the channeling problem it has been applied for the first time in [10].

The spectral method idea is the following. Every solution of the time-dependent Schrödinger equation could be expressed as the superposition

$$\Psi(x,y,t) = \sum_{n,j} A_{n,j}\psi_{n,j}(x,y)\exp(-iE_n t/\hbar) \tag{3}$$

of the Hamiltonian eigenfunctions $\psi_{n,j}(x,y)$,

$$\hat{H}\psi_{n,j}(x,y) = E_n\psi_{n,j}(x,y) \quad ,$$

where the j index is used to distinguish the degenerate states corresponding to the energy E_n. Computation of the correlation function

$$P(t) = \int_{-\infty}^{\infty}\int_{-\infty}^{\infty} \Psi^*(x,y,0)\Psi(x,y,t)\,dxdy \tag{4}$$

for the wave function of the form (3) gives

$$P(t) = \sum_{n,n',j,j'} \exp(-iE_{n'}t/\hbar)A^*_{n,j}A_{n',j'}\int_{-\infty}^{\infty}\int_{-\infty}^{\infty} u^*_{n,j}(x,y)\psi_{n',j'}(x,y)dxdy =$$

$$= \sum_{n,n',j,j'} \exp(-iE_{n'}t/\hbar)A^*_{n,j}A_{n',j'}\delta_{nn'}\delta_{jj'} = \sum_{n,j}|A_{n,j}|^2\exp(-iE_n t/\hbar) \quad . \tag{5}$$

Fourier transformation of (5) leads to the expression

$$P_E = \int_{-\infty}^{\infty} P(t)\exp(iEt/\hbar)\,dt = 2\pi\hbar\sum_{n,j}|A_{n,j}|^2\,\delta(E-E_n) \quad . \tag{6}$$

We see the Fourier transform of the correlation function looks like a series of δ-form peaks, whose positions indicate the energy eigenvalues. So, the method is based on the numerical simulation of the time evolution of initial wave function $\Psi(x,y,0)$ according to time-dependent Schrödinger equation.

Of course, the numerical integration of (1) could be performed for the finite time interval T only, so the Fourier transform of the correlation function takes the form

$$P(E) = \int_0^T P(t)\exp\left(\frac{i}{\hbar}Et\right)dt = \sum_{n,j}|A_{n,j}|^2\frac{\sin\left[(E_n - E_n')T/2\hbar\right]}{(E_n - E_n')/2\hbar}\exp\left[\frac{i}{2\hbar}(E-E_n)T\right] \quad . \tag{7}$$

The $\sin(xT)/x$ function in the right part of (7) has the main maximum at $x = 0$, whose magnitude is increased and width is narrowed with T interval increasing. Its other maxima and minima at both sides from the main maximum have the lower magnitudes. So, instead of δ-like peaks set we have the superposition of maxima, whose widths are inverse proportional to

T. The required resolution of the computation dictates the T value, while the expected minimal distance ΔE between the neighbor energy levels dictates the resolution, and condition for T is

$$T > 2\pi\hbar/\Delta E \ . \tag{8}$$

To avoid the superposition of side maxima of $\sin(xT)/x$ function from neighbor energy levels this requirement should be enhanced. In our calculations we use

$$T = 16\pi\hbar/\Delta E \ . \tag{9}$$

The numerical procedure based on these ideas has been developed and successfully used for transverse energy levels search with high precision in [11]–[13].

The stationary wave functions could be obtained with the spectral method also [9, 10]. The superposition (3) multiplied by $\exp(iE_n t/\hbar)$ and integrated over the time interval T provides the eigenfunctions $\psi_{n,j}(x,y)$ of the eigenvalue E_n:

$$\int_0^T \Psi(x,y,t)\exp\left(\frac{i}{\hbar}E_n t\right)dt = \sum_{n',j} A_{n',j}\,\psi_{n',j}(x,y)\int_0^T \exp\left[\frac{i}{\hbar}(E_n - E'_n)t\right]dt = \tag{10}$$

$$= \sum_{n',j} A_{n'}\,\psi_{n',j}(x,y)\,\frac{\sin\left[(E_n - E'_n)T/2\hbar\right]}{(E_n - E'_n)/2\hbar}\exp\left[\frac{i}{2\hbar}(E_n - E'_n)T\right] \ .$$

For the large enough T the left part integral in (10) is proportional to either eigenfunction $\psi_n(x,y)$ (without state degeneration), or $\psi_{n,j}(x,y)$ superposition (if n-th energy level is degenerated) with a good accuracy.

3. Wave functions in the regular case

The field of single atomic string has the axial symmetry $U(x,y) = U(r)$, if the neighbor strings influence is neglected. This symmetry allows us to classify the electron quantum states. In the polar coordinates

$$\hat{H} = -\frac{\hbar}{2E_\parallel/c^2}\left[\frac{1}{r}\frac{\partial}{\partial r}\left(r\frac{\partial}{\partial r}\right) + \frac{1}{r^2}\frac{\partial^2}{\partial\varphi^2}\right] + U(r) \ . \tag{11}$$

The equation for the eigenfunctions and eigenvalues of the operator (11)

$$-\frac{\hbar}{2E_\parallel/c^2}\left[\frac{1}{r}\frac{\partial}{\partial r}\left(r\frac{\partial}{\partial r}\right) + \frac{1}{r^2}\frac{\partial^2}{\partial\varphi^2}\right]\psi(r,\varphi) + U(r)\psi(r,\varphi) = E_\perp\psi(r,\varphi) \tag{12}$$

allows us to separate the variables, and their solutions are

$$\psi_{n_r,m}(r,\varphi) = \frac{1}{\sqrt{2\pi}}e^{im\varphi}\rho_{n_r,|m|}(r) \ , \tag{13}$$

where the $\rho_{n_r,|m|}(r)$ function is the solution of

$$\frac{1}{r}\frac{\partial}{\partial r}\left(r\frac{\partial}{\partial r}\right)\rho_{n_r,|m|}(r) + \frac{2E_\parallel/c^2}{\hbar}\left(E_\perp - U(r) - \frac{\hbar}{2E_\parallel/c^2}\frac{m^2}{r^2}\right)\rho_{n_r,|m|}(r) = 0 \ . \tag{14}$$

The particle quantum states in the potential $U(r)$ with axial symmetry are characterized by two quantum numbers: the radial one n_r and projection m of the orbital momentum to the symmetry axis of the field. The n_r is equal to the number of zeroes of the function $\rho_{n_r,|m|}(r)$ at finite r

(except the possible zero at $r = 0$). The states with $m = 0$ are non-degenerated, while ones with non-zero m are doubly degenerated by the signum of m. The eigenvalues of Hamiltonian (11), i.e. the eigenvalues of the transversal motion energy of the channeling electron, depend in general on both quantum numbers: $E_\perp = E_{n_r,|m|}$.

First of all let's consider the low energy ($E_\parallel = 5$ MeV) electron's motion near direction of the atomic string [110] of the Si crystal. The continuous potential could be represented by the modified Lindhard potential [1]

$$U_1(x, y) = -U_0 \ln\left(1 + \frac{\beta R^2}{x^2 + y^2 + \alpha R^2}\right) , \tag{15}$$

where $U_0 = 60$ eV, $\alpha = 0.37$, $\beta = 3.5$, $R = 0.194$ Å (Thomas–Fermi radius).

In the upper left plot in figure 1 the wave function of the ground state is shown. It has single maximum in center ($r = 0$) and has not zeroes at all (except the asymptotic zero at $r = \infty$).

The other wave functions are represented by black (for $\psi(x, y) < 0$) and white (for $\psi(x, y) > 0$) regions on the plane (x, y) (see figure 1). The irregularity on the periphery of some plots is due to numerical uncertainties. This representation allows us to classify the wave functions using the quantum numbers n_r and m easily, by the wave function zeroes counting.

Indeed, at the $m = 0$ (upper row in figure 1) the wave functions $\psi_{n_r,0}(r, \varphi)$ depend on radial coordinate r only, so the $\psi(x, y) = 0$ zero lines are concentric circles.

Each next solution $\rho_{n_r,0}(r)$ of the equation (14) at $m = 0$ in the solution sequence, where the corresponding eigenenergy is increased, has more zeroes than previous by exactly one. So we can enumerate the corresponding quantum eigenstates by the quantum number n_r which is equal to the number of zeroes of the radial wave function $\rho_{n_r,0}(r)$. Practically n_r is equal to number of borders between the concentric regions.

For the $m \neq 0$ case the situation is analogous. The degeneration by signum of m means the our method provides for each $E_\perp$ eigenvalue the superposition of $\psi_{n_r,m}(r, \varphi)$ and $\psi_{n_r,-m}(r, \varphi)$ functions with equal weights instead of eigenfunction (13). This superposition looks as follows:

$$\rho_{n_r,|m|}(r) \cos\left[|m|\varphi + \alpha_m\right] . \tag{16}$$

So, the m value for calculated wave function equals to number of white (or black) sectors on the plot in figure 1. The energy levels of the considered system classified by quantum numbers n_r and m are presented in figure 2.

The character feature of the calculated stationary wave functions is the crossing of $\psi(x, y) = 0$ lines, which forms the chessboard-like picture. The counting of these zero lines allows us to find the quantum number values easily. The crossings presence is due to the considered system integrability. Indeed, this two-dimensional system has two degrees of freedom and two integrals of motion — transversal energy $E_\perp$ and projection m of orbital momentum to the field symmetry axis. So, the variables separation is possible in the motion equation, which appears integrable in quadratures. The presence of the zero lines (surfaces for dimensions number more than two) crossings (or thin quasicrossings) appears a general property of quantum systems, whose classical counterparts demonstrate the regular dynamic [4]–[7].

4. Wave functions in the chaotic case

The qualitatively different picture is observed for the non-integrable systems, for example, the electron motion in the field of two neighbor atomic strings [110] of Si crystal. This field (the contribution from other atomic string pairs could be neglected) could be written as

$$U(x, y) = U_1(x, y + a/8) + U_1(x, y - a/8) , \tag{17}$$

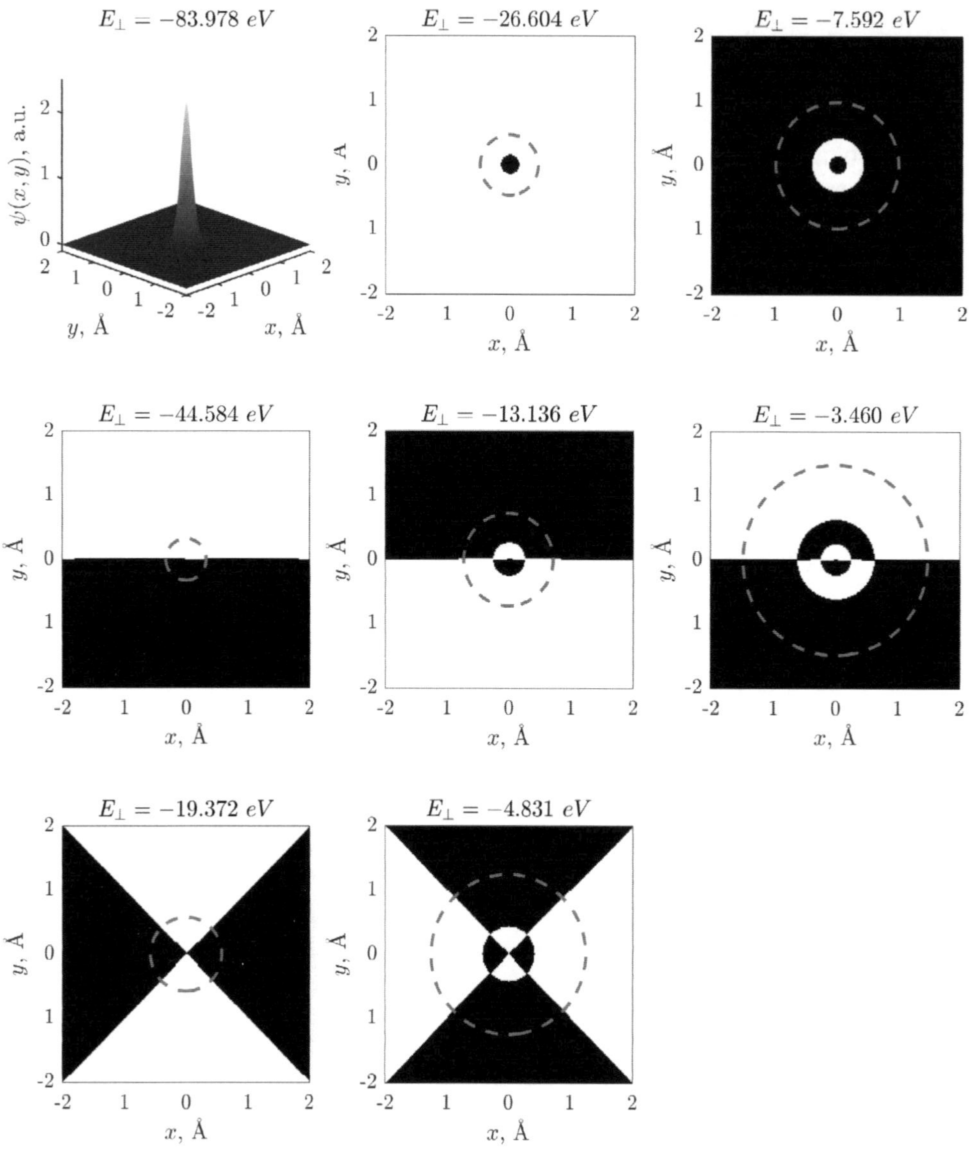

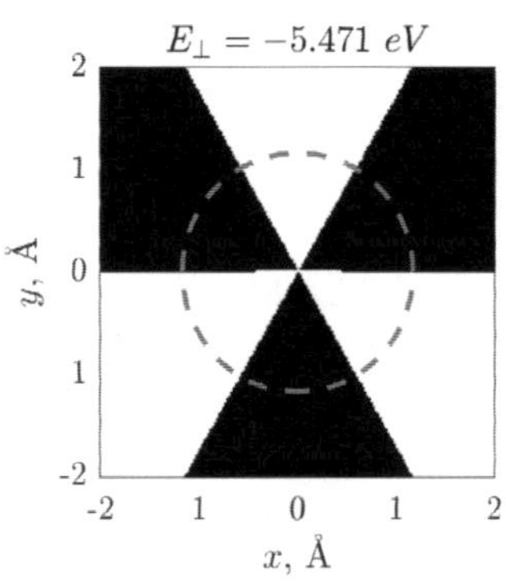

Figure 1. Eigenfunctions of the electron with $E_\parallel = 5$ MeV transversal motion in the field of atomic string [110] of Si crystal for the $m = 0$ (upper row), $m = 1$ (second row), $m = 2$ (third row), and $m = 3$ (lower row) cases. The red dashed curve is the border of the classically allowed region $U(x,y) < E_\perp$.

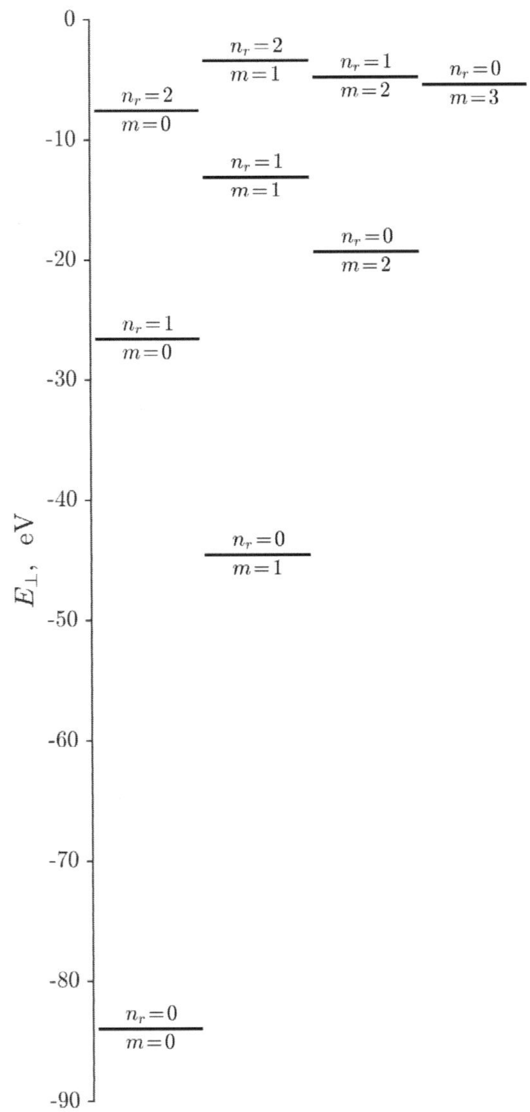

Figure 2. The energy levels classification of the electron with $E_\parallel = 5$ MeV transversal motion in the field of atomic string [110] of Si crystal.

$E_\perp$, eV	n_r	m
−83.978	0	0
−44.5838	0	1
−26.604	1	0
−19.3717	0	2
−13.136	1	1
−7.59175	2	0
−5.4715	0	3
−4.83075	1	2
−3.46025	2	1

where $a = 5.431$ Å is the lattice period. Some calculated eigenfunctions of the electron with longitudinal energy $E_\parallel = 5$ MeV channeling in this field are shown in figure 3, while all the eigenvalues — in table 1. Without the field axial symmetry the single integral of motion (energy $E_\perp$) still exists for two-dimensional system. This leads to dramatical change of the wave functions nature: the zero lines $\psi(x, y) = 0$ have not crossings. However the low energy $E_\parallel = 5$ MeV case not allows us to see this change clearly, so some wave functions of the electron channeling with $E_\parallel = 500$ MeV are presented in figure 4. The latter energy is high enough to ensure big total number of energy levels and small inter-level distances [12]. Namely this situation, similar to one in the classical mechanics with its continuous energy spectra, is needed for the quantum chaos manifestation studies [5, 6, 8]. We can see bizarre picture of black and white "islands" (see figure 4) instead of chessboard-like one for the axially symmetrical field. This behavior is general for the non-integrable systems [4]–[7].

In the classical mechanics the integrability (or not) of the system is closely related with regularity (or chaoticity) of motion of this system [5, 6]. The dynamical chaos in the classical case is the essential sensitivity of the system motion to the initial conditions, which leads to

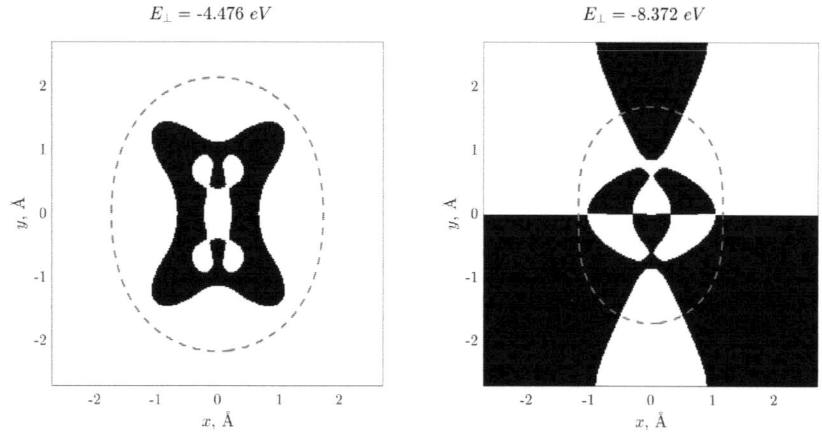

Figure 3. The eigenfunction samples for the electron with longitudinal energy of $E_{\parallel} = 5$ MeV channeling in the field of two neighbor atomic strings [110] of Si crystal.

Table 1. Full set of eigenenergies $E_{\perp}$ (eV) of the electron with $E_{\parallel} = 5$ MeV in the field of two neighbor atomic strings for each of four symmetry classes of the eigenfunction.

$\Psi(-x,y) = \Psi(x,y)$ $\Psi(x,-y) = \Psi(x,y)$	$\Psi(-x,y) = \Psi(x,y)$ $\Psi(x,-y) = -\Psi(x,y)$	$\Psi(-x,y) = -\Psi(x,y)$ $\Psi(x,-y) = \Psi(x,y)$	$\Psi(-x,y) = -\Psi(x,y)$ $\Psi(x,-y) = -\Psi(x,y)$
−49.0303	−48.7077	−48.711	−48.982
−31.6285	−23.7033	−24.3553	−30.8612
−24.894	−16.7748	−17.8317	−23.529
−19.98	−8.5155	−12.2515	−16.036
−12.3253	−6.3875	−9.101	−9.967
−11.214	−4.696	−6.39	−8.37175
−7.95275		−3.583	−6.071
−5.254		−3.22625	−3.36075
−4.4755			
−3.25275			
−2.84325			

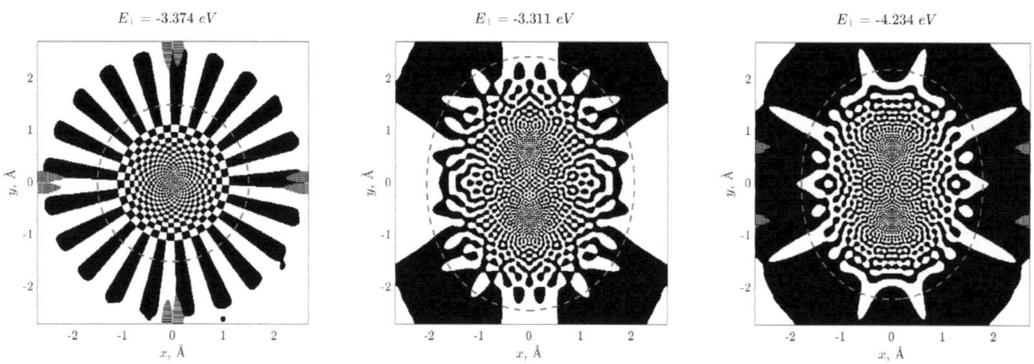

Figure 4. The eigenfunction samples for the electron with longitudinal energy of $E_{\parallel} = 500$ MeV channeling in the field of one (left plot) and two neighbor (center and right plots) atomic strings [110] of Si crystal.

exponential divergence of the initially close trajectories. The trajectories still deterministic (because the noise or accidental forces are assumed absent in the system) however appear very like to random ones.

Table 2. The spectral method algorithm parameters.

longitudinal energy $E_{\parallel}$, MeV	space lattice	time step value dt, eV^{-1}	number of time steps N_t	energy step value dE, eV
5	256×256	9.2×10^{-5}	27324375	2.5×10^{-4}
500	384×384	9.2×10^{-5}	136621875	5.0×10^{-5}

In the quantum mechanics case the exponential dependence on the initial conditions is absent. However the behavior of systems, which are regular and chaotic in the classical limit, is qualitatively different by the number of features. The search and investigation of these differences is the quantum chaos topic contents [8]. The observed difference of the wave functions structure, namely, zero lines $\psi(x, y) = 0$ picture of these functions is the one of manifestations of the quantum chaos.

5. Conclusion

The quantum mechanical problem of the fast electron motion in the axially-symmetric field of single atomic string [110] and the field of two neighbor ones of Si crystal is considered. The spectral method suitability for numerical solving, i.e. obtaining the eigenfunctions and energy eigenvalues, of the Schrödinger equation for this system in the axial channeling mode is demonstrated. This approach could be used for investigation of the quantum chaos manifestations in the electron axial channeling through the crystal.

Acknowledgments

The work was supported by the Ministry of Education and Science of the Russian Federation (the State assignment N 3.500.2014/K).

Appendix

First of all the spectral method algorithm was implemented on the MATLAB for debugging. After the code quality was proven the algorithm was implemented on C language using FFTW (for fast Fourier transforms) and GSL (for complex values calculations) libraries, because the MATLAB FFT performance and memory constraints appear unsatisfactory. The production calculations were performed under UNIX–like operating systems on computers of `amd64` multi-processor architecture. The same algorithm with minor changes allows us to calculate both energy levels and (in the separate run) eigenfunctions of the considered system. The algorithm parameters for present data calculations at $E_{\parallel} = 5$ and 500 MeV are summarized in table 2. For field of both single and two neighbor atomic strings the full set of eigenenergies and eigenfunctions at $E_{\parallel} = 5$ MeV was computed.

References

[1] Akhiezer A I and Shul'ga N F 1996 *High-Energy Electrodynamics in Matter* (Amsterdam: Gordon and Breach)
[2] Akhiezer A I, Shul'ga N F, Truten' V I, Grinenko A A and Syshchenko V V 1995 *Physics-Uspekhi* **38** 1119
[3] Uggerhøj U I 2005 *Rev. Mod. Phys.* **77** 1131
[4] Stratt R M, Handy N C and Miller W H 1979 *J. Chem. Phys.* **71** 3311
[5] Gutzwiller M C 1990 *Chaos in Classical and Quantum Mechanics* (New York: Springer-Verlag)
[6] Schuster H G 2005 *Deterministic Chaos. An Introduction* (Weinheim: WILEY-VCH Verlag)
[7] Berezovoj V P, Bolotin Yu L and Cherkaskiy V A 2004 *Phys. Lett.* A **323** 218
[8] Berry M V 1987 *Proc. R. Soc. Lond.* A **413** 183
[9] Feit M D, Fleck J A Jr and Steiger A 1982 *J. Comput. Phys.* **47** 412
[10] Dabagov S and Ognev L I 1988 *Nucl. Instrum. and Methods in Phys. Res.* B **30** 185

[11] Shul'ga N F, Syshchenko V V and Neryabova V S 2013 *Nucl. Instrum. and Methods in Phys. Res.* B **309** 153
[12] Shul'ga N F, Syshchenko V V and Isupov A Yu 2014 *Problems Atom. Sci. Technol.* **63** 5 (93) 120
[13] Shul'ga N F, Syshchenko V V, Neryabova V S and Isupov A Yu 2014 *J. Phys.: Conf. Series* **517** 012030

Influence of absorption on the X-ray diffraction by the trapeziform model of superlattice with a stacking fault

H M Manukyan[1] **and S M Manukyan**[2]
[1]Faculty of Physics, Yerevan State University, 1Alex Manoogian, Yerevan 0025, Armenia
[2]Faculty of Informatics and Applied Mathematics, Yerevan State University, 1Alex Manoogian, Yerevan 0025, Armenia

E-mail: hasmikm@ysu.am

Abstract. The problem of a dynamic diffraction of a plane monochromatic X-ray wave at the absorbing superlattice with a stacking fault between the layers for the trapeziform model of superlattice is considered. It is shown, that presence of stacking fault reduces the anomalous absorption. The interference absorption depends on the shift vector and depth of the stacking fault and the structural factor of superlattice.

1. Introduction

When a plane monochromatic X-ray wave is incident on a superlattice (SL) double diffraction occurs, first by atomic planes causing the formation of a modulated X-ray wave, and then by diffraction of the latter by the interfaces between layers, since the layer structure acts as a diffraction grating. Consequence of this diffraction is appearance of satellites around the principal maximum of X-ray diffraction in SL.

SL based on heterojunction are applied widely in microelectronics and computer techniques [1–3]. Dynamic diffraction of X-ray radiation has been applied in [4] for harmonic SL. Work [5] develops dynamical theory of X-ray diffraction on one-dimensional perfect SL of arbitrary model. Obtained results are employed for different models [6]. Being important in applications, superlattice perfection has been investigated by various methods of X-ray analysis [7–11].

One of the possible defects impairing superlattice parameters is a stacking fault. Work [12] develops the theory of X-ray dynamic diffraction on one-dimensional SL with a stacking fault between the layers if a absorption by the medium is neglected. In [13] X-ray dynamic diffraction on a SL with a stacking fault between the layers is considered theoretically in the case of interdiffusion of SL components if a absorption by the medium is neglected. In [14] we obtained the reflectivity in the exact directions of diffraction maxima for SL with a stacking fault between the layers when absorption is taken into account. The results are used for square-wave model of SL.

In this paper we develop the theory of X-ray diffraction on trapeziform model of SL with a stacking fault between the layers when absorption is taken into account.

2. Absorption in the superlattice with a stacking fault

Let us consider a superlattice of thickness Nz_0 (z_0 period of SL, N number of identical layers). Let a stacking fault be located at the depth N_1z_0 in SL (figure 1). In this case, the fault plane divides the

Content from this work may be used under the terms of the Creative Commons Attribution 3.0 licence. Any further distribution of this work must maintain attribution to the author(s) and the title of the work, journal citation and DOI.
Published under licence by IOP Publishing Ltd

crystal into two SLs of thicknesses N_1z_0 and N_2z_0 ($N_1 + N_2 = N$). Let a plane monochromatic X-ray wave of unit amplitude be incident at this SL. The waves emerging from the first SL are incident on the second one and for reflection amplitude of SL one obtains [15]

$$\Phi_{h,N} = r_{hN_1} t_{\bar{0}N_2} + e^{-i\alpha} t_{0N_1} r_{hN_2}, \tag{1}$$

where r_{hN_1}, r_{hN_2} and t_{0N_1}, $t_{\bar{0}N_2}$ are, respectively, the reflection and transmission amplitudes of two parts of SL; $\alpha = 2\pi\mathbf{hu}$ (**h** is the diffraction vector, **u** is the shift vector characterizing the displacement of second SL lattice with respect to the lattice of the first part of SL) and the dash corresponds to incidence from the back of the reflecting planes.

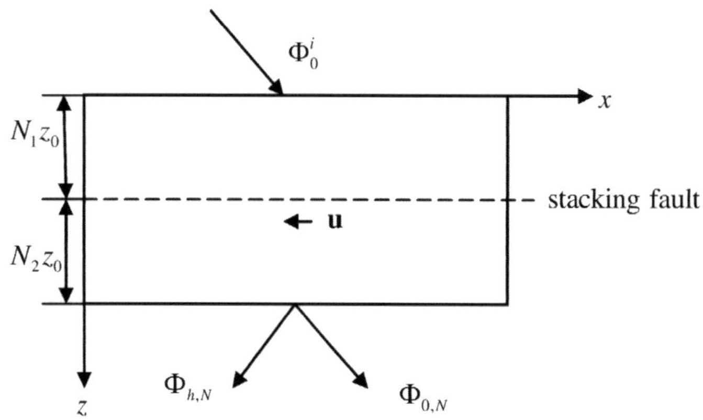

Figure 1. The geometry of diffraction.

As known, consequence of X-ray diffraction for SL with short period $z_0 \ll \bar{\Lambda}$ ($\bar{\Lambda}$ the mean extinction length of the crystal) is appearance of non-overlapping satellites around the principal diffraction maximum, the location of which is defined by an average over the superlattice period parameters. It is shown in [5], that for short-period SL within the limits of the *m*th reflection one may consider the SL as an ideal crystal with the modified Fourier component of the crystal polarizability:

$$\chi_{hm} = M_m \bar{\chi}_h, \tag{2}$$

where $\bar{\chi}_h$ is the Fourier component of the polarizability of crystal averaged over SL period and M_m the SL model-dependent structural factor of SL.

Substituting (2) in dynamical formulas of the reflection and transmission amplitudes of ideal crystal for symmetrical Laue case [16], we can write:

$$r_{hN_i,m} = i(\bar{\chi}_h/\bar{\chi}_{\bar{h}})^{1/2} M_m \frac{\sin(\pi Z_0 N_i (p^2 + M_m{}^2)^{1/2})}{(p^2 + M_m{}^2)^{1/2}}, \tag{3}$$

$$t_{0N_i,\bar{0}N_i,m} = \cos(\pi Z_0 N_i (p^2 + M_m{}^2)^{1/2}) \mp ip \frac{\sin(\pi Z_0 N_i (p^2 + M_m{}^2)^{1/2})}{(p^2 + M_m{}^2)^{1/2}}, \tag{4}$$

where

$$Z_0 = z_0/\bar{\Lambda} = \frac{C|\bar{\chi}_h|}{\lambda \cos\bar{\theta}_B} z_0, \tag{5}$$

$$p = (\sin 2\bar{\theta}_B / C|\bar{\chi}_h|)(\theta - \theta_m) \tag{6}$$

is a dimensionless parameter proportional to the deviation from the *m*th reflection direction θ_m, λ the wavelength of an incident wave, *C* the polarization factor, $\bar{\theta}_B$ the Bragg angle, averaged over SL period, and $i = 1;2$.

In [12] we develop expressions for reflectivity within the limits of the m th satellite for non-absorptive SL with a stacking fault between the layers, depending on the shift vector and depth of the stacking fault. We show, that the existence of a stacking fault is reducing satellite intensity. In [13] X-ray dynamic diffraction on a SL with a stacking fault between the layers is considered theoretically in the case of interdiffusion of superlattice components if a absorption by the medium is neglected.

The procedure of taking absorption into account based on the method used in the classical theory of dispersion, i.e. on the introduction of complex parameters of polarizability and its Fourier components:

$$\overline{\chi} = \overline{\chi}_r + i\overline{\chi}_i, \quad \overline{\chi}_h = \overline{\chi}_{hr} + i\overline{\chi}_{hi}, \quad \overline{\chi}_0 = \overline{\chi}_{0r} + i\overline{\chi}_{0i}. \tag{7}$$

Since for X-rays the dynamic absorption is usually small, i.e. $|\chi_{hi}| << |\chi_{hr}|$, neglecting terms of the order of $(|\overline{\chi}_{hi}|/|\overline{\chi}_{hr}|)^2$, for symmetrical reflection one can write:

$$Z_{0r} = C|\overline{\chi}_{hr}|z_0/\lambda\cos\overline{\theta}_B, \qquad Z_{0i} = Z_{0r}æ\cos\nu_h, \tag{8}$$

$$p_r = (\sin 2\overline{\theta}_B/C|\overline{\chi}_{hr}|)(\theta - \theta_m), \qquad p_i = -p_r æ\cos\nu_h, \tag{9}$$

where $æ = |\chi_{hi}|/|\chi_{hr}|$ and for crystals with a centre of symmetry $\cos\nu_h = \pm 1$.

In [14] we obtained the reflectivity in the direction of m th satellite $\theta = \theta_m$ (i.e. for $p = 0$) for SL with a stacking fault between the layers when absorption is taken into account. The results are used for square-wave model of SL.

In accordance with approach, mentioned above, after some transformations for the reflectivity within the limits of the mth reflection, depending on the deviation from the mth reflection direction θ_m, we obtain the following expression:

$$R_m = \frac{e^{-\mu D/\cos\overline{\theta}_B}}{(p_r^2 + M_m^2)}\Bigg|\sin(A_m + i\beta_m) + \frac{1 - \exp(-i\alpha)}{2}\Bigg\{i\frac{p_r[\cos(B_m + i\gamma_m) - \cos(A_m + i\beta_m)]}{(p_r^2 + M_m^2)^{1/2}}$$
$$+\sin(B_m + \upsilon_m + i\gamma_m) - \sin(A_m + \upsilon_m + i\beta_m)\Bigg\}\Bigg|^2, \tag{10}$$

where

$$A_m = \pi Z_{0r} N\sqrt{p_r^2 + M_m^2}, \quad \beta_m = æA_m/(p_r^2 + M_m^2), \tag{11}$$

$$B_m = \pi Z_{0r}(N_1 - N_2)\sqrt{p_r^2 + M_m^2}, \quad \gamma_m = æB_m/(p_r^2 + M_m^2), \tag{12}$$

$$\upsilon_m = æ/(p_r^2 + M_m^2)^{3/2}, \tag{13}$$

$D = Nz_0$ is the thickness of the superlattice, $\mu = -2\pi\overline{\chi}_{0i}/\lambda$ is the coefficient of linear absorption.

Since the p_r dependence of the reflectivity (10) is very lengthy, let us present it for a particular case, when $\alpha = \pi$:

$$R_m = \frac{e^{-\mu D/\cos\overline{\theta}_B}}{(p_r^2 + M_m^2)^2}\Big\{\Big[\sqrt{p_r^2 + M_m^2}\left(\sin A_m \operatorname{ch}\beta_m + \sin(B_m + \upsilon_m)\operatorname{ch}\gamma_m - \sin(A_m + \upsilon_m)\operatorname{ch}\beta_m\right)$$
$$+p_r\left(\sin B_m \operatorname{sh}\gamma_m - \sin A_m \operatorname{sh}\beta_m\right)\Big]^2 + \Big[p_r\left(\cos B_m \operatorname{ch}\gamma_m - \cos A_m \operatorname{ch}\beta_m\right)$$
$$+\sqrt{p_r^2 + M_m^2}\left(\cos A_m \operatorname{sh}\beta_m + \cos(B_m + \upsilon_m)\operatorname{sh}\gamma_m - \cos(A_m + \upsilon_m)\operatorname{sh}\beta_m\right)\Big]^2\Big\}. \tag{14}$$

According to (2) the interference absorption factor of m th diffraction maximum for ideal SL has the following form: $\sigma_m = \sigma_0 M_m$, where $\sigma_0 = 2\pi k C|\chi_{hi}|\cos\nu_h / (\cos\theta_B \sqrt{1+p_r^2})$ is the interference absorption factor for homogeneous crystal [16]. As is shown in [5], $M_m < 1$ for any SL model, therefore the interference absorption factor in SL decreases.

Presence of stacking fault reduces the effect of anomalous absorption. The interference absorption depends on the shift vector and depth of the stacking fault. For the maximum shift ($\alpha = \pi$) the wave field, weakly absorbable in the first part of SL, is strongly absorbable in the second one and vice versa.

3. Trapeziform model

Artificial SL crystals based on heterojunctions are one-after-another layers of different compositions with close interplanar spacings (like GaAs–AlAs). At early stage after fabrication when interdiffusion of semiconductor compounds entering the bilayer composition is absent, SL can be described by a rectangular (square-wave, if layers of different materials have the same thickness) model. Taking into account the interdiffusion of heteromaterials, one may describe SL by trapeziform model. If the SL layers are thin due to the interdiffusion of SL components different materials will be overlapped throughout the layer. One may describe such a SL by either sinusoidal or triangular model.

Assume that in some time after fabrication of SL, a diffusion layer of thickness z_d has been formed. Denote z_1 and z_2 the thicknesses of ideal layers. Then SL is described by a trapeziform model whose structural factor has the following form [6]:

$$M_m = \left| \frac{\sin(\pi p_{1m}\varepsilon_1)}{\pi p_{1m}\varepsilon_0} + (-1)^m \frac{\sin(\pi p_{2m}\varepsilon_2)}{\pi p_{2m}\varepsilon_0} - F_{1m} - (-1)^m F_{2m} \right|, \quad (15)$$

where the following notations are introduced:

$$p_{1m} = \frac{m - \varepsilon_2 - \varepsilon_d}{\varepsilon_0}, \qquad p_{2m} = \frac{m + \varepsilon_1 + \varepsilon_d}{\varepsilon_0}, \quad (16)$$

$$\varepsilon_j = 2kz_j \sin\bar{\theta}_B \mathrm{tg}\bar{\theta}\, \Delta d / \bar{d} \qquad (j = 0;1;2;d), \quad (17)$$

$$F_{jm} = \frac{\sqrt{\varepsilon_d}}{\varepsilon_0}(\sin(\pi p_{jm}\varepsilon_j)U_{3/2}(2q_{jm}^2,0) - (-1)^j \cos(\pi p_{jm}\varepsilon_j)U_{1/2}(2q_{jm}^2,0)), \quad (18)$$

$$q_{jm} = \sqrt{\pi\varepsilon_d p_{jm}} \qquad (j = 1,2), \quad (19)$$

$$U_\nu(2x,0) = \sum_{k=0}^{\infty} \frac{(-1)^k x^{\nu+2k}}{\Gamma(\nu+2k+1)}, \quad (20)$$

ε_0 is a parameter characterizing the degree of misfit of heteromaterials, Δd the difference in interplanar spacings of heterostructures, and the bar denotes the averaging over the SL period.

By inserting $z_d = 0$ into expressions (15)–(20) the structural factor of rectangular model is obtained. On the other hand, at full interdiffusion of semiconductor materials constituting SL, by substitution of $z_1 = z_2 = 0$ into (15)–(20), the structural factor of triangular model is obtained.

The numerical calculations are carried out for GaAs–AlAs SL, CuK_α radiation, when diffusion layer thickness is 20% of the bilayer thickness. With increasing interdiffusion of heteromaterials, redistribution of intensity of diffraction maxima takes place. In discussed case the magnitude of the principal maximum decreases, while oscillations increase.

Figure 2 shows dependence of reflectivity of the principal maximum ($m = 0$) on dimensionless parameter (p) of deviation from the direction of diffraction maximum at $\alpha = \pi$ and the stacking fault depth $N_1 z_0 = 0.1\ Nz_0$. As seen the plot on figure 2(b) becomes asymmetrical. If the depth of the stacking fault satisfies the condition $N_2 z_0 = 0.1\ Nz_0$, we obtained the reflectivity of absorbing SL similar to figure 2(b), mirrored with respect to Y-axis.

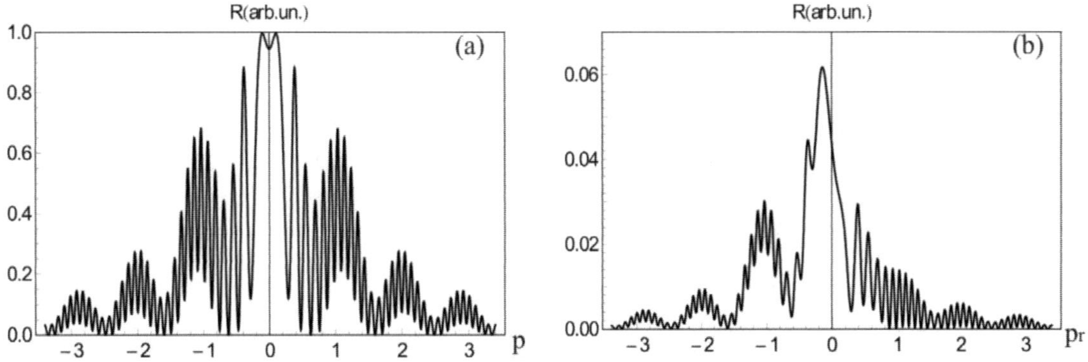

Figure 2. The reflectivity of the principal maximum for $\alpha = \pi$, $N_1 = 0.1N$ and $z_d = 0.2\ z_0$
(a) nonabsorbing SL, (b) absorbing SL

Figure 3 shows dependence of reflectivity of the principal maximum ($m = 0$) on dimensionless parameter (p) of deviation from the direction of diffraction maximum at $\alpha = \pi$ and the stacking fault depth $N_1 z_0 = N_2 z_0 = 0.5\ Nz_0$. In this case the plot on the figure 3(b) is symmetrical.

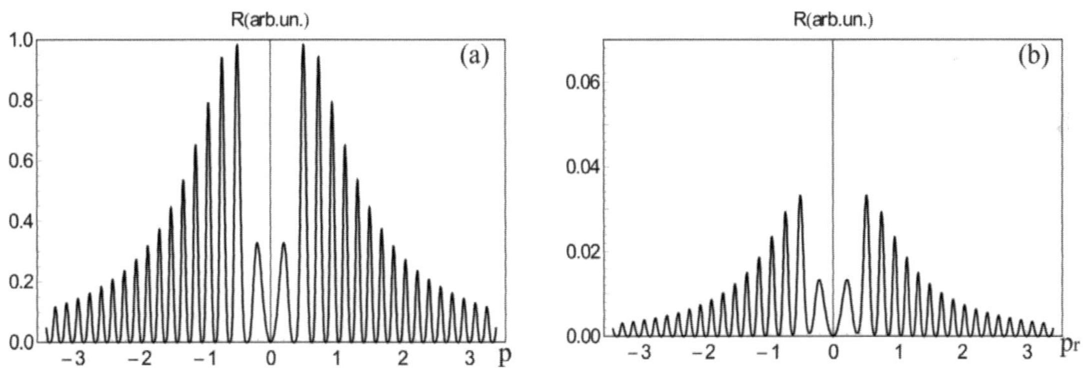

Figure 3. The reflectivity of the principal maximum for $\alpha = \pi$, $N_1 = N_2$ and $z_d = 0.2\ z_0$
(a) nonabsorbing SL, (b) absorbing SL

4. Conclusion

In this study we obtained, that in the case of a short-period SL with a stacking fault between the layers, taking absorption into account decreases the reflectivity of SL. Presence of stacking fault reduces the effect of anomalous absorption. The interference absorption depends on the shift vector and depth of the stacking fault and the structural factor of SL (which in turn depends on the interdiffusion degree of constituent heteromaterials).

References

[1] Shik A Ya 1974 *FTP* **8** 1841
[2] Golubev L V and Leonov E I 1977 *Sverkhreshetki* (Superlattices) Moscow: Znaniye
[3] Maslyuk V T and Fennich P A 1981 *Poluprovodnikoviye sverkhreshetki* (Semiconductor Superlattices) Moscow: Zarubezhnaya Elektronnaya Tekhnika
[4] Khapachev Yu P and Kuznetsov G F 1983 *Kristallografiya* **28** 27
[5] Vardanyan D M, Manoukyan H M and Petrosyan H M 1985 *Acta Cryst.* **A41** 212
[6] Vardanyan D M, Manoukyan H M and Petrosyan H M 1985 *Acta Cryst.* **A41** 218
[7] Afanasyev A M, Zaytsev A A and Imamov R M 1998 *Kristallografiya* **43** 1
[8] Engström C, Birch J, Hultman L, *et al* 1999 *J. Vac. Sci. Technol. A* **17** 2920
[9] Holy V, Stangl J, Springholz G, *et al* 2001 *J. Phys. D: Appl. Phys.* **34** A1
[10] Roch T, Holy V, Daniel A, *et al* 2001 *J. Phys. D: Appl. Phys.* **34** A6
[11] Zaytsev A A, Subbotin I A, Pashaev E M, *et al* 2004 *Plenki-2004 Materiali mezhdunarodnoi nauchnoi konferentsii* (Films-2004 Materials of International Conference) Moscow MIRE 49
[12] Manukyan H M 2007 *Poverkhnost'. Rentgen., sinkhrotron. i neytronniye issledovaniya* **10** 17
[13] Manukyan H M 2014 *Izvestiya NAN Armenii Fizika* **49** 360
[14] Manukyan H M, Manukyan S M 2015 *Proceedings of the Yerevan State University Physical and Mathematical Sciences* **3** 55
[15] Amelinckx S 1964 *The direct observation of dislocations* New York and London: Academic press
[16] Pinsker Z G 1982 *Rentgenovskaya Kristallooptika* (X-ray Crystal Optics) Moscow: Nauka

Planar channelling of relativistic electrons in half-wave silicon crystal and corresponding radiation

Y Takabayashi[1], V G Bagrov[2,3], O V Bogdanov[2,4], Yu L Pivovarov[5], T A Tukhfatullin[4]

[1]SAGA Light Source, 8-7 Yayoigaoka, Tosu, Saga 841-0005, Japan
[2]National Research Tomsk State University, 36 Lenin Avenue, 634050 Tomsk, Russia
[3]Institute of High Current Electronics, SB RAS, 2/3Academychesky Avenue, 634055, Tomsk, Russia
[4]National Research Tomsk Polytechnic University, 30 Lenin Avenue, 634050 Tomsk, Russia
[5]V.E. Zuev Institute of Atmospheric Optics SB RAS, 1 Zuev Square, 634034 Tomsk, Russia

E-mail: TTA@tpu.ru

Abstract. New experimental data on planar channeling of 255 MeV electrons in a 0.74 μm Si Half-Wave Crystal (HWC) obtained at SAGA-LS facility are presented. The computer simulation showed that the angular distribution of electrons after penetration through the HWC revealed the number of unknown before peculiarities is connected with specific electron trajectories in HWC. These specific trajectories lead to specific radiation, the properties of which are analyzed.

1. Introduction

The first prediction of charged particles deflection using a bent crystal was made in [1]. Since then, the extensive studies have been performed (see [2] and references therein). Recent works at CERN have been demonstrated the possibility to apply this effect to the beam collimation [3–5]. These early studies have been carried out using positively charged particles. Recently, these techniques has been proven to be useful even for negatively charged particles - relativistic electrons - at MAMI (855 MeV [6]) and at SLAC (3.35 and 6.3 GeV [7]). These experiments showed that there appear a large deflection angle in the channeling orientation and a smaller one in the volume-reflection orientation.

When the charged particles penetrate through the unbent crystal, the similar deflection effect may occur in a crystal of the thickness equal to a half wavelength of the planar channeling oscillation. In this case, the beam is deflected as if it is mirror-reflected by the crystal planes, as predicted in 1995 [8]. The mirroring effect for 2 MeV protons channeled in a Half-Wave (HWC) Si crystal was demonstrated for the first time in [9] in 2012. In 2014, the experimental studies were performed at ultra-relativistic energies using 400 GeV protons at CERN [10].

In 2014, the first experiments on channeling of 255 MeV electrons in a 1 μm-thick Si crystal were performed at SAGA LS [11]. Although the results were suggestive of a mirroring effect, the crystal thickness remained slightly greater than the optimum HWC thickness for mirroring. Here we present new theoretical and experimental results on angular distributions of 255 MeV electrons planar

Content from this work may be used under the terms of the Creative Commons Attribution 3.0 licence. Any further distribution of this work must maintain attribution to the author(s) and the title of the work, journal citation and DOI.
Published under licence by IOP Publishing Ltd

channeled in a thinner 0.74-μm Si crystal. Also we present estimations of spectral-angular properties of electromagnetic radiation appearing in mirroring process.

2. Channelling experiment in HWC Si

The experiments were performed at a linac at SAGA-LS, a Japanese synchrotron radiation facility [12]. The schematic of the setup is shown in figure 1. The electron beam energy was 255 MeV (corresponding to a Lorentz factor γ of 500). The details are described in [13]. The repetition rate of the beam acceleration was 1 Hz, and the average beam current was 7 nA. The horizontal and vertical beam sizes at the target were $\sigma_x \approx 0.3$ mm and $\sigma_y \approx 0.9$ mm, respectively (σ denotes one standard deviation). The horizontal and vertical angular divergences of the incident beam were $\sigma'_x \approx 0.1$ mrad and $\sigma'_y \approx 0.1$ mrad, respectively.

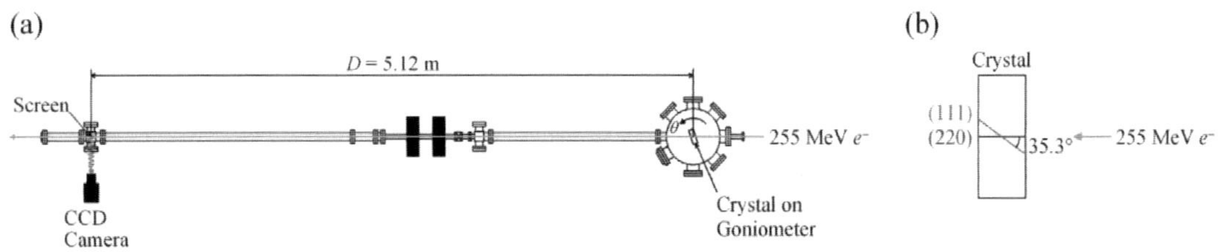

Figure 1. (a) Experimental setup (top view). (b) Orientation of the (220) and (111) planes.

A 0.74-μm-thick Si crystal was employed as a target. The crystal thickness has been selected based on simulations of the mirroring of 255 MeV electrons in an ultrathin crystal [11, 14]. The <001> crystallographic axis was normal to the crystal surface. The Si crystal was installed on a two-axis goniometer in a vacuum chamber. The (220) plane was aligned with the vertical plane. The profiles of the electron beam transmitted through the crystal were measured using a screen monitor, which consists of a 100-μm-thick alumina screen and a CCD camera. The distance between the crystal and the screen was D = 5.12 m. The (220) and (111) planes make an angle of 35.3°, and so the crystal was rotated by 35.3° from the normal incidence condition to observe (111) planar channeling (see figure 1(b)). Accordingly, the effective crystal thickness is calculated to be $0.74/\cos(35.3°) = 0.91\,\mu\text{m}$ for the (111) channeling experiment.

3. Experimental results and simulations

Simulations of the electron trajectories are carried out using the code "Basic Channeling with Mathematica" BCM–1.0 [15]. This code computes numerical solutions of the classical equations of motion using the Doyle-Turner potential for the crystallographic planes [16]. A series of trajectories for 255 MeV electrons in a Si HWC channeled along the (220) and (111) planes are depicted in figure 2. Because the potential for planar-channeled electrons is not harmonic, the electrons trajectories are not characterized by a single oscillation wavelength λ as they are for planar-channeled positive particles. Instead, for planar-channeled electrons the oscillation period depends on the point and angle of incidence, as shown in figure 2.

Typical electron trajectories channeled in (111) HWC Si are shown in figure 2(b). It can be clearly seen that some electrons change the sign of their perpendicular momentum (in the X-direction) for crystal thicknesses of about 0.8-0.9 μm (these trajectories are depicted by red solid lines). Due to this specific motion of channeled electrons in HWC, the mirror effect occurs for incidence angle $\theta = \theta_c / 2$ (figure 2(b) right column).

To simulate the angular spread of electrons the procedure described in [17] is used. For each point of entry, 20 values of the incident angle θ are generated randomly with a normal distribution. The standard deviation of the distribution matches the experimental value of 0.1 mrad. Thus, two thousand trajectories are calculated for different entry points within one period of the planar channeling potential. The calculated coordinates and angles of exit of these electrons are used to simulate the 2D beam profiles. Horizontal profiles are obtained by projecting the 2D beam profiles onto the horizontal plane.

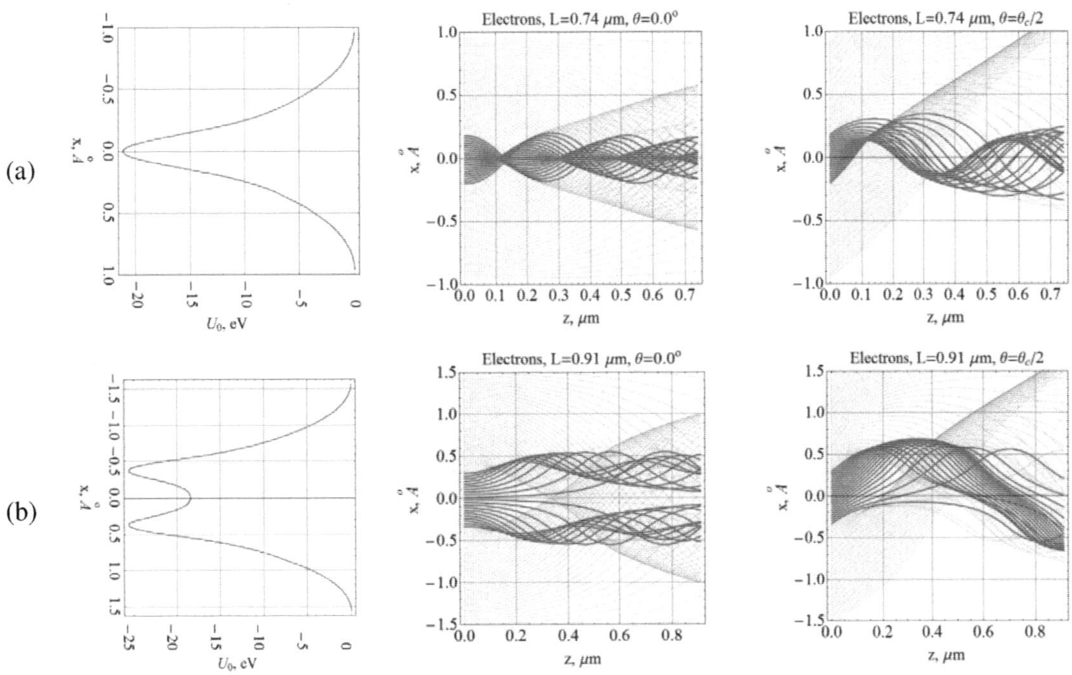

Figure 2. Simulated trajectories of 255 MeV electrons in a HWC Si and potential energy of electrons in the electric field of the crystal planes. (a) (220) planar potential and simulated trajectories for incidence angle $\theta = 0.0°$, $\theta = \theta_c / 2$. (b) (111) planar potential and simulated trajectories for incidence angle $\theta = 0.0°$, $\theta = \theta_c / 2$ (θ_c is the critical channeling angle). Solid red lines represent trajectories with entry points lying near the bottom of the potential well.

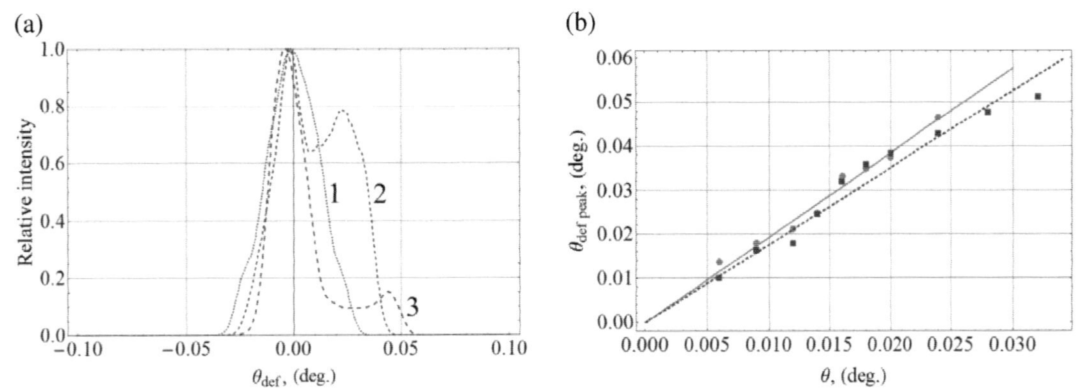

Figure 3. (a) Simulated angular distribution of 255 MeV electrons channeled in (111) HWC after penetrating trough 0.91-μm Si crystal for different angle of incidence θ: (1) $\theta = 0.0°$, (2) $\theta = \theta_c / 2$, (3) $\theta = \theta_c$. (b) The angle corresponding to the peak of angular distribution $\theta_{def\ peak}$ versus an angle of incidence θ (● simulation for the parallel beam, ■ simulation for the beam with angular spread).

Figure 3(a) shows simulated angular distribution of 255 MeV (111) channeled electrons after penetrating trough crystal for different angle of incidence. The calculation highlights that the peak of angular distribution lays in region $\theta_{\text{def peak}} = (1.75 \div 1.91)\theta$ (see in figure 3(b)). But the deflection efficiency decrease if incidence angle $\theta > \theta_c / 2$ (θ_c is the critical channeling angle). Thus the optimal angle of incidence for observation of mirroring effect for electron channeled in HWC is $\theta = \theta_c / 2$, the deflection angles $\theta_{def} = 0.023° = 0.4$ mrad which is comparable to that for planar channeling in bent crystals [6, 7] because the critical angle increases at lower energies. The fraction of deflected particles calculated by Gaussian gives 29% [14]. This result suggests that mirroring in HWC could be an effective method for beam deflection in that energy region, since multiple scattering is suppressed and a complex crystal bender is not required.

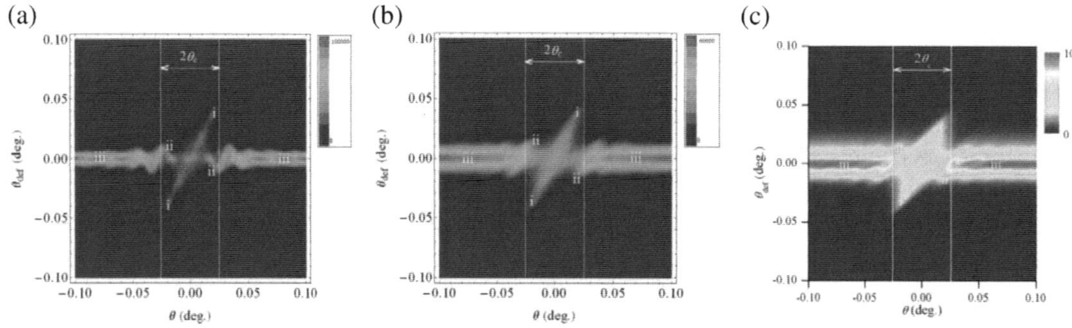

Figure 4. Beam intensity as a function of the beam deflection angle and of the crystal rotation angle θ relative to the (111) plane for (a) simulations for parallel beam, (b) simulations for beam with angular distribution and, (c) experimentally recorded distribution.

Figure 4(a, b) show calculated beam intensity distributions as functions of the crystal rotation angle θ and of the beam deflection angle θ_{def} for parallel beam (a), beam with angular divergence (b). Experimentally recorded distribution is presented in figure 4(c). The range $-\theta_c \leq \theta \leq \theta_c$ is indicated by the vertical lines. The experimental results are in good agreement with the simulations. Each graph can be divided into three regions. Region (i) corresponds to electrons channeled in the plane, lying along the line $\theta_{\text{def peak}} = 1.91\theta$ for parallel beam and $\theta_{\text{def peak}} = 1.75\theta$ for beam with angular spread, i.e., these electrons can be regarded as being mirror-reflected by the plane. Electrons in region (ii) are in the range of angles less than the critical angle. These electrons are slightly deflected into the direction opposite to the channeling direction, which is similar to the volume-reflection process [19]. Such a phenomenon has also been observed for 2 MeV protons [9] and for 400 GeV protons [10]. Region (iii) represents unchanneled (over-barrier) electrons whose deflection angles slightly oscillate with θ affected by the planar potential. A similar oscillation is observed in simulations for 400 GeV protons [10].

4. Simulation of spectral-angular distribution of radiation from electrons under channeling in in HWC Si

If the trajectory is known, the spectral-angular distribution of radiation from electron can be calculated using the well-known formula [18]:

$$\frac{d\varepsilon}{d\Omega d\omega}=\frac{e^2}{4\pi^2 c}\left|\int_0^{\tau}\frac{[\mathbf{n}[(\mathbf{n}-\boldsymbol{\beta})\dot{\boldsymbol{\beta}}]]}{(1-\mathbf{n}\boldsymbol{\beta})^2}\cdot e^{i(\omega t-\mathbf{kr})}dt\right|^2, \tag{1}$$

where $\mathbf{r}(t)$ is a radius-vector of electron, $\boldsymbol{\beta}(t)=\mathbf{v}(t)/c$, $\mathbf{v}(t)$ is an electron velocity and c is the speed of light, $\mathbf{k}=\mathbf{n}\cdot\omega/c$ the photon wave vector, $\mathbf{n}$ the unit vector determining the photon emission direction. Regarding the channeling radiation (CR), the formula (1) is used under assumption of a large number of oscillations of the channeled particle (the ideal trajectory). In this case it is possible to obtain relatively simple analytical formulas for the spectral-angular distributions of CR (see, for example [19-21]). In the case of channeling in a HWC one deals with approximately half of oscillation and one can expect the appearance of new features in radiation spectrum (let us call it CR-HWC).

The calculated CR-HWC spectrum in a forward direction from a 255 MeV electron planar channeled in (220) Si crystal is shown in figure 5 – only one selected trajectory. It is clearly seen the principal difference of CR-HWC spectrum (electron makes approximately a half oscillation in the transverse direction during its motion through the HWC crystal, figure 5(a) and that in a thicker crystal (electron makes more than a half of oscillation, see in figure 5(b,c)). With increasing the number of oscillations of electron in a crystal peaks in the CR spectrum become narrower and higher.

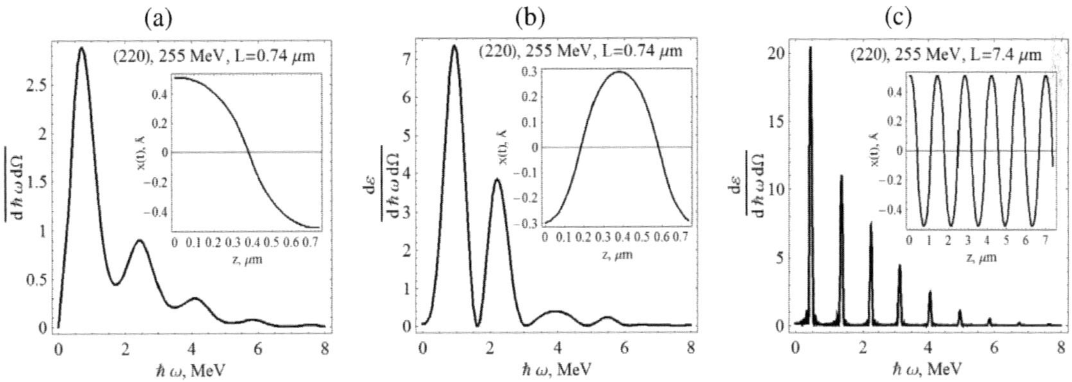

Figure 5. Calculated CR spectra in forward direction from an electron channeled in (220) Si crystal: (a) channeling in HWC – electron makes approximately a half of oscillation , (b) electron makes only one oscillation, (c) channeling in thick crystal.

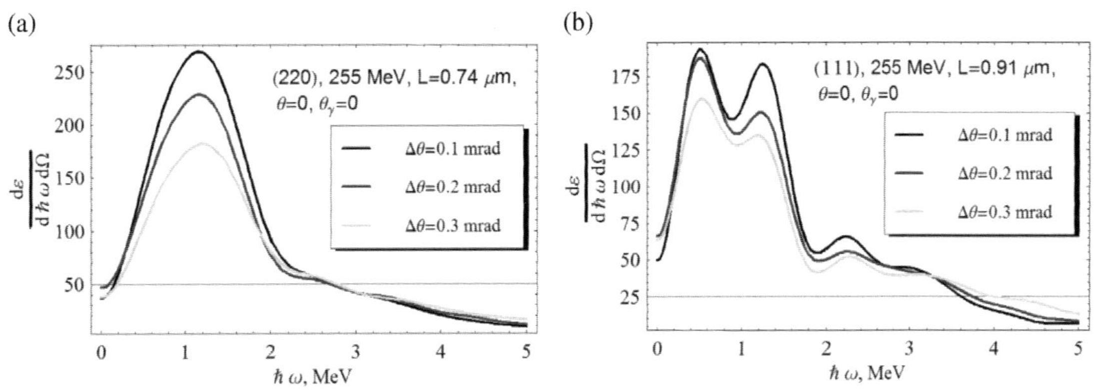

Figure 6. Calculated CR-HWC spectra from 255 MeV electrons in a forward direction ($\theta_\gamma=0.0°$) averaged over 500 trajectories and for different incident beam angular spread $\Delta\theta$. (a) electrons channeled in (220) HWC Si, (b) electrons channeled in (111) HWC Si. Here, the angle of incidence relative to crystallographic planes equals zero.

In a real experiment, one deal with a number of different electron trajectories in a crystal, the shape of every one depends on the point of incidence into a crystal and angle of incidence. Figure 6 shows calculated CR-HWC spectra from 255 MeV electrons in a forward direction ($\theta_\gamma = 0.0°$) averaged over 500 trajectories (in figure 2 only 100 trajectories are presented) and taking account beam angular spread $\Delta\theta$. The black line corresponds to the angular spread of the electron beam at SAGA-LS linac equals to $\Delta\theta = 0.1$ mrad. Whereas the angular spread increases three times, the height of the spectrum maximum decreases by 1.6 times for electrons channeled along (220) in HWC Si (see, in figure 6(a)). This is due to the major contribution of electrons involved in the quasi-periodic (over-barrier) motion.

The CR-HWC spectrum from electrons channeled in (111) HWC Si is more complicated. It contains of two peaks in the soft part (see, in figure 6(b)). This is due to the fact that the potential energy of electrons in an electric field of (111) planes has specific shape with two well. As a sequence, the trajectories of the electrons can be characterized by two specific wavelengths (see, in figure 2 (b)). With increase of the incident beam angular spread $\Delta\theta$ the second peak practically disappears. Obviously, the CR-HWC spectra will depend on the angle of incidence of electron beam relative to the channeling planes, i.e. will show the so-called orientation dependence. The calculated CR-HWC spectra (in a forward direction) from 255 MeV electrons channeled in HWC Si, for different incidence angles θ are shown in figure 7. For an incidence angle less than the critical channeling angle θ_c, the spectra contain the peaks in a soft part due to radiation from electrons involved in a bound motion (black and red lines in figure 7.). The heights of these peaks decrease with increase of the angle of incidence from $0.2\theta_c$ up to $0.5\theta_c$ for electrons channeled in (220) HWC, whereas for (111) channeling the heights of the peaks increase. For the angle of incidence $\theta = 2\theta_c$, all the electrons are involved into the quasi-periodic (over-barrier) motion, as a sequence the peaks of CR-HWC spectra are shifted to the region of greater photon energies (blue lines in figure 7(a, b)).

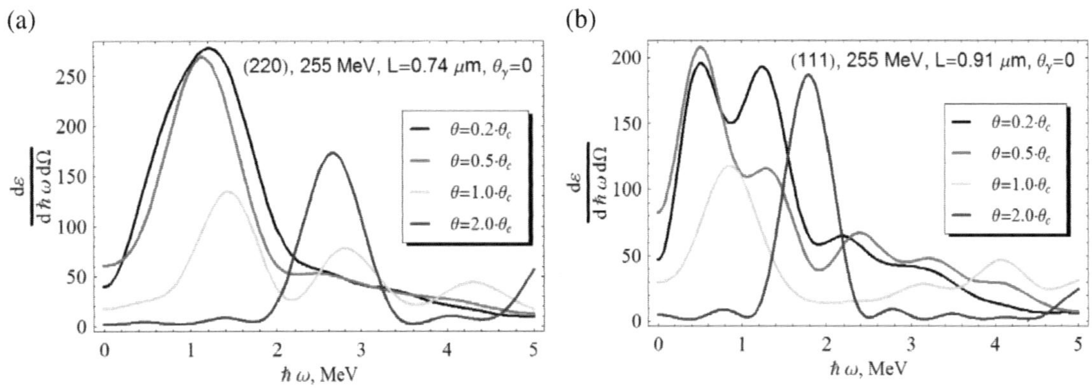

Figure 7. Calculated CR-HWC spectra from 255 MeV electrons channeled in HWC Si in a forward direction and for different incidence angle θ : a) (220) HWC Si, b) (111) HWC Si.

Since the efficiency of 255 MeV electron beam bending in a HWC Si crystal was estimated as 29%, it is worth to compare the intensities of gamma-beams emitted at the angles $\theta_\gamma = 0.0°$ and $\theta_\gamma = 0.5\,\theta_{def}$. The qualitative analysis can be made straightforward by comparison of the typical emission angle of relativistic particle $\theta_\gamma \approx 1/\gamma$ and deflection angle $\theta_{def} \cong \theta_c$. For parameters chosen (electron beam energy 255 MeV and (220) channeling) one obtains $\theta_\gamma \approx 1/\gamma = 2$ mrad and $\theta_{def} \cong \theta_c = 0.4$ mrad, respectively. Thus, $\theta_\gamma >> \theta_c$ and we may conclude, no sufficient difference in intensity spectra at $\theta_\gamma = 0 \div 1/\gamma$ and $\theta_\gamma = 0.5\,\theta_{\rm def}$ should appear. Indeed, the simulated CR-HWC spectra at $\theta_\gamma = 0.5\,\theta_{\rm def}$ presented in figure 8 together with ones at $\theta_\gamma = 0.0°$ show only small difference both in maxima positions and maximum values of CR-HWC intensity spectra at $\theta_\gamma = 0.0°$ and at $\theta_\gamma = 0.5\,\theta_{\rm def}$.

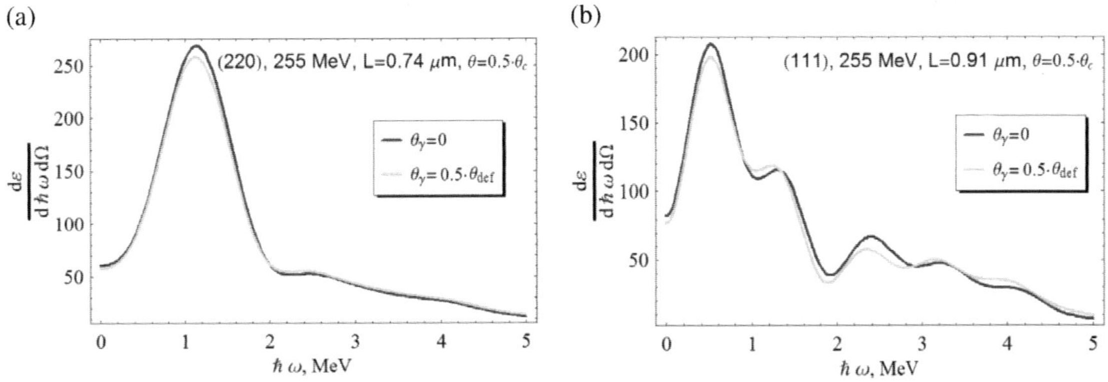

Figure 8. Calculated CR-HWC spectra from 255 MeV electrons channeled in HWC Si (incident angle $\theta = 0.5\theta_c$) at emission angles $\theta_\gamma = 0.0°$ and $\theta_\gamma = 0.5\,\theta_{\rm def}$: a) (220) HWC Si ; b) (111) HWC Si.

5. Conclusions

The results can be summarized:

New experimental data on angular distributions of planar channeled 255 MeV electrons in a 0.74 μm Si Half-Wave Crystal (HWC) obtained at SAGA - LS facility are presented. The data are explained well by computer simulations.

The computer simulations showed that the angular distributions of electrons after penetration through a HWC revealed the number of unknown before peculiarities and are connected with specific electron trajectories in HWC as well as with initial electron beam divergence and angle of incidence with respect to the channelling planes.

Third, it is closely connected with radiation spectrum from a particle moving in an arc [24-26] or in a short magnet. The experimental studies of so-called CER - coherent edge radiation (or "chicane radiation") have been carried e.g. in [27] to demonstrate the application of CER as a coherent THz radiation source and as a nondestructive beam diagnostic tool. The similarities and differences of CR-HWC considered in our paper and CER [27] are as follows: a) the bent non-periodic trajectories of electrons in both cases, but very different radii of curvature: 1.2 m in the case of CER [27] instead of only about 900 micrometers [11] in the case of CR-HWC in our case); b) the *coherent* radiation from a 61 MeV *compressed electron bunch* [27], but the radiation from a single 255 MeV electrons (our paper); c) the huge difference in spectral regions: THz or $10^{-5} - 10^{-4}$ eV in the case of CER [27], but MeV in the case of CR- HWC from 255 MeV electrons. In view of above, further theoretical and experimental studies of CR-HWC from relativistic electrons are necessary, aiming especially at polarization properties, angular distributions and search for possible applications.

Acknowledgements
The experimental work was supported in part by JSPS KAKENHI Grant Number 21740217. The theoretical work was supported by a grant from the President of the Russian Federation Number MK-5202.2015.2., by the Ministry of Education and Science of the Russian Federation (contract no. 1.676.2014/K, program "Nauka") and by RFBR grant No. 16-32-00464 mol_a.

References

[1] Tsyganov E N 1976 *Preprint TM-682, TM-684,* (Batavia: Fermilab)
[2] Biryukov V M, Chesnokov Y A, Kotov V I 1997 *Crystal Channeling and Its Application at High-Energy Accelerators, Springer*, (Berlin: Springer)
[3] Scandale W, Carnera A, Mea G A *et al* 2008 *Phys. Rev. ST Accel. Beams* **11** 063501
[4] Scandale W, Vomiero A, Baricordi S *et al* 2008 *Phys. Rev. Lett.* **101** 234801
[5] Scandale W, Vomiero A, Baricordi S *et al* 2009 *Phys. Rev. Lett.* **102** 084801
[6] Mazzolari A, Bagli E, Bandiera L *et al* 2014 *Phys. Rev. Lett.* **112** 135503
[7] Wienands U, Markiewicz T W, Nelson J *et al* 2015 *Phys. Rev. Lett.* **114** 074801
[8] Tsyganov E N, Taratin A 1995 *Nucl. Instr. & Meth. in Phys. Res*. A **363** 511
[9] Guidi V, Mazzolari A, De Salvador D, Bacci L 2012 *Phys. Rev. Lett.* **108** 014801
[10] Scandale W, Arduini G, Butcher M *et al* 2014 *Phys. Lett.* B **734** doi:10.1016/j.physletb.2014.04.062
[11] Takabayashi Y, Bagrov V G, Bogdanov O V, Pivovarov Yu L, Tukhfatullin T A 2015 *Nucl. Instr. & Meth. in Phys. Res*. B **355** 188
[12] Takabayashi Y, Kaneyasu T, Iwasaki Y 2011 *Nuovo Cimento* C **34** (4) 221
[13] Takabayashi Y, Pivovarov Yu L, Tukhfatullin T A 2014 *Phys. Lett.* A **378** 1520
[14] Takabayashi Y, Pivovarov Yu L, Tukhfatullin T A 2015 *Phys. Lett.* B **751** 453
[15] Bogdanov O V, Fiks E I, Korotchenko K B, Pivovarov Yu L and Tukhfatullin T A 2010 *J. Phys.: Conf. Ser.* **236** 012029
[16] Doyle P A, Turner P S 1968 *Acta Crystallogr.* A **24** 390
[17] Bogdanov O V, Pivovarov Yu L, Takabayashi Y, Tukhfatullin T A 2012 *J. Phys.: Conf. Ser.* **357** 012030
[18] Landau L D, Lifshitz E M 1980 *The classical theory of fields, Course of Theoretical Physics Series* vol 2 (Oxford: Butterworth-Heinman)
[19] Akhiezer A I, Shul'Ga N F 1996 *High Energy Electrodynamics in Matter* (Luxembourg: Gordon & Breach Publishers)
[20] Baier V N, Katkov V M, Strakhovenko V M 1998 *Electromagnetic Processes at High Energies in Oriented Single Crystals* (Singapoure: World Scientific Publishing)
[21] Bogdanov O V, Korotchenko K B, Pivovarov Yu L 2008 *J. Phys. B: At. Mol. Opt. Phys.* **41** 055004
[22] Bogdanov O V, Pivovarov Yu L 2008 *Nucl. Instr. & Meth. in Phys. Res*. B **266,** 3852
[23] Fomin S P, Shul'ga N F, Shul'ga S N 2003 *Phys. of At. Nucl* **66** 394
[24] Bagrov V G, Ternov I M, Fedosov N I 1982 *Journ. of Exper. and Theor. Phys* **55** 835
[25] Bagrov V G, Ternov I M, Fedosov N I 1982 *Doklady AN SSSR* **263** 1339 (in Russian)
[26] Bagrov V G, Razina G K, Sirota N N, Fedosov N I 1982 *Soviet Phys. Jour.* **25** 648
[27] Andonian G, Cook A, Dunning M *et al* 2009 *Phys. Rev. Spec. Top.-Accel. & Beams* **12** 030701

The source of THz radiation based on dielectric waveguide excited by sequence of electron bunches

A M Altmark[1] **and A D Kanareykin**[1,2]

[1]St. Petersburg Electrotechnical University, St. Petersburg, 5 Prof. Popov St., 197376, St. Petersburg, Russia
[2]Euclid Techlabs LLC 400 Professional Dr. Ste. 220 Gaithersburg, MD 20879, USA

E-mail: aaltmark@mail.ru

Abstract. We present a new method for excitation of THz Cherenkov radiation in a dielectric waveguide by relativistic electron bunches. A sequence of bunches generates monochromatic radiation. The frequency of radiation is defined by the distance between the bunches. The studies were carried by using the newly updated BBU-3000 code which permits taking into account a number of additional options: an external quadrupole focusing system, group velocity of the wakefield, and the dielectric material loss factor. In this paper, we present our algorithm for optimizing the number and sequential positions of bunches for generation of narrow band high power THz radiation.

1. Introduction

Many applications benefit from the imaging capabilities of high power THz radiation. Laser driven THz emitters, solid state oscillators (high frequency diodes), and QCL (quantum cascade lasers) are currently considered as viable THz sources [1,2]. At the same time, a high power narrow bandwidth THz source can also be implemented using coherent Cherenkov radiation (CCR) generated by an electron bunch train passing through an all-metal or dielectric slow wave structure. In [3] observation of coherent Cerenkov radiation in the terahertz regime emitted by a relativistic electron pulse train passing through a dielectric lined cylindrical waveguide was presented. With the proposed technique, selective excitation of modes beyond the fundamental were demonstrated by the use of the appropriate bunch spacing [3]. Use of bunch train with adjustable spacing to selectively excite high order modes in a multimode structure was demonstrated in [4]. Narrowband THz generation by a bunch train in a multimode dielectric loaded wakefield structure was studied experimentally. The structure parameters (geometry and permittivity) provided a wide spacing between the modes in the frequency domain and a number of different frequencies were individually excited by a tunable bunch train [4].

In this paper we consider THz radiation based on a dielectric lined waveguide (DLW). THz radiation is produced by sequence of relativistic electron bunches passing along the DLW, figure 1. This work was initiated by the experimental studies presented in [4]. The bunch train was produced following three major steps: energy modulation of long bunch by self-wakefield generation, energy modulation conversion into spatial modulation by using a chicane technique, and finally THz radiation by the electron bunch train passing through the DLW. This scheme allows tuning the bunch separation in the train providing coherent THz generation and mode selectivity as well. The spectrum of THz

Content from this work may be used under the terms of the Creative Commons Attribution 3.0 licence. Any further distribution of this work must maintain attribution to the author(s) and the title of the work, journal citation and DOI.
Published under licence by IOP Publishing Ltd

radiation is defined by both the DLW parameters and bunch separation of the train. In this paper, we present our results of numerical modelling of the THz radiation generation by the train consisting of 100 μm long electron bunches taking in account the group velocity effects and loss factor at the propagation distance of the wave envelope in the DLW. We also considered the selectivity of the first three TM modes generated by the bunch train of up to 10 bunches. Special attention has been paid to the generation of the TM_{03} mode in the 0.8 THz frequency range. Monochromaticity and field amplitude of the first three THz modes have been studied as well.

2. Problem formulation

We studied the wakefield created by Gaussian bunch train by using the Green function for DLW, figure 1. All bunches in train are ultrarelativistic so the phase velocities of each mode are close to speed of light $V_z = \beta c$, $\beta \cong 1$. If the radial offsets of the bunches are equal to zero, we can take into account only the *TM*-modes with a high value of the longitudinal component of electric field and no transverse deflecting force for relativistic electrons. The charge distribution in bunch train is described by the equation:

$$p(\zeta)=\frac{Q}{\sqrt{2\pi}\,\sigma_z}\sum_{n=1}^{N}\exp\left(-\frac{(\zeta-l_n)^2}{2\sigma_z^{\,2}}\right)$$

where *Q*-charge of single bunch, σ_z is the rms length of a single bunch, $\zeta = z - V_z t$ is the axial position behind the first bunch, l_n is the position of the nth bunch in the train.

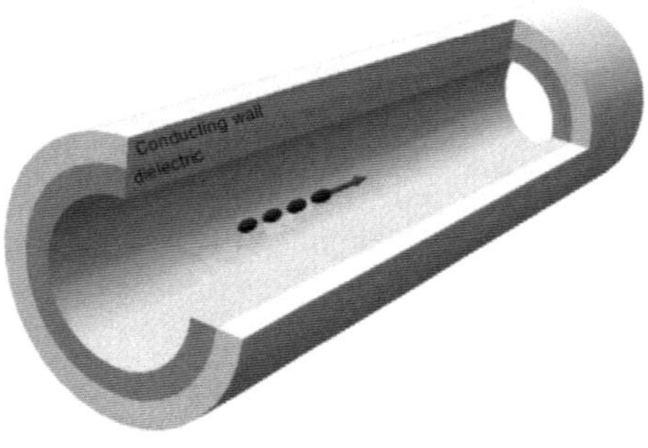

Figure 1. The bunch train passing along the axis of cylindrical dielectric waveguide.

We take into account the attenuation coefficient and group velocity for each TM mode. The Green function is defined by equation:

$$G_{Ez}(\zeta)=\sum_{m=0}^{\infty}\left(E_m\cdot\begin{cases}\cos(k_{zm}\zeta)\exp(-\alpha_m\zeta) & if\ \zeta < L\left(1-v_{gm}/c\right)\\ 0 & if\ 0\geq\zeta\geq L\left(1-v_{gm}/c\right)\end{cases}\right),$$

where E_m, $k_{zm}=\omega_m/V_z$, α_m, v_{gm} -amplitude, wave number, attenuation coefficient and group velocity of TM_{0m}-mode, L- way passed by bunch train in waveguide, $\beta = V_z/c$. The attenuation coefficient depends on losses in waveguide according to:

$$\alpha_m=\frac{\omega_m}{2Q_{wm}v_{gm}},\quad \frac{1}{Q_{wm}}=\frac{1}{Q_m^{diel}}+\frac{1}{Q_m^{wall}},$$

where Q_m^{diel}, Q_m^{wall} quality factors caused by losses for TM_{0m}-mode in dielectric volume and metallic wall respectively. The group velocities significantly limit the length of wave pocket:

$$v_{gm} = \left.\frac{d\omega}{dk_z}\right|_{\omega=\omega_m} = \frac{P_{zm}}{U_m},$$

where P_{zm} U_m - power flow through cross section, energy stored per length of waveguide for TM_{0m} - mode. The wakefield (resulted THz radiation) is calculated by a convolution of the Green function with the charge density of the bunch train:

$$E_z(\zeta) = \int_{-\infty}^{\zeta} G_{Ez}(s-\zeta)\,p(s)\,ds.$$

3. Calculation of THz radiation

We performed a calculation for a DLW with parameters from table 1 in the updated BBU-3000 code [5,6] with a multibunch module. BBU-3000 is used for calculation of wakefield and beam dynamics for different geometries of DLW based on the analytic for m of the Green function, figure 2.

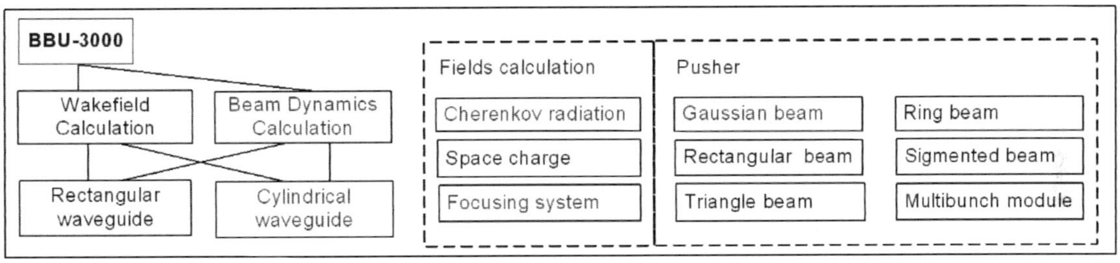

Figure 2. Capabilities of BBU-3000 code.

Table 1. Parameters of waveguide and bunch

Parameter	Value
Inner dielectric radius (um)	600
Outer dielectric radius (um)	850
Length (cm)	4
Epsilon	3.8
Loss tangent of dielectric	0.001
Conductivity of wall (S/m)	$5.7 \cdot 10^7$
Charge (nC)	1
σ_z (um)	100

For the THz band we chose a dielectric waveguide with a small aperture and thin dielectric layer (250 um), because the frequency intervals in the discrete spectrum of Green functions grow as the thickness of the dielectric layer is reduced, so that it is possible to adjust frequencies over a large band. The parameters of the TM modes are presented in table 2. The main purpose of using a bunch train is the selection of only one mode with the desired frequency without exciting any of the other modes. The most interesting case is the possibility to generate radiation consisting only of the TM_{03} mode.

Table 2. Properties of TM modes

Mode	frequency, GHz	Vg/c	Attenuation coefficient (1/cm)
TM_{01}	149.5	0.351	0.0466
TM_{02}	439.3	0.553	0.059
TM_{03}	768.0	0.616	0.0813

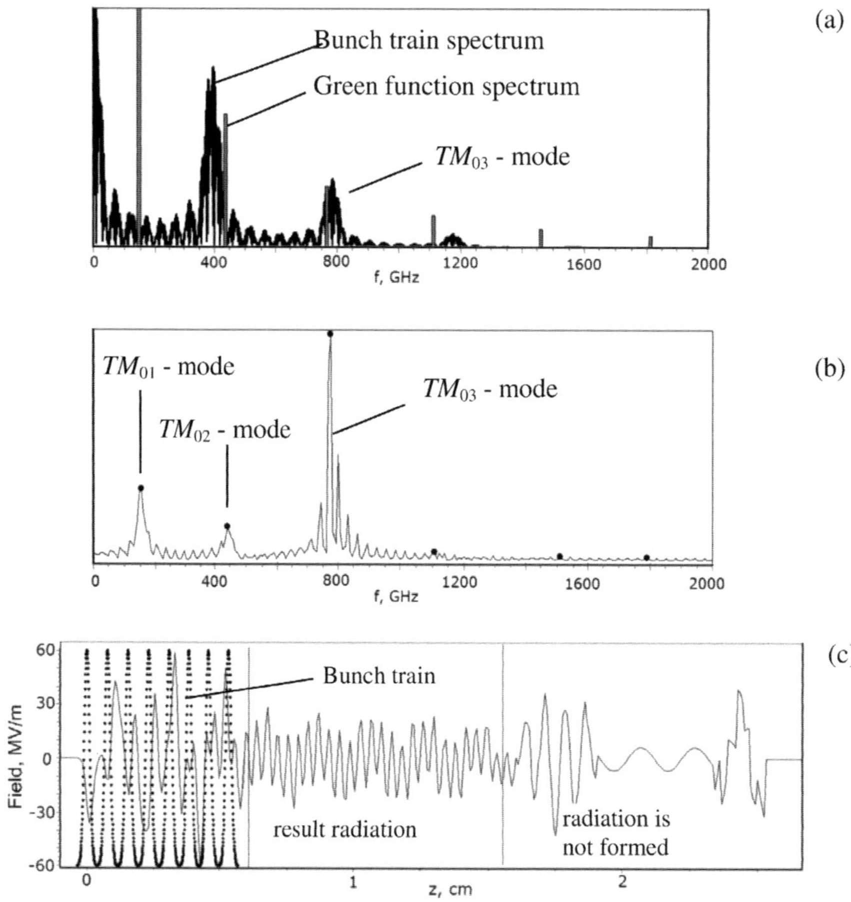

Figure 3. THz radiation based on TM_{03} mode: (a) Amplitude Spectrum distribution of Green-function and bunch train (a) and for result radiation (b), Wakefield created by bunch train (c), distances between bunches are equal 0.0765 cm.

The radiation is characterized by four main parameters: frequency, amplitude, monochromaticity and number of periods. Monochromaticity is defined as the ratio of the amplitude of the selected mode to the sum of the amplitudes of all radiated modes. In figure 3 the calculated results are shown for radiation with the frequency of TM_{03} mode. From figure 3 (a) we can see that amplitude spectrum distributions of the bunch train and the Green function are covered in frequency by this mode. Amplitude spectrum distribution of result radiation contain principally TM_{03} mode, figure 3 (b). It is possible only for a unique distance between bunches: 0.0765 cm. As we can see from figure 3 (c) the number of wavelengths of result radiation is limited by influence of the group velocities. The high value of group velocity means to the decrease of wave pocket for each mode. Therefore, the length of the result radiation is defined by the maximal value of the group velocity (0.616*c* for TM_{03} mode).

The figure 4 show dependence of radiation parameters of $TM_{01}, TM_{02}, TM_{03}$ modes as a function of number of bunches in train. It is shown that there is possibility to generate monochromatic radiations based on TM_{01}, TM_{02} and TM_{03} modes by adjusting distances between the bunches. The amplitude of wakefield depend on number of bunches only for TM_{01} mode, figure 4 (a). Monochromaticity, in particular, does not depend on bunch number in train for TM_{01} and TM_{03} modes, figure 4 (b). It is possible to excite TM_{01}, TM_{02} modes with different distances between the bunches, but the excitation of TM_{03} mode is possible only with train consisting 5, 7 and 8 bunches with single value of distance between bunches 0.0765 cm.

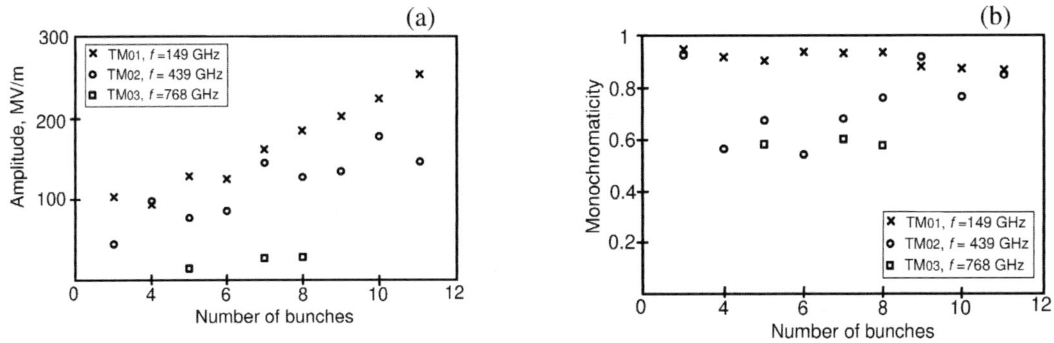

Figure 4. Calculations results for TM_{01}, TM_{02}, TM_{03} modes.

4. Conclusion

In conclusion, we have numerically modeled narrowband THz generation by a bunch train in a multimode dielectric loaded structure. We showed that for a bunch train for 8 bunches passing through the selected dielectric lined waveguide the length of the wave envelope at 768 GHz does not exceed 1 cm for the bunch separation of 765 μm. This limitation is defined by the group velocity value for that particular mode. It was also demonstrated that the THz radiation monochromaticity and field amplitude are reduced for higher frequencies. The radiation monochromaticity is in the range of 0.55 caused by the weak lower TM_{01} and TM_{02} mode generation, figure 4 (b).

Acknowledgments. This work was supported by grant 15-02-08745 from the Russian Foundation for Basic Research.

References

[1] Gallerano G P, Biedron S 2004 *Proc. of FEL* 216
[2] Lang M, Deninge A 2012 *Photonik Int* 36
[3] Andonian G, Williams O, Wei X et al 2011 *Applied Physics Letters* **98** 202901
[4] Antipov S, Jing C, Baryshev S et al 2015 *Proc. of IPAC* 1929
[5] Schoessow P, Jing C, Kanareykin et al 2011 *Proc. of PAC* **WEP111**
[6] Kanareykin A, Altmark A 2014 *Journal of Physics Conference Series* **517** 1-5

Corrugated capillary as THz Cherenkov Smith-Purcell radiator

K V Lekomtsev[1,2], A S Aryshev[1], A A Tishchenko[2], A A Ponomarenko[2], V M Sukharev[2], N Terunuma[1], J Urakawa[1] and M N Strikhanov[2]

[1]High Energy Accelerator Research Organisation (KEK), Tsukuba, Japan
[2]National Research Nuclear University (MEPhI), Moscow, Russia

E-mail: konstlek@post.kek.jp

Abstract. In this article we discussed Particle In Cell electromagnetic simulations and mechanical design of dielectric capillaries that produce THz Cherenkov Smith-Purcell radiation (ChSPR), arising when a femtosecond electron multi-bunch beam propagates through corrugated and non-corrugated dielectric capillaries with metallic radiation reflectors. We investigated the influence of the four-bunch beam on the SPR field spectrum and on the ChSPR power spectrum, and the influence of the non-central beam propagation on the ChSPR power spectrum. We also discussed the design and assembly of the capillaries, constructed as sets of cylindrical rings.

1. Introduction

In the past decade accelerator based sources of THz radiation have seen progress in generation of short femtosecond duration electron bunches as well as in THz radiators development. LUCX accelerator at High Energy Accelerator Research Organisation (KEK) produces short, hundreds of femtosecond duration electron multi-bunch beam with variable distance between bunches [1,2]. This type of multi-bunch beam may be used to generate tunable SPR. Capillary structures capable of producing powerful Cherenkov radiation (ChR) wakefields have been used for THz radiation generation [3] as well as for energy modulation of electron beams [4]. Characterization of coherent ChR sources and high gain antennas for such sources are currently being discussed as well [5]. Introduction of the corrugation into a dielectric capillary increases spectral tunability of THz radiation emitted due to the mechanism of SPR [6].

In this paper we discuss a corrugated capillary as a source of THz radiation based on coherent ChSPR mechanism, and a non-corrugated capillary as a source of THz radiation based on coherent ChR. We consider the geometries constructed as sets of cylindrical dielectric rings enclosed in metallic radiation reflectors. Designs of corrugated and non-corrugated capillaries, holders, and radiation reflectors are explained. We also present laser microscope scans of the outer surfaces of the corrugated and non-corrugated capillaries, which were used to check assembly accuracy. Particle In Cell (PIC) simulations are performed for the different distances between bunches and for an off-central propagation of the beam through the capillaries. The spectral response and the radiation power produced by both types of the geometries are compared for the different values of the distance between bunches and for off-central propagation of the four-bunch beam.

Content from this work may be used under the terms of the Creative Commons Attribution 3.0 licence. Any further distribution of this work must maintain attribution to the author(s) and the title of the work, journal citation and DOI.
Published under licence by IOP Publishing Ltd

2. PIC Simulations

The electromagnetic simulations were performed using Computer Simulation Technologies (CST) PIC solver that calculates propagation of charges and fields in a calculation domain [7]. We reported in [6] the spectral-angular distribution of THz ChSPR from a corrugated channel in infinite dielectric. In [8] we compared the theoretical model [6] with the PIC simulation of the electron bunch propagation in a corrugated channel in quasi-infinite medium. Because the spectral angular distribution could not be directly applied to the geometry of a capillary with limited dimensions and a reflector, PIC simulations were required to investigate the characteristics of ChSPR. In [9] we presented an extension of the study and showed a comparison between the simulation model and the theory for the SPR peak, as well as the optimisation of corrugation parameters, to achieve maximum radiation intensity for the capillary with a radiation reflector.

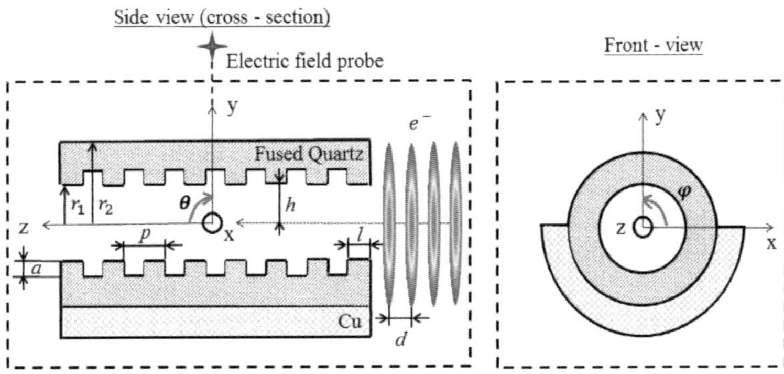

Figure 1. Geometry of PIC simulation of four-bunch beam propagation through corrugated capillary with radiation reflector.

Figure 1 schematically shows the corrugated capillary and the radiation reflector placed in the calculation domain. We considered propagation of four bunches separated by the distance d through the capillary. In figure 1 the solid lines show reflective borders and the dashed lines show outer non-reflective borders of the calculation domain. Values of the electric field were calculated at each discretization point in the calculation domain. The electric field values at the outer non-reflective borders were extrapolated in order to obtain electric field values in the wave zone at the location of the electric field probe. The method used for the extrapolation is described in [10]. The electric field probe measured all three component of electric field, including each component's real and imaginary part. Simulation parameters are shown in table 1. The values of r_2 and a were chosen based on the optimization discussed in [9]. The value of r_1 was chosen to accommodate electron beam with σ_{transv} up to 500 μm. The capillary material was assumed to be loss free Fused Quartz with the real part of dielectric permittivity $\varepsilon_{real} = 3.75$. As shown in [11] ε_{real} stays relatively constant throughout the frequency range 100 - 700 GHz.

Table 1. Simulation parameters.

Parameter name	Value
Beam Lorentz - factor, γ	16
Bunch charge	0.1 nC
Bunch long. dimension, σ_{long}	0.03 mm
Bunch transv. dimension, σ_{transv}	0.3 mm
Number of bunches	4

Distance between bunches, d	Variable
Simulated frequency range	up to 700 GHz
Capillary material	Fused Quartz (loss free)
Fused Quartz diel. perm. ε (real part)	3.75
Material of the Radiation reflector	Copper (pure)
Internal radius, r_1	2 mm
External radius, r_2	2.7 mm
Corrugation depth, a	0.2 mm
Number of corrugation periods	30
Corrugation width, l	0.5 mm
Corrugation period, $p = 2l$	1 mm
Bunch offset, h	Variable
Capillary length, L	30 mm

2.1. Smith-Purcell radiation characteristics for different distances between bunches

In this paper we discuss characteristics of the SPR emitted at 90 deg, which as a primary source of THz radiation at large angles θ. On the contrary, ChR is reflected by the outside surface of the corrugated capillary and directed in the forward direction at small angles $\theta \approx 10$ deg. Efficient extraction of the radiation produced by both mechanisms is a separate topic requiring thorough investigation, and it is left for future publications.

Figure 2 shows the spectra of the SPR field calculated at the electric field probe, which is located at $\theta = 90$ deg. at the distance $\approx \gamma^2\lambda$. The spectra were calculated for $\gamma = 16$ and $\lambda = 1$ mm. The contribution of the non-corrugated capillary with $r_1 = 2.2$ mm and $r_2 = 2.7$ mm was subtracted. The absolute value of the field at the probe location is first calculated as a function of time, and then by taking Fourier transform we obtain a frequency representation of the electric field. The spectra are shown for four different distances between bunches $d = 0.25, 0.5, 0.75, 1$ mm. In the case of the $d = 1$ mm (equal to the period of the corrugation) one may observe the SPR of (-1) and (-2) diffraction orders, however when the distance is $d = 0.5$ mm (equal to the half period of the corrugation) only the SPR of (-2) diffraction order may be observed. For the distances $d = 0.25$ mm and $d = 0.75$ mm no peaks of SPR are present in the considered frequency range.

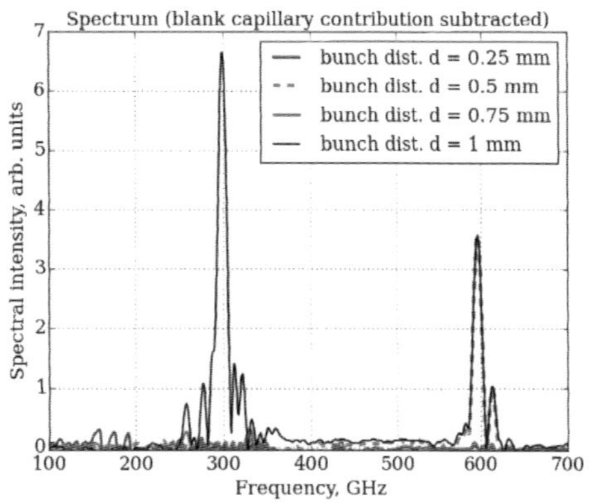

Figure 2. SPR spectra at probe location at $\theta = 90$ deg., $\varphi = 90$ deg. and distance to probe $\gamma^2\lambda$, calculated for $\gamma = 16$ and $\lambda = 1$ mm.

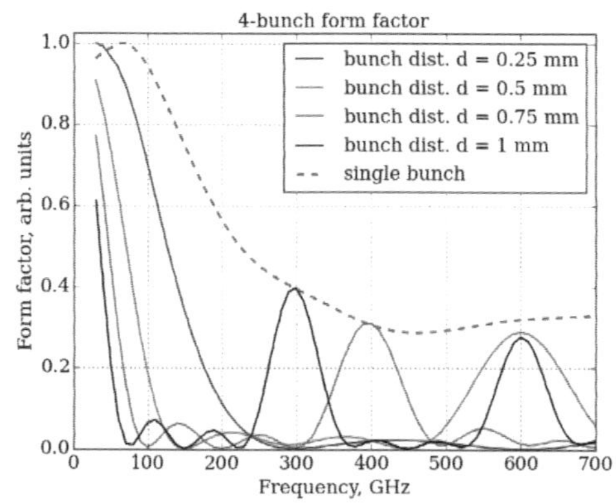

Figure 3. Form factors of four-bunch beam with different bunch spacing, and form factor of one bunch.

The fact that diffraction orders may be suppressed for different values of the bunch spacing can be explained by the power spectrum (form-factor) of four-bunch beam, shown in figure 3. The power spectra were calculated for four bunches with the bunch spacing $d = 0.25, 0.5, 0.75, 1$ mm and for one bunch in the calculation domain without a capillary. When the corrugated capillary is added in the calculation domain the radiation distribution and the spectrum are defined by the modulated power spectrum of the four-bunch beam and by the dispersion relation (1). For example, in the case of $d = 0.5$ mm above 200 GHz the power spectrum of the beam has only one peak around 600 GHz, therefore giving only one peak in the spectrum of the SPR in figure 2. In the case of $d = 1$ mm in both figures 2 and 3 the peaks are present at the frequencies 300GHz and 600GHz.

The diffraction orders of the Cherenkov peak and the SPR peaks satisfy the following dispersion relation:

$$\cos(\theta) = \frac{2\pi m}{kp} + \frac{1}{\beta\sqrt{\varepsilon(\omega)}}; \tag{1}$$

where θ is the polar angle in figure 1; β is the charge speed in terms of the speed of light; k is the wave number in the dielectric; p is the corrugation period; and m is a diffraction order. The value of $m = 0$ corresponds to the Cherenkov peak, and the values of $m = \pm n$; $n = 1,2,3 \ldots$ correspond to the diffraction orders of SPR. This dispersion relation is only valid for the radiation propagation inside the dielectric material, since the theory was developed for the case of a corrugated channel in infinite dielectric. At a capillary border the radiation follows Snell's law of refraction.

Figure 4 shows the power radiated through the surface of the outside boundary A of the calculation domain during the simulation time Δt. The calculated power for each frequency is given by the following expression:

$$P(\omega) = \left|\int_0^{\Delta t} \oiint \boldsymbol{S}(\omega) * \boldsymbol{n}\, dA\, dt\right|; \tag{2}$$

where $\boldsymbol{S}(\omega)$ is the Poynting vector, $\boldsymbol{n}$ is a unity vector in the outward normal direction from the boundary A of the calculation domain. The radiated power was calculated for the non - corrugated capillary (black dashed curve) and for the corrugated one (red solid curve).

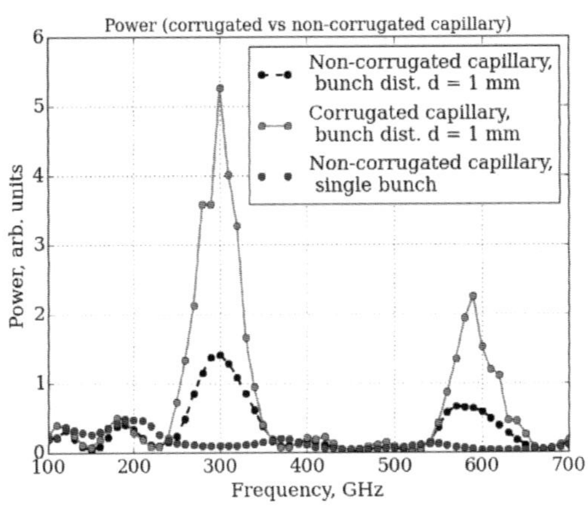

Figure 4. Power spectrum of ChSPR radiated in all directions outside of calculation domain, for corrugated and non-corrugated capillaries (four bunches and one bunch).

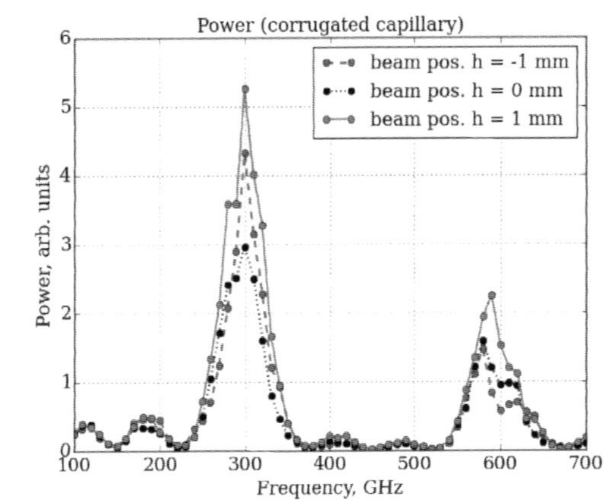

Figure 5. Power spectrum of ChSPR radiated in all directions outside of calculation domain, for corrugated capillary and off-centre beam propagations: $h = -1, 0, 1$ mm.

The distance between bunches was equal to the corrugation period $p = d = 1$ mm, and the beam was propagating at the offset $h = 1$ mm (figure 1). When these equalities are satisfied, the power radiated from the corrugated capillary is approximately three times more intense than from the non-corrugated one. The power increase occurs at 300GHz, corresponding to the SPR of (-1) order, and at 600 GHz, corresponding to the SPR of (-2) order (figure 4). In the case of the non-corrugated capillary two less intense radiation peaks also present, however it is important to clarify that they originate not due to SPR, but due to the ChR from the four-bunch beam. To prove this point figure 4 also shows the power spectrum of one bunch (blue dotted curve) that demonstrates relatively flat response at all frequencies.

2.2. Smith-Purcell radiation characteristics for different beam positions
In addition to the investigation of the influence of the bunch spacing on the radiation characteristics, it is also interesting to understand how the radiation characteristics change if the beam propagates not through the centre of a capillary, but at an offset from it. As a measure of the offset we can introduce parameter h (figure 1). When the beam propagates through the centre of a capillary, h equals zero, when the beam is shifted towards the positive values of the axis y, h is positive, and when the beam is shifted towards the negative values of the axis y, h is negative. Figure 5 shows the comparison of the SPR power spectra for $h = -1, 0, 1$ mm. When the beam travels at $h = 1, -1$ mm SPR is more intense than for the central propagation.

2.3. ChSPR power patterns
Using the capability of CST Particle Studio to calculate radiation power patterns based on the fields in the calculation domain, we obtained 3D power patterns of the ChSPR for fixed frequencies.

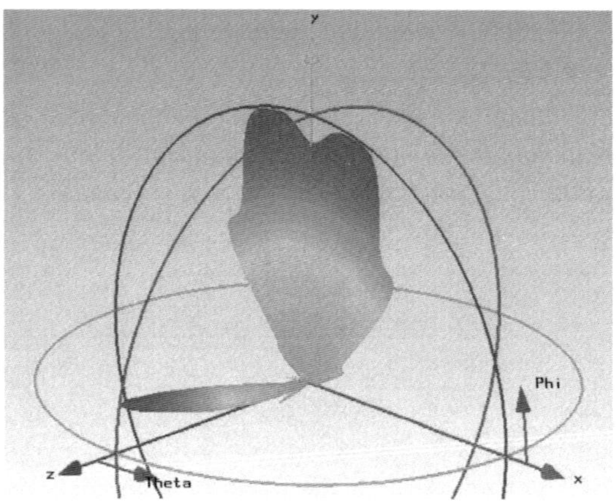

Figure 6. 3D power pattern of ChSPR at 300GHz for off-central propagation of four-bunch beam at $h = 1$mm.

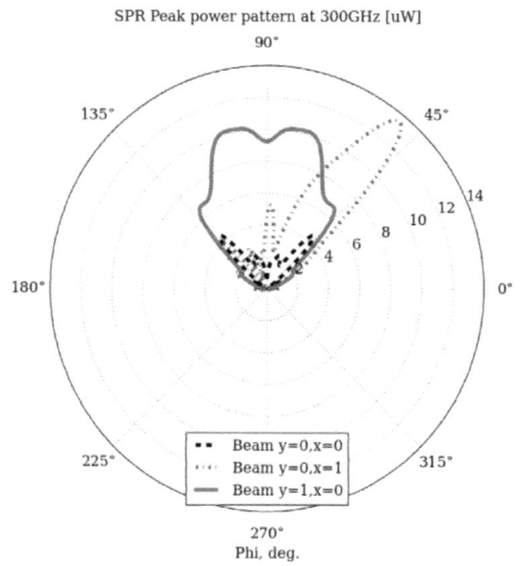

Figure 7. Polar cross-sections of SPR peak power patterns at $\theta = 90$ deg, for three different non-central beam propagations.

Figure 6 demonstrates the power pattern of the ChSPR from the corrugated capillary at 300GHz for the beam propagation at the offset $h = 1$ mm. There are two effects that may be separated: Cherenkov radiation is reflected by the outside boundaries of the corrugated capillary and directed at the small angles $\theta \approx 10$ deg., for the chosen capillary parameters SPR of (-1) order is emitted at $\theta = 90$ deg. The power pattern changes when the beam propagates off-centre. Figure 7 demonstrates this behaviour, it shows the cross-sections of the SPR power pattern for the two different beam offsets and for the central beam position (x, y) = (0, 0). While θ is fixed at 90 deg., φ is varied in the range (0; 360) deg. The values of the peak power were obtained from the expression (2) as $P_{peak} = P(\omega)/\Delta t$, where $\Delta t = 0.13$ ns is the simulation time. When the beam is shifted along the axis y by 1 mm, the radiation has the pattern depicted by the red curve. When the beam is shifted along the axis x by 1 mm, the maximum of the radiation pattern is shifted by approximately 40 deg. towards the smaller values of the angle φ (green dash dot-line). For the beam shift along the axis x, changes in the patterns are not necessarily the same for all frequencies. Depending on the considered frequency the radiation can experience reflections inside the capillary in different ways, subsequently changing the power pattern considerably.

3. Capillary design

The capillaries, both corrugated and non-corrugated, were constructed as sets of cylindrical rings manufactured from Fused Quartz. A composite design was used to allow for greater flexibility and for potential modifications. For example, the cylindrical rings may be produced from different materials, the internal radius of the rings can be designed to be gradually decreasing or increasing along the capillary length. These changes may be used to achieve higher radiation intensities, or to investigate their influence on electron beam dynamics.

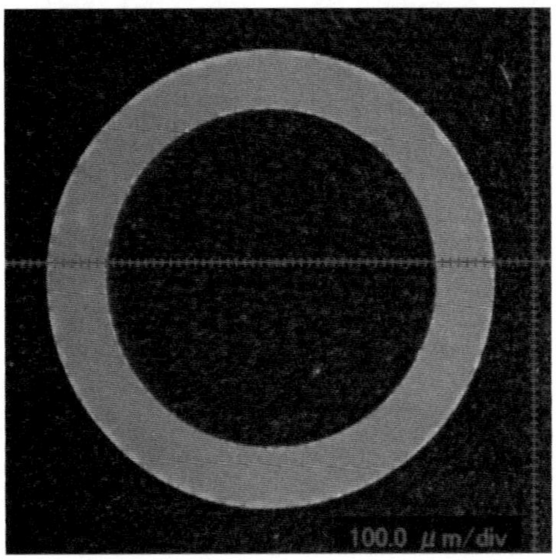

Figure 8. Image of Fused Quartz cylindrical ring obtained using electronic microscope.

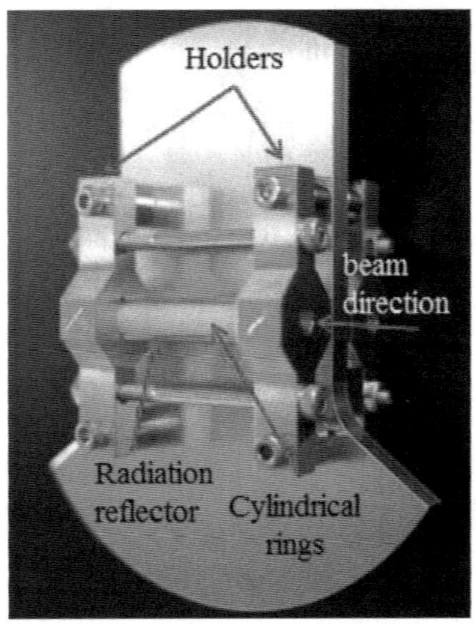

Figure 9. Corrugated capillary assembly mounted on frame.

Figure 8 shows the image of one of the dielectric rings, taken using an electronic microscope. The image demonstrates some material chipping in the order of 50 to 100 micron at the ring edges (white areas on the light blue background). For both corrugated and non-corrugated capillaries the holders are designed to be identical. Figure 9 shows the assembly of the holders, the radiation reflector, and the corrugated capillary. Using the radiation reflector as an alignment tool, the cylindrical rings were assembled and then secured by two holders on both sides of the reflector.

3.1. *Measurements of assembly accuracy*

We used a laser scanning setup to check the assemblies' accuracies. A laser beam was scanned along the outer surfaces of each capillary. The light reflected from the surfaces was detected by an array detector to obtain the vertical off-set of the laser beam. We obtained a dependence of the vertical offset as a function of the horizontal travel range. The manufacturing accuracies of the inside and outside surfaces of the rings were assumed to be roughly the same.

Figure 10 shows the laser scan for the non-corrugated capillary, the top plot shows the scan for 25 mm long section, the bottom plot shows a close-up of the scan for the range of the horizontal scanning position from 15 to 20 mm. The bottom plot shows that the width of each cylindrical ring corresponds to 500 ± 50 μm, the resolution of 50 μm is defined by the minimum step of the scan. The data points in the top plot were fitted with a line, the tilt of the line is -0.045 deg., which shows that the beginning and the end of the capillary were horizontally levelled within a fraction of degree. Similar scan over the outside surface was performed for the corrugated capillary. It was horizontally levelled with the tilt of 0.21 deg. over the length of 25 mm.

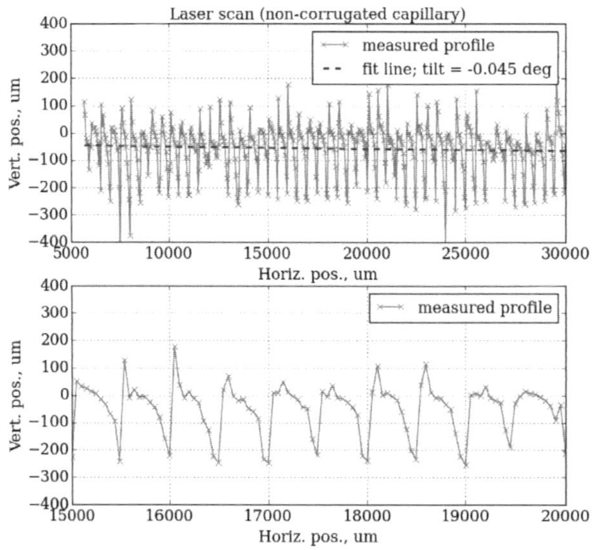

Figure 10. Scan of outside surface of non-corrugated capillary.

4. Conclusions

In this paper we considered the capillary structures as THz radiators optimised for multi-bunch femtosecond electron beams. Using CST PIC solver we investigated the change in the spectrum of the SPR radiation at $\theta = 90$ deg for the four different values of bunch spacing. For (-1) and (-2) diffraction orders of SPR it was shown, that the corrugated capillary emits more intense radiation than the non-corrugated one. For the given corrugated capillary parameters we identified that more efficient generation of SPR was achieved for the non-central propagation of the four-bunch beam.

We discussed a composite design of the capillaries constructed as sets of cylindrical rings, and showed the assembly accuracy checks, which were performed using the laser-scanning setup. Both capillaries were horizontally levelled with the accuracy of less than 0.3 deg. over the length of 25 mm.

As was pointed out earlier, current design allows generating only SPR at large angles θ. After initial experimental studies we will consider other designs of the radiation reflector and the cylindrical rings in order to achieve extraction of hybrid ChSPR at large angles θ. The hybrid mechanism should provide more intense radiation, compared to only SPR mechanism.

Acknowledgments

This work was supported by Photon and Quantum Basic Research Coordinated Development Program from the Ministry of Education, Culture, Sports, Science and Technology, Japan, and also by the Leverhulme Trust International Network Grant IN-2015-012.

References

[1] Aryshev A, Shevelev M, Honda Y et al 2015 Femtosecond response time measurements of a Cs2Te photocathode *arXiv:1507.03302v1* (physics.acc-ph)

[2] Shevelev M, Aryshev A, Araki S et al. 2015 Coherent radiation spectrum measurements at KEK LUCX facility *Nucl. Instr. Meth.* A **771**, 126

[3] Cook A M, Tikhoplav R, Tochitsky S Y et al 2009 Observation of Narrow-Band Terahertz Coherent Cherenkov Radiation from a Cylindrical Dielectric-Lined Waveguide *Phys. Rev. Lett.* **103** 095003

[4] Antipov S, Jing C, Fedurin M et al 2012 Experimental Observation of Energy Modulation in Electron Beams Passing through Terahertz Dielectric Wakefield Structures *Phys. Rev. Lett.* **108** 144801

[5] Smirnov A V 2015 Characterization of coherent Cherenkov radiation source *Nucl. Instr. Meth.* A **771** 147

[6] Ponomarenko A A, Ryazanov M I, Tishchenko A A et al 2013 Terahertz radiation from electrons moving through a waveguide with variable radius, based on Smith-Purcell and Cherenkov mechanisms *Nucl. Instr. Meth.* B **309** 223

[7] https://www.cst.com/Products/CSTPS

[8] Ponomarenko A A, Lekomtsev K V, Tishchenko A A et al 2015 CST simulations of THz radiation from a channel with periodically variable radius *Nucl. Instr. Meth.* B **355** 160

[9] Lekomtsev K, Ponomarenko A A, Tishchenko A A et al 2015 CST simulations of THz Cherenkov Smith-Purcell radiation from corrugated capillary *Proc. of Int. Particle Accelerator Conf. IPAC15 (Richmond, Virginia, USA, 3-8 May 2015)* TUPWA070 1594

[10] Yee K S, Ingham D and Shlager K 1991 Time-Domain Extrapolation to the far field based on FDTD calculations *IEEE Trans. of Ant. and Prop.* **39** 410

[11] Shu Chen, Kim N. Nguyen and Mohammed N. Afsar 2006 Complex Dielectric Permittivity Measurements of Glasses at Millimeter Waves and Terfahertz Frequencies *Proc. of the 36th European Microwave Conference (Manchester, UK, September 2006)* 384

Design, development and use of the spectrometer for investigating coherent THz radiation produced by micro-bunching instabilities at Diamond Light Source

Aiveen Finn[1,2], **Pavel Karataev**[1], **Guenther Rehm**[2]

[1] JAI at Royal Holloway University of London, Egham, UK

[2] Diamond Light Source, Oxfordshire, UK

E-mail: aiveen.finn.2013@live.rhul.ac.uk

Abstract. Schottky barrier diodes (SBDs) are known for their low noise, ultra-fast response and excellent sensitivity. They are often implemented as detectors in the millimetre wavelength regime. Micro-bunch instabilities (MBI) have been detected at many light sources around the world including the Diamond Light Source, UK. These MBI can result in bursts of coherent synchrotron radiation (CSR) with millimetre wavelengths. More research needs to be carried out with regards to the dynamics of MBI in order to confirm the simulations and to eventually harness the power of the CSR bursts. A single shot spectrometer has been designed and is under operation at the Diamond Light Source (DLS). It is composed of eight SBDs ranging from 33-1000 GHz. Unlike previous measurements carried out, each of the SBDs has been individually characterised thus making the results obtained comparable to simulations. In this paper, we present the assessment of each SBD in the spectrometer and the first results of the spectrometer's use in the beam.

1. Introduction

MBI occur when the single bunch charge exceeds a threshold current and the bunch begins to filament [1]. The filaments proceed to emit CSR which is of mm-wavelengths at DLS. CSR only occurs at wavelengths longer than the bunch unless their emission is shielded. However, when the CSR is caused as a result of MBI, the wavelength of CSR can occur at wavelengths shorter than the bunch length. It is important to note that the wavelength of CSR is limited by the longitudinal electron bunch dimension and must be longer than the beam filament emitting it (coherent emission), but the wavelength is usually much shorter than the overall bunch length.

DLS along with other light sources across the globe have observed micro-bunch instabilities [2–7]. Correspondingly, MBI and their resulting CSR bursts are monitored and investigated worldwide utilising different methods and instruments. The approach undertaken here is to investigate these instabilities in a turn-by-turn regime in order to measure the spectral characteristics of the CSR and compare with simulations. A single shot spectrometer has been developed with each of the detectors it contains characterised in order to thoroughly analyse the data.

Content from this work may be used under the terms of the Creative Commons Attribution 3.0 licence. Any further distribution of this work must maintain attribution to the author(s) and the title of the work, journal citation and DOI.
Published under licence by IOP Publishing Ltd

Journal of Physics: Conference Series **732** (2016) 012039 doi:10.1088/1742-6596/732/1/012039

2. Schottky Detector Diodes

Schottky Barrier Diodes (SBDs) are able to detect mm-waves at room temperature and hence are often used as detectors within this wavelength range. Their main attributes are low noise, excellent sensitivity and fast responses [8]. SBDs are versatile and can be operated so as to fit a specific application, when terminated with a low impedance (typically 50 Ω) the SBDs operate at a much faster speed but at the cost of lower sensitivity. In the set-up as described in this paper, the SBDs are connected to a high impedance (10 kΩ in this case) and thus achieve a higher sensitivity at a compromise of slower speed.

Of the eight detectors chosen, seven are housed within waveguides and fed signal via horn antennas. The eighth detector is a broadband quasi-optical detector (QOD) with a silicon lens. All eight SBDs are in a high impedance set up (10 kΩ) to take full advantage of the high sensitivity. Each detector and its corresponding horn antenna (where applicable) have been individually characterised on the test bench. Consequently, a spectrometer has been designed using eight SBDs, with each detector covering a specific frequency band from 33-1000 GHz. The three detectors with the lower frequency ranges are supplied by Millitech Inc, USA, while the remaining five detectors are from Virginia Diodes Inc, USA. Table 1 shows the properties of the chosen SBDs.

Table 1. SBD Specifications

Detector Model	Stated Range	Waveguide Range
DXP-22 [9]	33 - 50 GHz	26.36 - 52.73 GHz
DXP-12 [9]	60 - 90 GHz	48.41 - 96.81 GHz
DET-08 [9]	90 - 140 GHz	73.82 - 147.64 GHz
WR5.1ZBD [10]	140 - 220 GHz	115.79 - 231.59 GHz
WR3.4ZBD [10]	220 - 330 GHz	173.69 - 347.38 GHz
WR2.2ZBD [10]	330 - 500 GHz	295.28 - 590.55 GHz
WR1.5ZBD [10]	500 - 750 GHz	393.70 - 787.40 GHz
QOD [10]	100 - 1000 GHz	n/a (silicon lens)

3. Test Bench Set Up

A Thomas Keating (TK) Power Meter was used to characterise both the source and thus the individual detectors. It is a broadband detector requiring modulated signal. The main benefit of the TK power meter is that it can be calibrated by injecting a known amount of power electrically into the instrument. As a result of this, the TK power meter can then be used to quantify the source (on the test bench) and therefore the SBDs too.

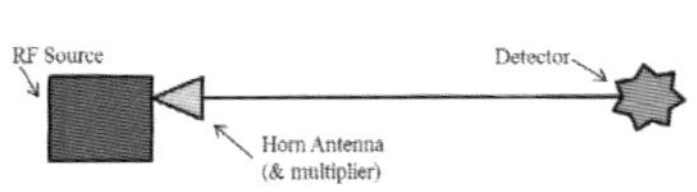

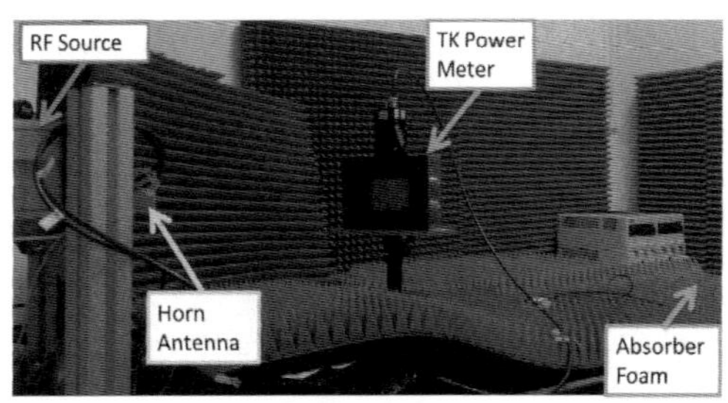

Figure 1. A schematic and photograph of the Test Bench set-up.

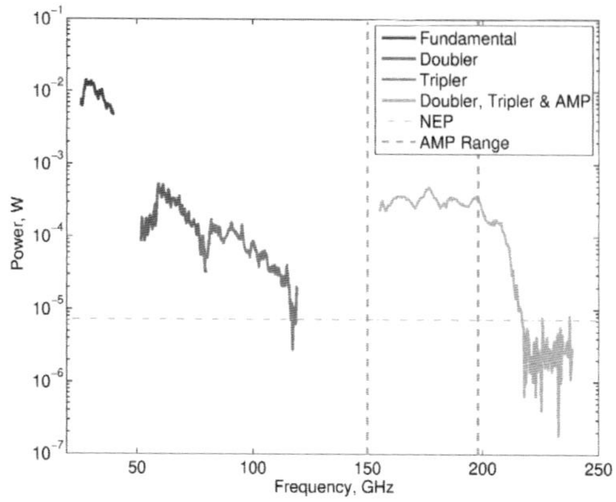

Figure 2. Power of signal received by the TK power meter for 26 - 240 GHz at a distance of 50 cm from the source with amplification on the doubler-tripler combination. NEP is the calculated Noise Equivalent Power of the TK within the described set-up.

The Ka band emitter (26.5 - 40 GHz) is the radio frequency (RF) source used on the test bench (figure 1) to produce a signal with which the detectors are quantified. The emitter employs a Gunn diode to produce the output frequency. In order to achieve the desired frequency ranges, various multipliers were utilised. The properties of the multipliers are show in table 2. Due to the nature of multipliers, the strength of the signal unceasingly declines when passing through one (or more). To counteract this continual decrease in signal, an amplifier was used to boost the output power. The amplifier utilised is an AMP-15 from Millitech Inc, USA with an input frequency of 50 - 66 GHz and a gain of 22 dB.

Table 2. Frequency Multipliers

Model	Type	Output Frequency
MUD-15 [9]	Doubler	50 - 75 GHz
MUT-10 [9]	Tripler	75 - 110 GHz
WR5.1X3 [11]	Tripler	140 - 220 GHz

First, the TK power meter was calibrated and using this information the RF source was quantified over the entire range of frequencies using the multipliers and amplifier (26 - 240 GHz). In figure 2, the power of the signal received by the TK head over the range of frequencies can be seen. In order to minimise reflections, the area in and around the test bench was lined with pyramidal absorber and the TK head was placed at Brewster's angle with respect to the source, 55.5°. The distance of 50 cm was decided upon to ensure the SBDs would receive sufficient signal but would not be saturated. Despite the source and combination of multipliers achieving frequencies higher than 240 GHz, due to the limited range of the amplifier approximately 30 GHz of the 240 GHz occurs below the NEP. Therefore, this cannot be used with regards the characterisation and thus the enforced upper limit of the set-up becomes 217 GHz.

4. Characterisation of Detectors

A power density approach was taken in order to determine the responsivities of each detector and its particular horn antenna combination. The aperture of the TK is large, especially in contrast to the physical apertures of the individual horn antennas, thus by calculating sensitivities with respect to area a more applicable result was obtained. Furthermore, the sensitivities determined from the power densities have been calculated for the detector and horn antenna combination in the forward direction. For every detector and horn antenna combination displayed here, the fundamental (i.e. TE_{10}) mode cut off of the waveguide is depicted as well as the lower cut off of the frequency of the next mode.

From figure 3, it can be seen that the sensitivity across the frequency range of DXP22 is quite constant. After the lower cut off of the next mode, the detector is still able to observe signal albeit erratically. This is to be expected because once dealing with higher modes it is difficult to predict what will happen with respect to the power in the waveguide. With regards DXP12

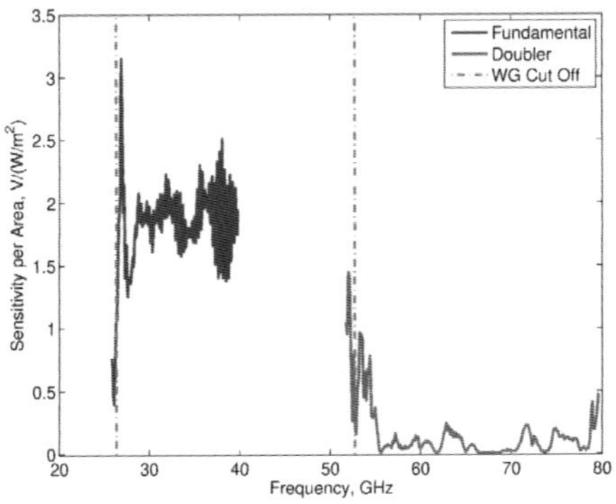

Figure 3. The sensitivity per unit area for DXP22 (33 - 50 GHz).

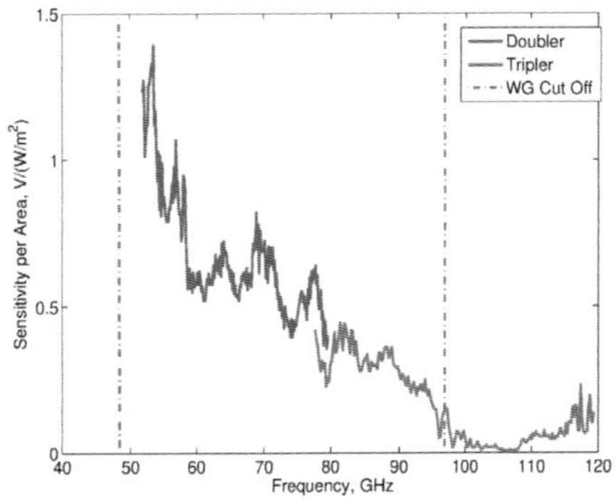

Figure 4. The sensitivity per unit area for DXP12 (60 - 90 GHz).

shown in figure 4, the sensitivity per area decreases with an increase of frequency. There is a clear overlap between the signal observed with the doubler and the tripler. Though signal is detected after the waveguide boundary, it is with a lower sensitivity per area. The shape of the curve of DET08 in figure 5, contrasts greatly with the previous figures from fellow detectors by Millitech Inc. DET08 has a peak sensitivity per area at 100 GHz.

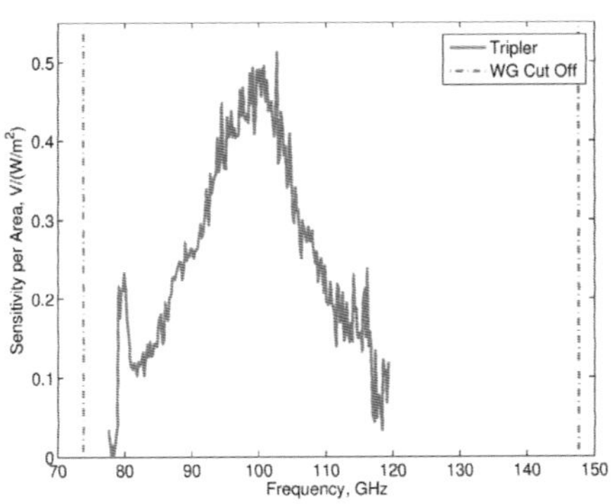

Figure 5. The sensitivity per unit area for DET08 (90 - 140 GHz).

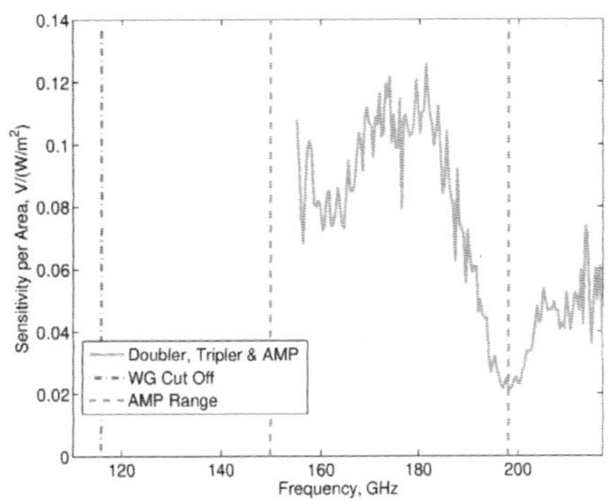

Figure 6. The sensitivity per unit area for WR5.1 (140 - 220 GHz).

As seen in figure 7, the sensitivity of WR3.4 is extremely fluctuating. There are two main points of change, the first is at and after the lower cut off of the waveguide and then again after the fall off from the power amplifier.

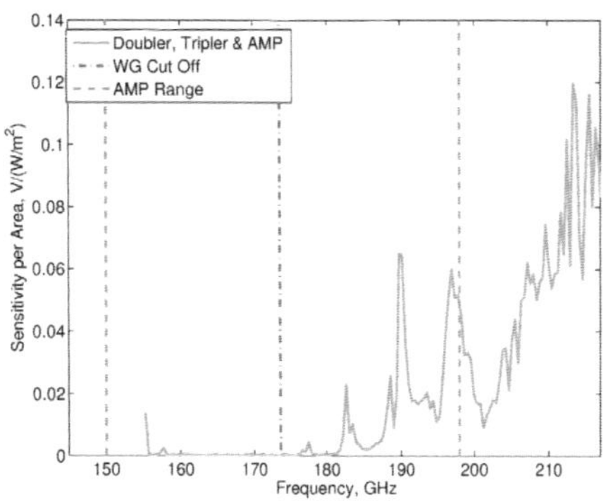

Figure 7. The sensitivity per unit area for WR3.4 (220 - 330 GHz).

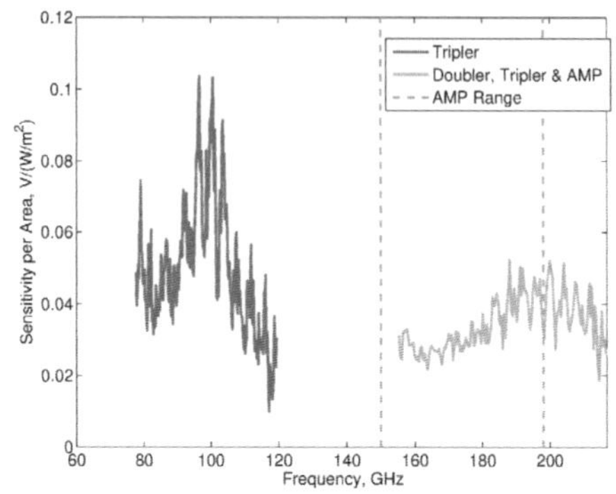

Figure 8. The sensitivity per unit area for Quasi-Optical Detector (100 - 1000 GHz).

The QOD is not housed within a waveguide, but rather a silicon lens to guide the signal into an antenna. The range as stated by the manufacturer of the QOD is 100 - 1000 GHz and it is clear that from probing the lower end of the frequencies the sensitivity is not uniform. As shown in figure 8, the sensitivity fluctuates a lot within the frequency ranges. The variability in the sensitivities can be attributed to the silicon lens being of the same or similar size as the wavelength thus causing standing waves.

The remaining detectors (WR2.2 and WR1.5) and the higher frequencies of already mentioned detectors will be further quantified using a THz laser.

5. Spectrometer/Detector Array

The spectrometer has been created in such a manner that all eight SBDs are as close together as possible to observe the most signal simultaneously (figure 9). In order to eliminate the risk of shadowing all entrance apertures are in one plane. Due to the nature of waveguides, the horn antennas and hence detectors must be aligned correctly in order to observe the E-field.

Figure 9. Layout of detector array plate.

Within the Diamond Storage Ring tunnel, there is a viewport that has been dedicated to the investigations of CSR from MBI [12]. The viewport transports the synchrotron radiation from dipole B06 to the silica viewport window (figure 10). The distance travelled is approximately

3.5 metres, whereby the radiation travels parallel but vertically lower than the electron beampipe. In order to alter the path of the desired radiation, copper mirrors are brought into play. These water-cooled mirrors not only guide the mm-radiation through the viewport but also absorb the x-ray radiation which is created by the dipole magnet and usually required by beamlines at DLS.

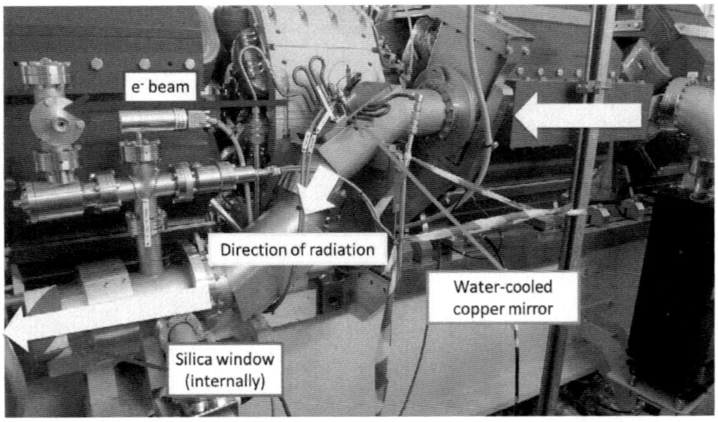

Figure 10. The path the radiation takes within the viewport from just after B06 to the silica window.

As shown in figure 11, the detector plate is installed in open air in front of the silica window, not further than 1 metre away. The detector array plate is attached to a three-way motion set up in order to move the spectrometer in the x-, y- and z-plane and thus to optimise the signal obtained in the different channels. Directly after the viewport window is a sheet of styrofoam, which is used to block unwanted frequencies that may affect the measurements, namely visible wavelengths of light and infrared while allowing THz and mm-wavelengths to pass through.

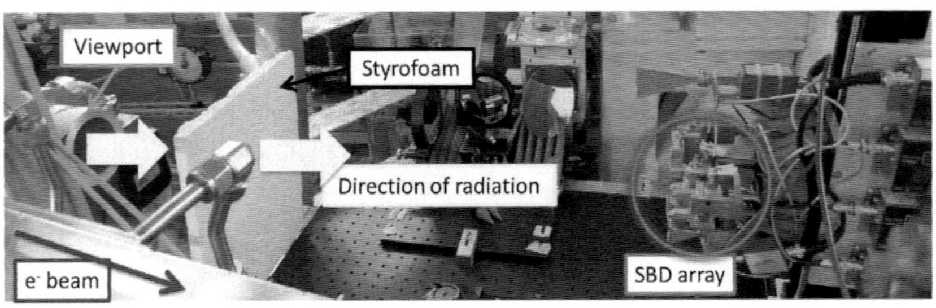

Figure 11. The detector array installed in the tunnel with respect to the viewport.

6. Data Analysis

The signals from the eight channels are individually amplified before being carried out of the tunnel via coaxial cables to be analysed. The amplifiers (FEMTO, Germany HVA-S Switchable Gain) are used to boost the signal above the noise, because otherwise the distanced travelled by the signals is great (less 30 m) and the signals would become engulfed by noise.

The amplifiers connected to each of the SBDs have a high input impedance so that the SBDs achieve a high sensitivity rating. The input impedance of the amplifiers is 10 kΩ and because we are working in the high impedance regime, the capacitance of the cables becomes very important.

The effect of capacitive loading of the cables can be altered by choosing shorter cables and using cables with a lower capacitance. In the set-up described here, cables with a lower capacitance were employed, but the primary contribution was that the amplifiers are mounted close to the detector array, to minimise cable length.

The signals are amplified and then travel along approximately 30 metres of cables until they reach the simultaneous multichannel digitiser. The ADC is a d-tAcq ACQ425ELF with a sampling rate of 1 MSPS and the possibility for 16 input channels. It constantly streams out the data where it is fed into a local computer. This computer carries out FFTs of the data, selects the region of interest and executes a power average over every ten data sets. The region of interest is focussed around the revolution frequency (533.820 kHz) of the synchrotron storage ring, as opposed to DC. SBDs produce flicker noise, which has a 1/f power spectrum, so the signal to noise on their readout will be better when at higher frequency. As Nyquist sits at 500 kHz, the frequency range is imposed to be 33.820 kHz symmetrically either side of the central frequency.

7. MBI Observations

Since its installation in the storage ring, the SBD array has been capturing data. The data presented here was obtained when the storage ring was set to user optics with a single bunch fill [13]. When the RF cavities had a combined voltage of 2.2 MV, a single bunch was injected into the storage ring with a low current of just 1.07 mA. Slowly the current was increased to 4.15 mA. It was not possible to inject more charge into the single bunch, as this caused the bunch to become unstable and the beam was kicked out of the storage ring. During the current ramp of 1.07 mA to 4.15 mA, data was continually obtained and the bursting of MBI observed. In figure 12, the signal obtained at the revolution frequency (533.820 kHz) is displayed for each

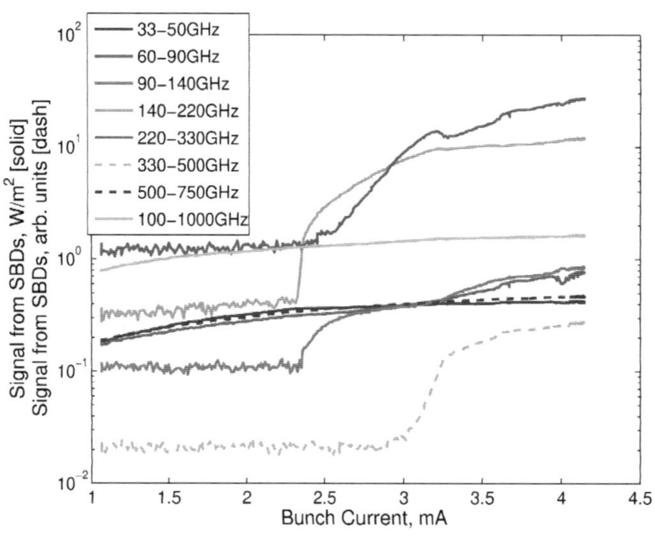

Figure 12. The signal obtained by the SBDs and depicted as the power for a unit area.

of the eight detectors. As only six of the eight detectors were characterised, the remaining two detectors (WR2.2 (330 - 500 GHz) and WR1.5 (500 - 750 GHz)) have dashed lines to represent the different units. Figure 13 depicts the signal observed solely in the sidebands i.e. within the $\pm$ 33.820 kHz range excluding the central frequency for each detector. The activity occurring either side of the revolution frequency was termed the 'variation' or 'noise' because it only occurs with bursting or disturbances in the ring. The variation in most cases remains quite constant.

Finally, in figure 14 the relationship between the signal at revolution frequency and its variation can be observed.

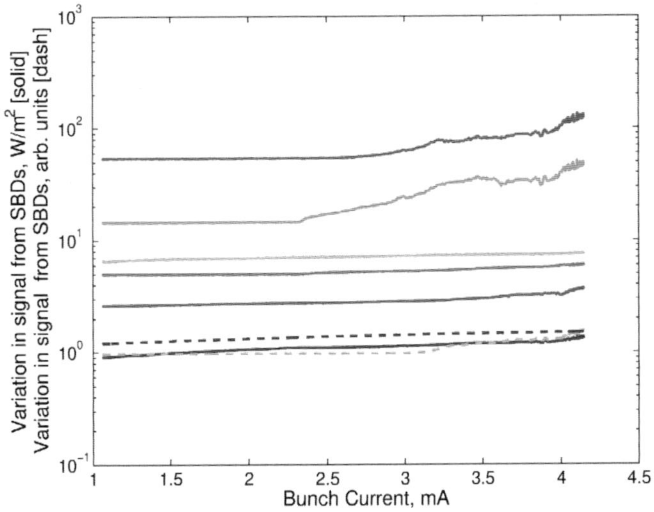

Figure 13. The variation in each of the channel's signals, whereby the variation could be interpreted as the standard deviation.

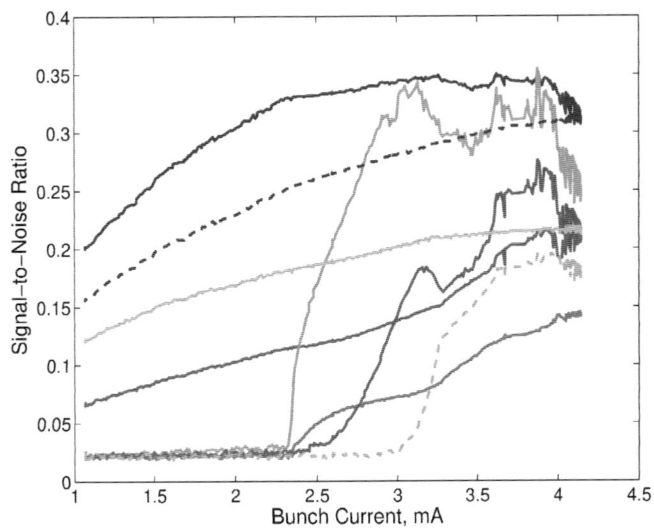

Figure 14. The ratio between figure 12 and figure 13 is depicted.

DXP22 (33 - 50 GHz) portrays a smooth signal and variance, this is due to its low(er) frequency view where there is less change than for the higher frequency detectors to observe. Furthermore, the broadband QOD can probe the higher frequencies but because of the strong and constant signal from the DXP22, the QOD becomes overpowered by this lower frequency signal and thus they achieve similar curves shapes. It is clear to see from the data obtained, WR5.1 (140 - 220 GHz) and WR3.4 (220 - 330 GHz) observed the most amount of signal. This is extremely obvious in figure 12 where the power they observe is at least an order of magnitude above the rest. These two detectors also have the most variance (figure 13) and thus incur an erratic ratio (figure 14). In figure 14, there is a sudden change at 2.3 mA representing the lengthening of the bunch. This value of 2.3 mA for these machine parameters could thus be declared as the bursting threshold.

8. Conclusions & Future Plans

The data presented here show that six of the eight detectors were characterised on the test bench, thus allowing for quantitative analysis when using the detector array in the beam at DLS. As the source and accompanying multipliers can only generate frequencies with sufficient power up to 217 GHz, the SBDs requiring higher frequencies will be quantified using a terahertz laser.

The initial results from experiments taken at DLS are depicted here and for the first time the results are presented as real units (W/m^2) and not arbitrary units.

A full analysis of raster scans will be undertaken so that a location favourably suitable for all eight detectors can be determined for its placement in the storage ring tunnel with respect to the viewport silica window. Following on from this determination, further data will be taken at DLS and thus analysis carried out. Future experiments at DLS include obtaining data under various beam condition in particular 'Low Alpha modes', when the momentum compaction factor is set to be small [14].

References

[1] Venturini M, Warnock R, Ruth R and Ellison J 2005 *Phys. Rev. ST Accel. Beams* **8** 014202
[2] Abo-Bakr M, Feikes J, Holldack K, Wüstefeld G and Hübers H W 2002 *Phys. Rev. Lett.* **88** 254801
[3] Rehm G, Morgan A F, Bartolini R, Martin I P S and Karataev P 2009 *Proc. of DIPAC* (Basel, Switzerland) pp 369–371
[4] Podobedov B, Carr G L, Kramer S L and Murphy J B 2001 *Proc. of Particle Accelerator Conf.* vol 3 (Chicago, IL, USA) pp 1921–1923
[5] Byrd J M, Leemans W P, Loftsdottir A, Marcelis B, Martin M C, Mckinney W R, Sannibale F, Scarvie T and Steier C 2002 *Phys. Rev. Lett.* **89** 224801
[6] Mochihashi A, Hosaka M, Katoh M, Shimada M and Kimura S 2006 *Proc. of EPAC* (Edinburgh, Scotland) pp 3380–3382
[7] Kuske P 2009 *Proc. of Particle Accelerator Conf.* (Vancouver, BC, Canada) pp 4682–4684
[8] Smith R J and Dorf R C 1992 *Circuits, Devices and Systems* 5th ed (John Wiley & Sons Inc.)
[9] Millitech Inc. *Detector Specifications* URL `http://www.millitech.com/`
[10] Virginia Diodes Inc. *VDI User Guide* URL `http://vadiodes.com/index.php/en/`
[11] Virginia Diodes Inc. *VDI User Guide WR5.1 WR3.4 WR2.2* URL `http://vadiodes.com/index.php/en/`
[12] Shields W, Boorman G and Karataev P 2012 *Proc. of IPAC* (New Orleans, Louisiana, USA) pp 3114–3116
[13] CCLRC 2002 *Diamond Synchrotron Light Source: Report of the Design Specification*
[14] Martin I P S, Rehm G, Thomas C and Bartolini R 2011 *Phys. Rev. ST Accel. Beams* **14** 040705

Sapphire decelerating capillary channel integrated with antenna at frequency 0.675 THz

I A Ashanin and S M Polozov

Department of Electrophysical Facilities, National Research Nuclear University - Moscow Engineering Physics Institute, Kashirskoe shosse, 31, Moscow, 115409, Russia

E-mail: ilya.ashanin@mail.ru; SMPolozov@mephi.ru

Abstract. In recent years, there has been an increasing interest in THz-radiation for application in medicine (THz tomographs), in pharmaceutics (composition analysis for medicines), in introscopy of large-scale objects (ships, trains, containers) and others. THz-radiation can be generated by relativistic electron bunches passing through the Cherenkov decelerating capillary channel (circular waveguide with dielectric filling) with horn extraction. Relativistic electron beams having ~100 μm in diameter and pulse durations of 1 ps or less (as in photoinjectors) are capable of producing substantial power of THz-radiation. High-peak power coherent Cherenkov radiation can be produced by a properly modulated high-brightness electron beam or by a single, high-density bunch having sub-wavelength dimension. The aperture of a Cherenkov decelerating structure should be comparable with the mm or sub-mm wavelength (0.1-3 mm). Different dielectric materials for the internal surface coating of the capillary channel of mm-sub-mm cross-section can be used. As is known, a frequency of 0.675 THz corresponds to the atmospheric window with high transparency. This report presents the results of electrodynamics study of the metallized sapphire decelerating Cherenkov capillary. A horn antenna attached to the metallized sapphire capillary channel at the 0.675 THz resonant frequency will be considered.

1. Introduction

THz-frequency range radiation sources are necessary for huge variety of applications. Many types of THz radiation generation principles are known: orotrons, TWT's, nonlinear convertors based on semiconductors or gas filled tubes, gyrotrons [1]. But only large scale facilities such as FEL's can generate coherent THz radiation. The other possible method of coherent THz-radiation generation is based on high brightness relativistic electron bunches passing through the Cherenkov decelerating capillary channel (circular waveguide with dielectric filling) or the corrugating channel with horn extraction [2]. Relativistic electron beams having ~100 μm in diameter and pulse durations of 1 ps or less (as in photoinjectors) are capable of producing substantial power of THz radiation.

Besides THz radiation generation, the problem of transportation and output of so powerful radiation into the atmosphere is not less difficult to develop such structures. The aperture of a Cherenkov decelerating structure should be comparable with the mm or sub-mm wavelength (0.1-3 mm). Different dielectric materials for the internal surface coating of the capillary channel of mm or sub-mm cross-section can be used. This paper describes estimation for electrodynamical characteristics of a decelerating capillary structure without radiation power estimation.

THz radiation in a broad band of frequencies possesses low penetration because of its high absorption in water. Based on data obtained in the observatory of Mauna Kea for THz radiation

Content from this work may be used under the terms of the Creative Commons Attribution 3.0 licence. Any further distribution of this work must maintain attribution to the author(s) and the title of the work, journal citation and DOI.
Published under licence by IOP Publishing Ltd

transmission in atmosphere, transmission windows are defined in ranges of frequencies 0.6-0.7; 0.8-0.98 THz and some over [3].

A frequency of 0.675 THz corresponds to the atmospheric window with high transparency and was chosen for the simulation. This paper presents the results of electrodynamics study of the metallized sapphire decelerating Cherenkov capillary. A horn antenna attached to such capillary channel at the 0.675 THz resonant frequency will be considered.

2. Geometrical and electrodynamical characteristics of dielectric loaded capillary

CST Microwave Studio was used for simulation of a decelerating capillary structure with dielectric filling [4]. Sapphire was chosen as a material of dielectric filling, because it has high durability, dielectric constant ε~9.9, thermal stability up to 1600 °C and there is almost no heat conductivity. The electrodynamical model of a decelerating sapphire capillary structure at a resonance frequency of 0.675 THz is shown in the figure 1. Geometrical and electrodynamics characteristics of the decelerating capillary channel with sapphire coating at a resonance frequency of 0.675 THz are presented in tables 1 and 2.

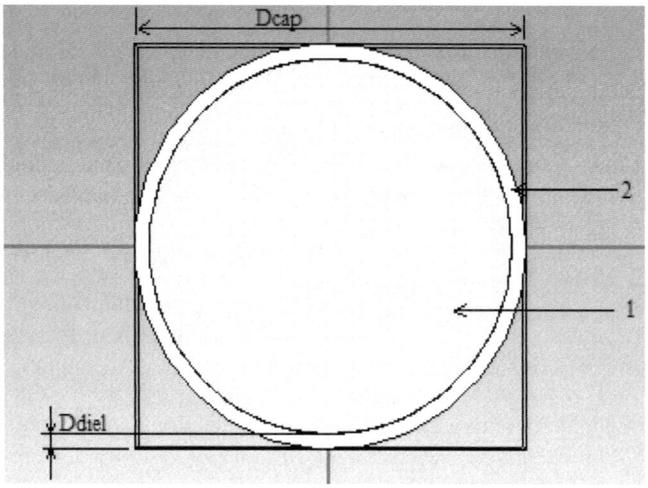

Figure 1. Electrodynamical model (front view) of a decelerating sapphire capillary structure at a resonance frequency of 0.675 THz. 1 – vacuum; 2 – dielectric coat; D_{cap} – capillary diameter; D_{diel} – dielectric coat thickness.

Table 1. Geometrical characteristics of the decelerating capillary channel with sapphire coating at a resonance frequency of 0.675 GHz.

Diameter of vacuum channel D_{vac} (μm)	Dielectric coat thickness D_{diel} (μm)	Structure diameter D_{cap} (μm)	Structure length L_{cap} (mm)
312.6	24.8	337.4	0.5

Table 2. Electrodynamical characteristics of the decelerating capillary channel with sapphire coating at a resonance frequency of 0.675 GHz.

Material	ε	Q-factor	Shunt impedance R_{sh} (MOhm/m)	R_{sh}/Q (Ohm/m)	Group velocity β_{group}	Resonant frequency F (THz)
Sapphire	9.9	1531	50.17	32.77	0.43	0.675

3. The radiating horn antenna integrated with the capillary channel

It is necessary to match a radiation source output with an open space for more efficient THz radiation. There is a need to obtain an effective matching of the Cherenkov capillary with an open space, as well as to provide the necessary directivity. The capillary channel radiated by the open end does not provide the high directivity of radiation. It can be seen from the diagrams shown in the figure 2. Furthermore, the reflection of the wave from the open end will be observed. The reflection coefficient of the open end is equal to -10.1 dB for the capillary having parameters which are given in tables 1 and 2. Higher radiation bands are far from the operating frequency and the next ones have a resonant frequency of 0.885 THz, which differs by 30% from the chosen frequency and the width of this band is 0.031 THz (3.5%), the reflection coefficient is equal to -21.6 dB.

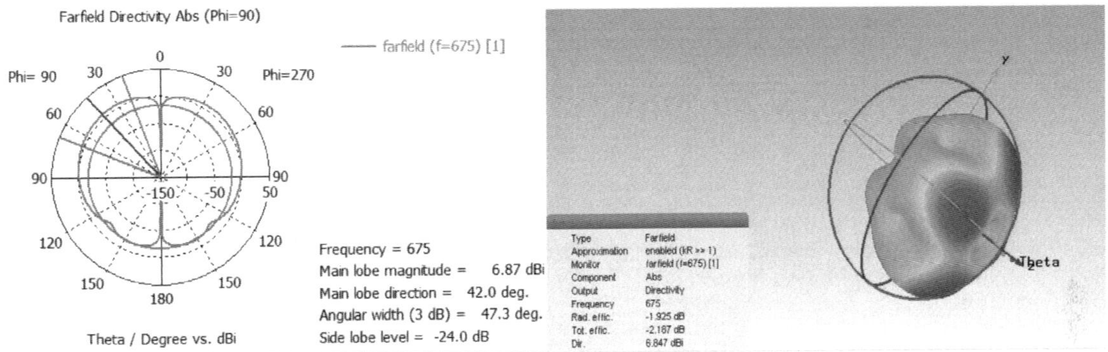

(a) (b)

Figure 2. Circular (a) and three-dimensional (b) patterns of the decelerating sapphire capillary at an operating frequency of 0.675 THz without the extractor.

An antenna must be used for increasing the radiation efficiency. The capillary has a circular cross section, therefore it is clear that the solution will be a horn antenna which will be directly attached to the end of a capillary structure. The main parameters of the antenna are directivity, angle between the axis of the antenna and direction of the main lobe, angular width of the main lobe and reflection coefficient of the horn, radiation reflectivity from the horn and width of the operating frequency band.

Further electrodynamics model of the copper antenna matched with the sapphire capillary was tuned to a resonant frequency of 0.675 THz (figure 3). Geometrical characteristics and simulation results of the radiating horn matching with the decelerating sapphire capillary channel are presented in the table 3. Directivity patterns of such extractor and variation characteristics are shown in the figure 4. The length of horn was variated to achieve the optimal directivity. Then the horn aperture angle was variated at the obtained fixed length of horn L = 0.168 mm. The S-parameter's analysis results are shown in the figure 5. Here the S11-parameter is interpreted as a wave reflection coefficient from the output side into a horn antenna.

Table 3. Geometrical characteristics of the radiating horn and matching simulation results of the decelerating sapphire with copper capillary tuned at a frequency of 0.675 THz (parameters are shown in figure 3).

D_{cap} (μm)	D_{diel} (μm)	D_{met} (μm)	D_{capmet} (μm)	D_{rup} (μm)	L_{cap} (mm)	$L_{caphorn}$ (mm)	Directivity (dB)	Angular width (°)	Reflectivity (dB)	Operating frequency band width (THz (%)
60	24.9	49.6	334.5	480	0.3	0.468	9.94	57.8	-33.52	0.031 (4.6)

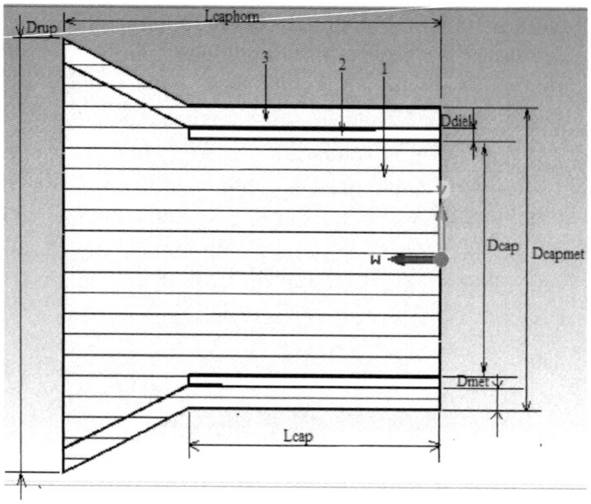

Figure 3. Electrodynamic model of the circular horn matched with the decelerating channel: 1 – vacuum; 2 – dielectric coat; 3 – copper; D_{cap} – diameter of vacuum channel; D_{diel} – dielectric coat thickness; D_{met} - copper thickness; D_{capmet} – capillary diameter; D_{rup} – horn diameter; L_{cap} – capillary length; $L_{caphorn}$ – length capillary with horn.

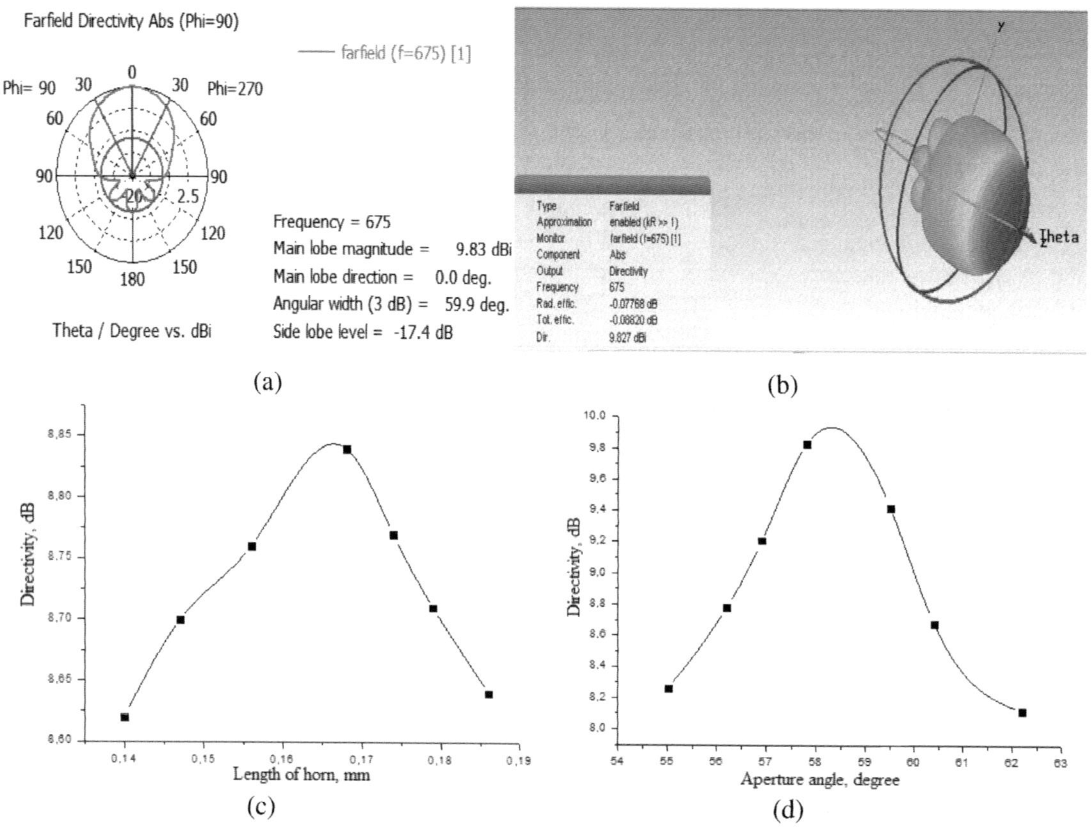

Figure 4. Circular (a) and three-dimensional (b) patterns of the decelerating sapphire and copper capillary at an operating frequency of 0.675 THz matching with the extractor.

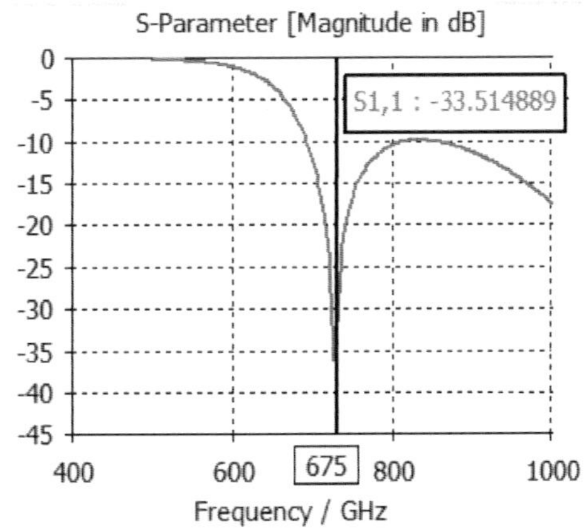

Figure 5. S11-parameters analysis results.

The directivity and operating frequency band width are reduced by increasing the horn beam angle. The width of the main lobe of radiation and its angle, on the contrary, increase. An increase of the length of the horn and a decrease of the beam angle lead to increasing the directivity, but this is accompanied by a withdrawal of the selected resonant frequency of 0.675 THz. In this case, a minimum reflectance is equal to -33.52 dB, and the frequency corresponds to 0.677 THz i.e. only 0.3% of the offset from the considered frequency of 0.675 THz.

4. Conclusions

A model of the radiating Cherenkov capillary and horn antenna with the coating of sapphire and copper at a resonant frequency of 0.675 THz was designed. The group velocity is in the range of 0.4-0.5 c, the effective shunt impedance R_{sh}/Q = 30-35 kOhm/m. THe calculated horn antenna directivity provides ~ 10 dB, the reflection coefficient of the capillary output is less than -25dB, the width of the operating frequency band ~ 5%.

The capillary without antenna has the radiation directivity of ~ 6 dB and the reflection coefficient from the capillary output is more than -12 dB and worse other parameters. Cherenkov capillaries with the integrated radiating antenna allows to achieve a wide range of operating parameters for different geometrical dimensions. It should be noted that these capillaries are difficult for production because of small sizes and high manufacturing tolerances requirements.

References

[1] Smirnov A V 2008 A high performance FIR radiator based on a laser driven e-gun *Photonics Research Developments* **8** 247-269
[2] Polozov S M, Bondarenko T V and Smirnov A V 2012 *Problems of Atomic Science and Technology Series: Nuclear Physics Investigations* **3** (79) 92-95
[3] Pardo J R, Wiedner M C and Serabyn E 2004 Side-by-side comparison of Fourier transform spectroscopy and water vapour radiometry as tool for calibration of millimeter/submillimeter ground-based observatories *The astrophysical journal supplement series* **153** 363-367
[4] CST GmbH – Computer Simulation Technology URL: http://www.cst.com

Simulation of variation characteristics at thermostabilization of 27 GHz biperiodical accelerating structure

Y D Kluchevskaya and S M Polozov
Department of Electrophysical Facilities, National Research Nuclear University - Moscow Engineering Physics Institute, Kashirskoe shosse, 31, Moscow, 115409, Russia

E-mail: kluchevskaia@mail.ru; SMPolozov@mephi.ru

Abstract. It was proposed to develop the biperiodical accelerating structure with operating frequency of 27 GHz to assess the possibility of design a compact accelerating structure for medical application. It is necessary to do the more careful simulation of variation characteristics this case because of decrease of wavelength 3-10 times in comparison with conventional structures 10 and 3 cm ranges. Results of such study are presented in the article. Also a combination of high electromagnetic fields and long pulses at a high operating frequency leads to the temperature increase in the structure, thermal deformation and significant change of the resonator characteristics, including the frequency of the RF pulse. Development results of three versions of system of temperature stabilization also discuses.

1. Introduction

In the accelerator it is proposed to use the conventional biperiodical accelerating structure (BAS) based on the disk–loaded waveguide (DLW) operating on the standing wave. Results of beam dynamics simulation and electrodynamics study was discussed early in [1-3]. The RF feeding of accelerator is planned to be realized using a gyrotron to be designed at the Institute of Applied Physics of Russian Academy of Sciences [4].The combination of high electromagnetic fields and long pulse at the high operating frequency leads to the increase of temperature patterns, thermal deformations and a noticeable change in the characteristics of the resonator, including frequency during an RF pulse [5]. Three variants of the thermal stabilization system of the accelerator were considered. To determine the frequency shift depending on power supply and temperature of the cooling system analysis of variation curves was carried out. The simulation results are defined using CST code [6]. Frequency range of 27 GHz is not traditional for the BAS, therefore the variation characteristics depending on the geometric parameters of the structure cells were calculated.

2. Simulation of variation characteristics at thermostabilization of 27 GHz BAS

Three variants of the cooling system with tubes, tubes and jacket, double jacket were considered. Results for three versions of temperature stabilization system simulation are presented in the table 1.

Table 1. Comparison of the BAS cooling system variants.

	Cooling system:		
Parameters:	tubes	tubes & jacket	double jacket
Water flux (l/s)	0,33	0,35	0,35
Maximal temperature (°C)	34	36	37
Frequency shift (MHz)	80	110	98

Content from this work may be used under the terms of the Creative Commons Attribution 3.0 licence. Any further distribution of this work must maintain attribution to the author(s) and the title of the work, journal citation and DOI.
Published under licence by IOP Publishing Ltd

Journal of Physics: Conference Series **732** (2016) 012041 doi:10.1088/1742-6596/732/1/012041

Three variants of the cooling system are presented in the figures 2-4.

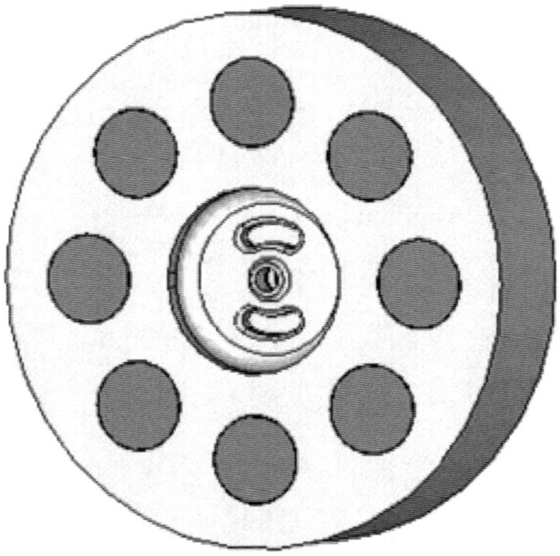

Figure 2. The cooling system with tubes.

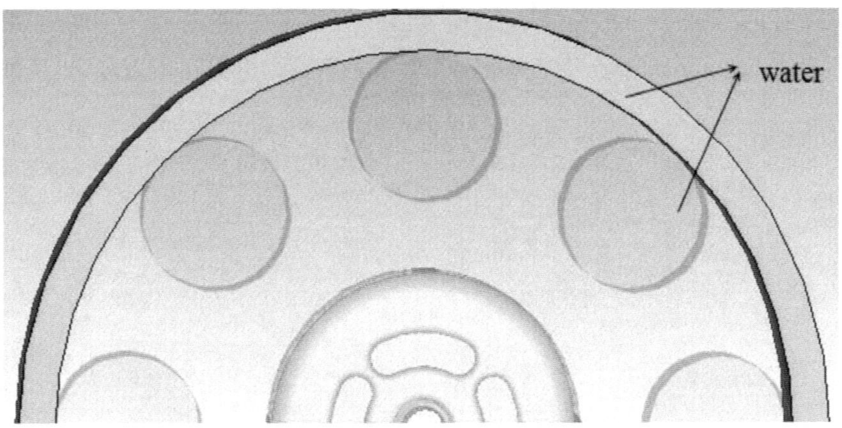

Figure 3. The cooling system with tubes and jacket.

Figure 4. The cooling system with double jacket.

To determine the frequency shift vs. changes of geometrical parameters of the structure as radius of the coupling cell, radius of the accelerating cell and radius of the nose on diaphragm tube drift are shown in the figure 5.

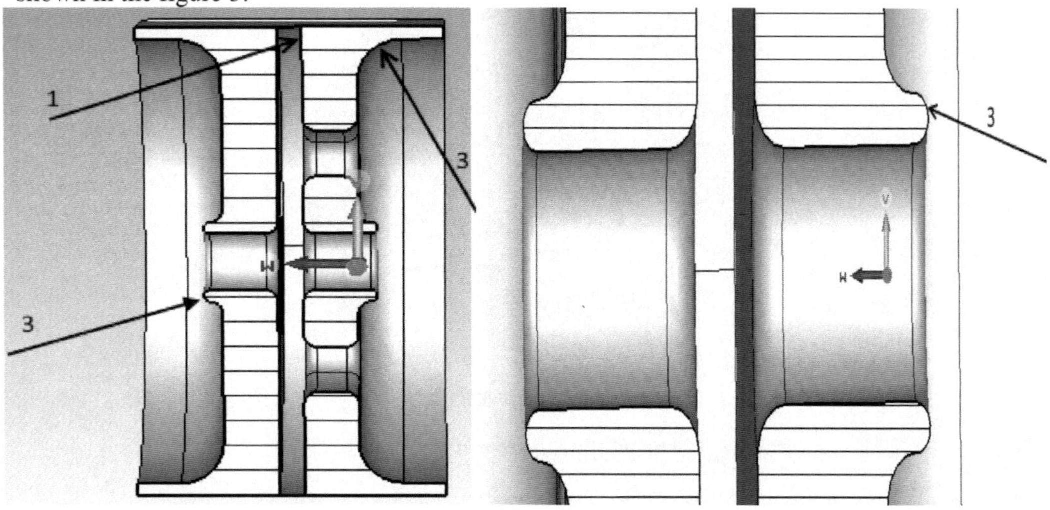

Figure 5. Variable parameters (1 – radius of the coupling cell; 2 – radius of the regular cell; 3 – radius of the nose).

The most significant effect on the resonant frequency has a radius of half-tube nose. Analysis of the variation curves shows that frequency shifts are $dF/dR_{coupling\ cells}$ = 80 MHz/mm, $dF/dR_{regular\ cells}$ = 130 MHz/mm, dF/dR_{nose} = 180 MHz/mm, where F - frequency, R - radius. Variation characteristics are presented in figures 6-8.

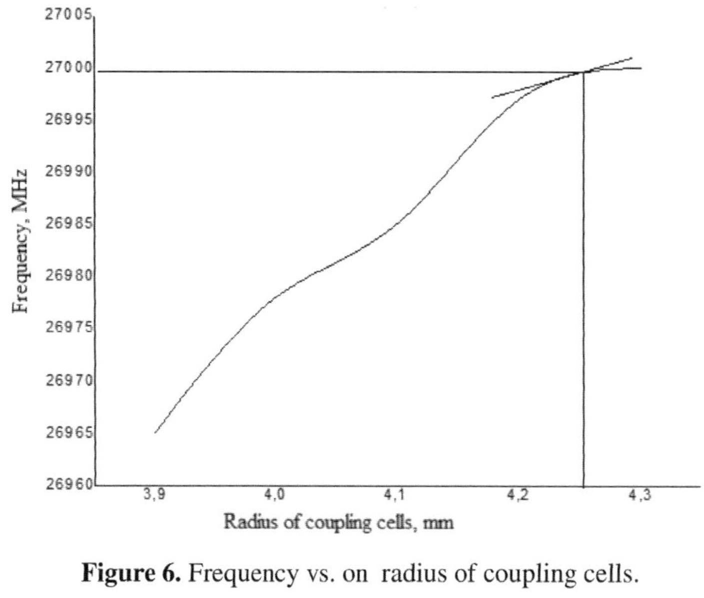

Figure 6. Frequency vs. on radius of coupling cells.

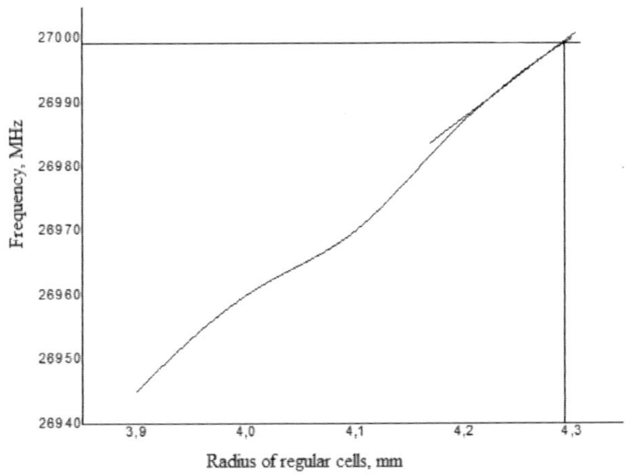

Figure 7. Frequency vs. on radius of regular cells.

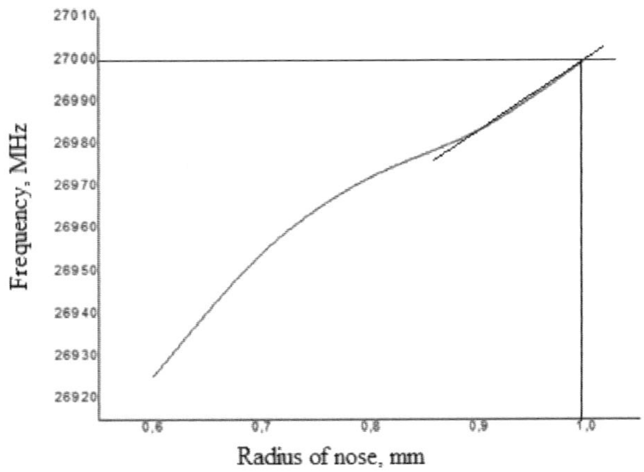

Figure 8. Frequency vs. on radius of nose.

3. Conclusion

According to the results of thermal simulations the necessary parameters of the cooling system, allowing to limit the maximum temperature in the structure to 34 °C was determined these in case the frequency shift of the RF field is negligible. Three variants of the cooling system were considered. Possibility of the BAS thermal stabilization problem solving with acceptable requirements for fluid flow was shown. Variant of double cover, which should be recommended for use, is a simplest constructional solution. Variation curves analysis revealed that it's need to produce structure with tolerances of 10 microns. Which requires the application of electroerosion technology.

References

[1] Bondarenko T, Kliuchevskaia Yu, Polozov S and Rashchikov V 2014 *Proc. of IPAC* 2191-2192
[2] Kliuchevskaia Yu, Polozov S and Bondarenko T 2014 *Proc. of XXVII Linear Accelerator Conference* 1096-1098
[3] Kliuchevskaia Yu and Polozov S 2014 Thermal simulations of the biperiodical accelerating Structure with the operating frequency 27 GHz *Proc. of XXIVth Russian Particle Accelerators Conference* **13** 237-239
[4] Zapevalov S A, Ilyakov E V and Kornishin S Yu 2008 *Proc. of the XXIth Russian Particle Accelerator Conference* ***1*** 339.
[5] Gusarova M A, Isaev I V and Kostin R A 2013 *Journal of Applied Physics* **83** 4
[6] CST GmbH – Computer Simulation Technology URL: http://www.cst.com

Generation of THz-radiation in the Cherenkov decelerating structure with planar geometry at frequency ~ 0.675 THz

I A Ashanin and S M Polozov
Department of Electrophysical Facilities, National Research Nuclear University - Moscow Engineering Physics Institute, Kashirskoe shosse, 31, Moscow, 115409, Russia

E-mail: ilya.ashanin@mail.ru; SMPolozov@mephi.ru

Abstract. One of the ways to generate THz-radiation is by the relativistic electron bunches travelling through Cherenkov decelerating dielectric filled capillary channel. Sapphire or other dielectric materials can be used for the internal surface coating of the capillary. Relativistic electron bunches of ~100 μm in diameter and pulse durations of 1 ps or shorter are capable to produce substantial power of THz-radiation. The aperture of Cherenkov decelerating structure should be comparable with the sub-mm wavelength (0.05-3 mm). Such type of decelerating system allows providing of the wide range of operating parameters at the various geometrical sizes. But it is necessary to consider that such capillaries are difficult in production as there is a requirement to drill a small aperture in a long crystal of high hardness but brittle. In this regard it would be desirable to offer transition option from the axial to the planar geometry. Furthermore the ribbon beam has some advantages as focusing at low energies and possessing smaller expansion in the drift space. The authors present design and results of electrodynamics study of the decelerating planar dielectric filling Cherenkov channel at frequency 0.675 THz in this article. It is also delivered characteristic comparison with axial geometry channel. A horn antenna attached to such channel at 0.675 THz resonant frequency is considered.

1. Introduction

At present facilities for THz-radiation can be used in such areas as introscopy of large objects, medicine (THz tomography for survey and determination of cancer neoplasm), pharmacy (definition of medical product composition). Such facilities can be realized based on sufficiently powerful radiation source.

As it is known Cherenkov decelerating capillary channel or corrugating channel with dielectric filling and horn antenna can be used for THz-radiation generation with relativistic electron bunches passing through [1, 2]. RF photo injectors can provide high-brightness relativistic electron bunches with pulse durations of 1 ps or less and are capable to produce substantial power of THz-radiation. The basic components of a RF photo injector are RF-gun with photocathode (Cu, Mg, Cs2Te, GaAs, Ba etc.), laser and optical system producing the desired pulse structure, a RF source, timing and synchronization system [3].

High-brightness relativistic electron bunch of mm-sub-mm length and 100 μm in diameter travelling through decelerating structure can produce a coherent Cherenkov radiation. The aperture of such Cherenkov decelerating structure shall be compared with the mm or sub-mm wavelength (0.1-3 mm). Different dielectric materials as quartz, sapphire, diamond, barium tetratitanate can be used for the internal surface coating of the dielectric capillary channel of mm-sub-mm cross-section. There isa need to use a dielectric material with higher dielectric permittivity (5 and higher), such as diamond or sapphire for obtaining the best results on the radiation generation. THz-radiation in a broad band of

Content from this work may be used under the terms of the Creative Commons Attribution 3.0 licence. Any further distribution of this work must maintain attribution to the author(s) and the title of the work, journal citation and DOI.
Published under licence by IOP Publishing Ltd

frequencies possesses low penetration because of its high absorption in water. In this case the frequency of 0.675 THz was considered as one of the high transparency window for THz-radiation transmission in atmosphere [4].

In the article [5] it was suggested to use sapphire as dielectric filling material for decelerating Cherenkov capillary structure, because of its high durability, ε~9.9, thermal stability up to 1600° C and with almost no conductivity. Electrodynamic and geometrical characteristics of capillary channel were analyzed and simulation results of main parameters of such structure were obtained.

It is necessary to consider that such capillaries are very difficult to produce as a small aperture must be drilled in a long crystal of high hardness. In this regard it would be desirable to offer transition option from the axial to the planar geometry and to use a ribbon beam. Planar structure is easier for manufacturing, for preparation and it is still of high dielectric permittivity. Moreover using ribbon beams has such features as smaller expansion in the drift space, slight impact of spatial charge and more simple focusing at low energies.

For transformation of bunch symmetry from axial to the ribbon quadruple lens can be used. Axial beam passing through contracts in one, and stretches in another plane and acquires a tape form.

In this paper design and results of electrodynamics study of the planar metallized sapphire decelerating Cherenkov channel are presented. Horn antenna integrated to such decelerating channel at 0.675 THz resonant frequency is considered. Electrodynamic characteristics comparison with axial geometry channel is presented as well.

2. Geometrical and electrodynamics characteristics of dielectric loaded planar structure

CST Microwave Studio was used to simulate the decelerating planar structure with sapphire filling and to calculate the main electrodynamics characteristics [6]. Electrodynamic model of decelerating sapphire planar structure at resonance frequency 0.675 THz is shown in Figure 1. Geometrical and electrodynamic characteristics of decelerating planar channel with sapphire coating at resonance frequency 0.675 THz are presented in Table 1.

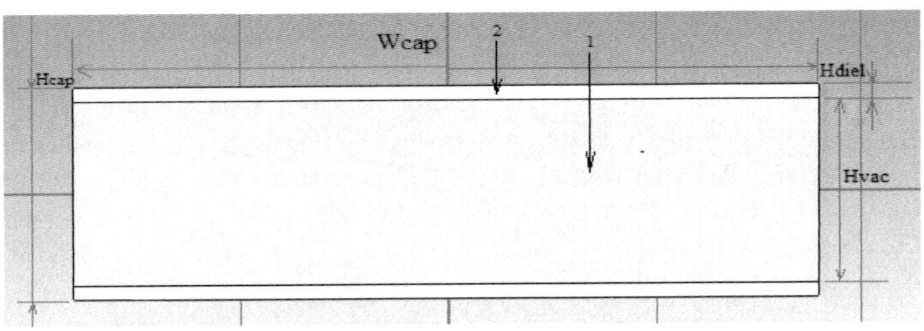

Figure 1. Electrodynamic model (front view) of decelerating sapphire planar structure at resonance frequency 0.675 THz. 1 – vacuum; 2 – dielectric coat; H_{cap} – channel height; H_{diel} – dielectric coat height; H_{vac} – vacuum channel height; W_{cap} – channel width.

Table 1. Geometrical and electrodynamic characteristics of decelerating planar structure with sapphire coating at resonance frequency 0.675 GHz (geometrical characteristics are shown in Figure 1).

Material	ε	H_{cap} (μm)	W_{cap} (μm)	*Q*-factor	*Shunt impedance* R_{sh} (MOhm/m)	R_{sh}/Q (kOhm/m)	Group velocity β_{group}
Sapphire	9.9	228	500	1133	61	53.8	0.62

3. The radiating horn antenna integrated into the sapphire planar channel

As it is known an axial or planar decelerating channel radiating at the end into open space does not provide the high directivity of radiation, which was shown on the diagrams in the article [5]. Rectangular or circular cross section horn antenna can be used for increasing the radiation efficiency. This type of decelerating channel has rectangular cross section, therefore the most relevant variant shall be the rectangular antenna which will be directly integrated into the end of the planar channel. The main parameters of the antenna is its directivity, the angle between its axis and the direction of the main lobe, the angular width of the main lobe and the reflection coefficient of the horn, the radiation reflectivity of the horn and width of the operating frequency band.

Electrodynamic model of a copper antenna matched with the sapphire planar channel was tuned to the resonant frequency of 0.675 THz (fig. 2). Electrical field distribution simulated at 1 J of storage power in the structure is shown in Figure 3. This condition was used for parameters simulation. Geometrical characteristics and simulation results of radiating antenna matching with decelerating sapphire planar channel are submitted in Table 2. Directivity patterns of such extractor are shown in Figure 4. S11-parameters analysis results are shown in Figure 5. Here S11-parameter is interpreted as a wave reflection coefficient from the output side into the horn antenna.

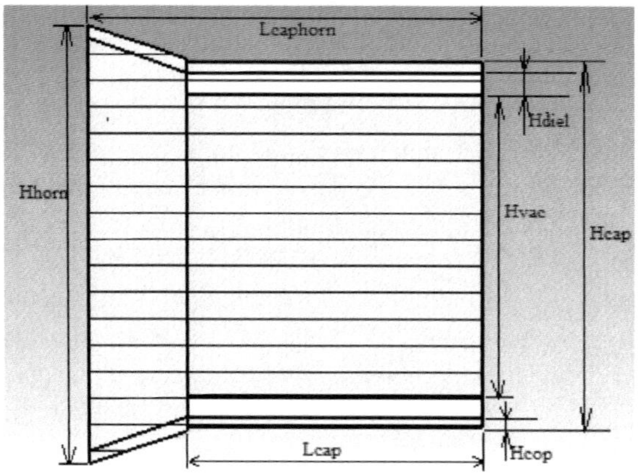

Figure 2. Electrodynamic model (side view) of the rectangular horn matched with decelerating channel: 1 – vacuum; 2 – dielectric coat; 3 – copper; H_{cap} – channel height; H_{vac} – vacuum channel height; H_{diel} – dielectric coat height; H_{cop} – copper part height; H_{horn} – horn height; L_{cap} – channel length; $L_{caphorn}$ – length channel with horn.

Table 2. Geometrical characteristics and simulation results of radiating antenna matching with decelerating sapphire and copper capillary tuned at frequency 0.675 THz (geometrical characteristics are shown in Figure 2).

H_{cap} (μm)	H_{vac} (μm)	H_{diel} (μm)	H_{cop} (μm)	H_{horn} (μm)	L_{cap} (mm)	$L_{caphorn}$ (mm)	Directivity (dB)	Angular width (°)	Reflectivity (dB)	Operating frequency band width (THz (%)
260	210	15	7.5	305.9	0.2	0.266	9.1	75.7	-38.3	0.042 (6.2)

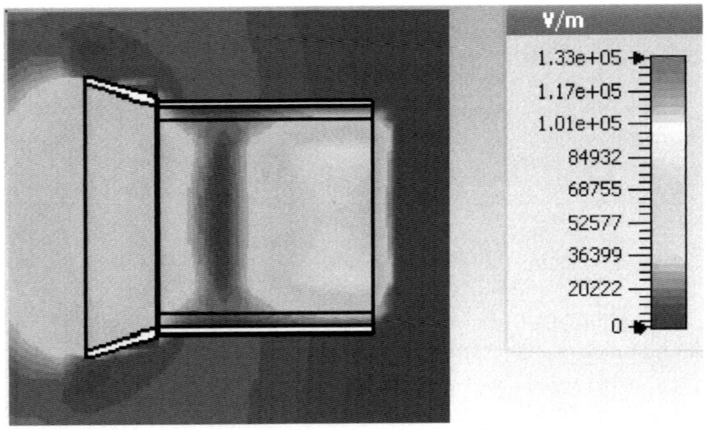

Figure 3. Electrical field distribution of decelerating planar channel with radiating antenna at frequency of 0.675 THz.

As in the case of axial structure angle reduces both directivity and operating frequency band width by increasing the horn beam. The width of the main lobe of the radiation and its angle contrary increase. Increasing the length of the horn and decreasing the beam angle lead to directivity increase. It is also worth to note that using the ribbon beam will allow to obtain higher directivity pattern values as the beam travelling through the planar dielectric structure will interact more with top and bottom of structure. Consequently it means a higher efficiency of THz-radiation output.

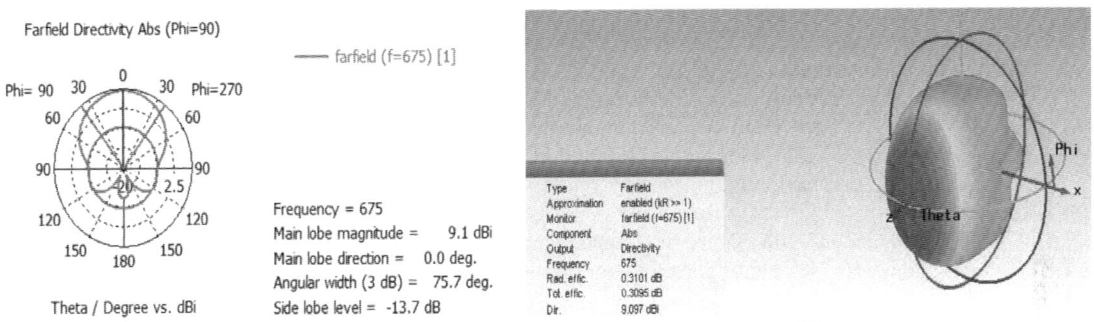

(a) (b)

Figure 4. Circular (a) and three-dimensional (b) patterns of decelerating sapphire and copper planar channel at operating frequency of 0.675 THz matching with the horn extractor.

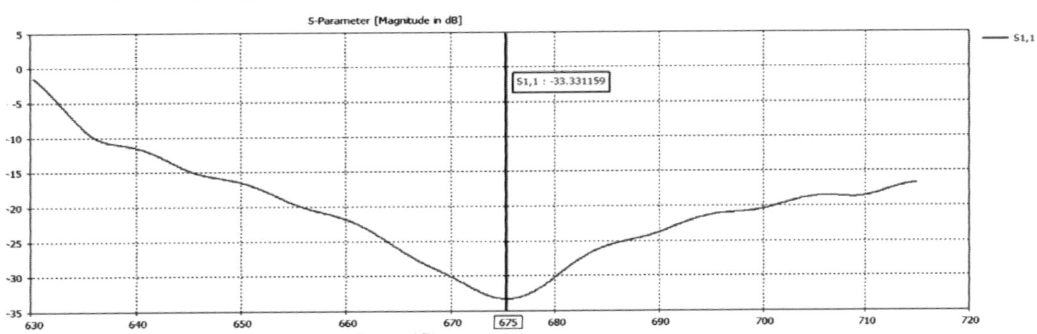

Figure 5. S11-parameters analysis results.

Comparison of electrodynamic characteristics of axial and planar Cherenkov structures are presented in Table 3.

Table 3. Comparison of electrodynamic characteristics of axial and planar structures.

Geometry	ε	R_{sh} (MOhm/m)	R_{sh}/Q (kOhm/m)	Q	β_{gr}	F (THz)
Axial	9.9	50.2	32.7	1531	0.43	0.675
Planar	9.9	61.0	24.8	1398	0.57	0.675

4. Conclusion

A model of planar Cherenkov radiating channel with horn antenna coated with sapphire and copper at the resonant frequency of 0.675 THz was designed. The group velocity is in the range of 0.5-0.6 c, the effective shunt impedance is R_{sh}/Q = 20-25 kOhm/m. Calculated horn antenna directivity provides < -10 dB, the reflection coefficient of the output of the capillary is less than -25dB, the width of the operating frequency band is ~ 6%. Apparently as it follows from the results electrodynamic characteristics of axial and planar structure geometry do not differ significantly which allow to considering the planar Cherenkov decelerating structure as accepted for THz-radiation output. Planar Cherenkov decelerating channel with integrated radiating antenna allows achieving a wide range of operating parameters for different geometrical dimensions with axial structure.

References

[1] Smirnov A V 2008 A high performance of FIR radiator based on a laser driven e-gun *Photonics Research Developments* **8** 247

[2] Polozov S M, Bondarenko T V and Smirnov A V 2012 *Problems of Atomic Science and Technology Series: Nuclear Physics Investigations* **3** (79) 92

[3] Feuerlein S, Roessl E, Proksa R et al. 2008 *Radiology* **249** 1010

[4] Pardo J R, Wiedner M C and Serabyn E 2004 Side-by-side comparison of Fourier transform spectroscopy and water vapour radiometry as tool for calibration of millimeter/submillimeter ground-based observatories *The astrophysical journal supplement series* **153** 363

[5] Ashanin I A and Polozov S M 2015 Report presented at *XI International Symposium RREPS-2015* Sapphire Decelerating Capillary Channel Integrated with Antenna at Frequency 0.675 THz

[6] CST GmbH – Computer Simulation Technology URL: http://www.cst.com